# Applied Biological Psychology

**Glen E. Getz, PhD, ABN,** is a board-certified clinical neuropsychologist through the American Board of Professional Neuropsychology. He is employed within the neuropsychology section in the department of psychiatry at Allegheny General Hospital and is an adjunct assistant professor at the Drexel University College of Medicine and Temple University School of Medicine. His responsibilities at Allegheny General include evaluating both inpatient and outpatient adult and child populations, providing cognitive behavioral therapy (CBT) interventions with psychiatric patients, and performing cognitive-based research within the Allegheny Singer Research Institute. He is the clinical training director of the neuropsychology postdoctoral program and also supervises clinical psychology interns, psychiatry residents, and graduate students. He has served a 3-year term on the Scientific Committee of Division 40 of the American Psychological Association (Clinical Neuropsychology). He has published over 30 full-length articles, abstracts, and book chapters, and he coauthored the third edition of *Screening for Brain Impairment* with Michael Franzen (Springer Publishing Company, 2010). He has taught applied biological psychology courses over a dozen semesters at Chatham University since 2007, where he serves as an adjunct professor.

Copyright © 2014 Springer Publishing Company, LLC

All rights reserved.

No part of this publication may be reproduced, stored in a retrieval system, or transmitted in any form or by any means, electronic, mechanical, photocopying, recording, or otherwise, without the prior permission of Springer Publishing Company, LLC, or authorization through payment of the appropriate fees to the Copyright Clearance Center, Inc., 222 Rosewood Drive, Danvers, MA 01923, 978-750-8400, fax 978-646-8600, info@copyright.com or on the Web at www.copyright.com.

Springer Publishing Company, LLC
11 West 42nd Street
New York, NY 10036
www.springerpub.com

*Acquisitions Editor*: Nancy S. Hale
*Composition*: Amnet

ISBN: 978-0-8261-0922-4
e-book ISBN: 978-0-8261-0923-1
*Instructor's Manual ISBN*: 978-0-8261-2723-5
*Instructor's Test Bank ISBN*: 978-0-8261-2724-2
*Instructor's PowerPoint Slides ISBN*: 978-0-8261-2722-8

*Instructor's Materials: Instructors may request supplements by emailing textbook@springerpub.com*

14 15 16 17 / 5 4 3 2 1

The author and the publisher of this Work have made every effort to use sources believed to be reliable to provide information that is accurate and compatible with the standards generally accepted at the time of publication. The author and publisher shall not be liable for any special, consequential, or exemplary damages resulting, in whole or in part, from the readers' use of, or reliance on, the information contained in this book. The publisher has no responsibility for the persistence or accuracy of URLs for external or third-party Internet websites referred to in this publication and does not guarantee that any content on such websites is, or will remain, accurate or appropriate.

**Library of Congress Cataloging-in-Publication Data**

Getz, Glen.
  Applied biological psychology / Glen E. Getz, PhD.
      pages cm
  Includes bibliographical references and index.
  ISBN 978-0-8261-0922-4 — ISBN 978-0-8261-0923-1 (e-book) 1. Neurobiology. 2. Psychobiology. 3. Biological psychiatry. I. Title.
  QP360.G48 2014
  612.8—dc23

2013036809

Special discounts on bulk quantities of our books are available to corporations, professional associations, pharmaceutical companies, health care organizations, and other qualifying groups. If you are interested in a custom book, including chapters from more than one of our titles, we can provide that service as well.

**For details, please contact:**
Special Sales Department, Springer Publishing Company, LLC
11 West 42nd Street, 15th Floor, New York, NY 10036-8002
Phone: 877-687-7476 or 212-431-4370; Fax: 212-941-7842
E-mail: sales@springerpub.com

Printed in the United States of America by Bradford & Bigelow.

# Applied Biological Psychology

GLEN E. GETZ, PhD, ABN

SPRINGER PUBLISHING COMPANY
NEW YORK

To Dana, my incredible wife, and to Maia and Anna, my amazing children.

# Brief Contents

*Preface*   xvii
*Acknowledgments*   xix

## Section I: Foundations of Neurobiology
1. History of Neurobiology   1
2. Research and Clinical Methods   13
3. Nervous System and Brain Structure   27
4. Major Neurobiological Brain Systems   45

## Section II: Application of Neurobiology in Psychology
5. Childhood Disorders   61
6. Schizophrenia   79
7. Mood Disorders   93
8. Anxiety Disorders   113
9. Eating Disorders   129
10. Sleep Disorders   145
11. Substance Disorders   159
12. Medical Disorders   173
13. Traumatic Brain Injury   193
14. Personality Disorders   207

*Glossary*   219
*References*   243
*Index*   277

# Contents

*Preface*   xvii
*Acknowledgments*   xix

**Section I: Foundations of Neurobiology**
1. History of Neurobiology   1
    Clinical Case Reports   2
        The Case of Phineas Gage   2
        The Case of Mr. Tan   3
        The Case of H.M.   4
    SPECIAL TOPICS: Decision-Making Capacity   5
        Current Case Examinations   6
    Important Methodological Advancements   6
        Development of Neuropsychology Testing   7
        Creation of Computed Tomography   7
        Creation of MRI and Beyond   7
    The History of Treatment Technique   8
        Lobotomy and Shock Treatment   8
    ETHICS: Safety of Patients   9
        Medication   10
        Psychotherapy   10
        Cognitive Behavioral Therapy   10
    Neuroscientists   11
    Conclusions   12
    Summary Points   12

2. Research and Clinical Methods — 13
   Microscopic Evaluations — 14
   Ablation — 14
   ETHICS: Animal Research—Necessity in Neuroscience? — 15
   Electrophysiology Recordings — 15
      Single-Cell Technique — 15
      Electroencephalogram — 16
      Event-Related Potentials — 17
   Neuroimaging Techniques — 17
      Structural Neuroimaging — 17
   SPECIAL TOPICS: Safety and the MRI Scanner — 20
      Functional Neuroimaging — 21
   Neuropsychological Evaluations — 24
   Behavior Genetic Testing — 25
   Conclusions — 26
   Summary Points — 26

3. Nervous System and Brain Structure — 27
   ETHICS: Competency and Neurobiology — 28
   Overview of the Nervous System — 29
      Peripheral and Central Nervous Systems — 29
   Neurotransmitters and Psychiatry — 33
      Acetylcholine — 33
      Serotonin — 34
      Catecholamines — 34
   Central Nervous System Development — 35
   Brain Division — 36
      Gray Versus White Matter — 36
      Left and Right Hemispheres — 37
      Cerebellum — 37
      Corpus Callosum — 38
      Glial Cells — 38
      Meninges and Cerebral Spinal Fluid — 38
   The Lobes of the Brain — 39
      The Occipital Lobe — 39
      The Parietal Lobe — 40
      The Temporal Lobe — 40
      The Frontal Lobe — 40
   SPECIAL TOPICS: Errors in Analyzing MRI for Understanding the Brain — 41
   Conclusions — 42
   Summary Points — 43

| | |
|---|---|
| **4.** Major Neurobiological Brain Systems | 45 |
|     ETHICS: Collaboration and Cooperation | 47 |
|     Primary Neurobiological Systems in Psychiatric Disorders | 48 |
|         The Limbic System | 48 |
|         The Basal Ganglia | 51 |
|         The Prefrontal System | 53 |
|         The Cingulate System | 54 |
|         The Fusiform Gyrus | 54 |
|     Secondary Neurobiological Systems in Psychiatric Disorders | 55 |
|         Cranial Nerves | 55 |
|         Mirror System | 56 |
|         Circle of Willis | 56 |
|     Application of the Neural Systems | 57 |
|     Conclusions | 57 |
|     SPECIAL TOPICS: The Application of the Concept of Neuroplasticity | 58 |
|     Summary Points | 59 |

### Section II: Application of Neurobiology in Psychology

| | |
|---|---|
| **5.** Childhood Disorders | 61 |
|     SPECIAL TOPICS: Psychotropic Medication in Children | 62 |
|     Neurodevelopment Overview | 63 |
|     ADHD | 65 |
|         Genetics and ADHD | 65 |
|         Neurotransmitters and ADHD | 66 |
|         Neuroimaging and ADHD | 66 |
|         Cognition and ADHD | 67 |
|         Treatment for ADHD | 67 |
|     Autism Spectrum Disorders | 68 |
|     ETHICS: The Harmful, Long-Lasting Effects of Falsifying Neurobiological Data | 69 |
|         Genetics and ASD | 70 |
|         Neurotransmitters and ASD | 70 |
|         Neuroimaging and ASD | 70 |
|         Cognition and ASD | 71 |
|         Treatment for ASD | 71 |
|     Intellectual Disability | 72 |
|         Learning Disorders | 73 |
|     Childhood Trauma | 74 |
|         Childhood Maltreatment | 74 |
|         Genetics and Childhood Maltreatment | 75 |
|         Neurotransmitters and Childhood Maltreatment | 75 |

Neuroimaging and Childhood Maltreatment ... 75
  Cognition and Childhood Maltreatment ... 76
  Therapy for Childhood Maltreatment ... 77
 Conclusions ... 77
 Summary Points ... 78

6. **Schizophrenia** ... 79
 Clinical Symptoms of Schizophrenia ... 80
 Onset and Course of Schizophrenia ... 81
 Genetics and Schizophrenia ... 82
 Neurochemistry and Schizophrenia ... 83
 Neuroimaging and Schizophrenia ... 83
 SPECIAL TOPICS: Clinical Imaging in Schizophrenia? ... 86
 Cognition and Schizophrenia ... 87
 Treatment for Schizophrenia ... 88
  Medication ... 88
 ETHICS: Decision-Making Capacity in Schizophrenia ... 89
  Therapy ... 90
 Conclusions ... 91
 Summary Points ... 91

7. **Mood Disorders** ... 93
 Depression ... 94
  Clinical Symptoms of Depression ... 94
  Subtypes of Depression ... 95
 Major Depressive Disorder ... 95
  Genetics and MDD ... 96
  Neuroimaging and MDD ... 96
  Cognition and MDD ... 98
  Treatment for MDD ... 98
 SPECIAL TOPICS: Neuroscience in Prediction of Success: MRI/
  Genetics/Cognitive ... 103
 Bipolar Disorder ... 104
  Genetics and BD ... 104
  Neuroimaging and BD ... 105
  Neurochemistry and BD ... 106
  Cognition and BD ... 107
  Treatment for BD ... 108
 Conclusions ... 109
 ETHICS: Suicidal Behavior: When to Break Confidentiality ... 110
 Summary Points ... 111

8. **Anxiety Disorders** ... 113
 ETHICS: Stress in Clinical and Research Environments ... 114

| | |
|---|---:|
| Basic Neurobiological Response to Stress: Fight or Flight? | 115 |
|     Physical Effects of SNS Activation | 116 |
| SPECIAL TOPICS: Why People Like to Be Frightened | 117 |
| Panic Disorder | 118 |
|     Genetics and PD | 118 |
|     Neurochemistry and PD | 119 |
|     Neuroimaging and PD | 119 |
|     Cognition and PD | 119 |
|     Treatment for PD | 119 |
| Agoraphobia | 120 |
| Social Anxiety | 121 |
|     Genetics and SA | 121 |
|     Neurochemistry and SA | 121 |
|     Neuroimaging and SA | 121 |
|     Cognition and SA | 122 |
|     Treatment for SA | 122 |
| Specific Phobias | 122 |
|     Genetics and Phobias | 123 |
|     Neuroimaging and Phobias | 123 |
|     Treatment for Phobias | 123 |
| Posttraumatic Stress Disorder | 123 |
|     Genetics and PTSD | 124 |
|     Neurochemistry and PTSD | 124 |
|     Neuroimaging and PTSD | 125 |
|     Cognition and PTSD | 126 |
|     Treatment for PTSD | 126 |
| Generalized Anxiety Disorder | 126 |
|     Genetics and GAD | 127 |
|     Neurochemistry and GAD | 127 |
|     Neuroimaging and GAD | 127 |
|     Cognition and GAD | 127 |
|     Treatment for GAD | 128 |
| Conclusions | 128 |
| Summary Points | 128 |
| **9. Eating Disorders** | **129** |
| Anorexia Nervosa | 130 |
|     Genetics and AN | 131 |
|     Neuroimaging and AN | 131 |
|     Cognition and AN | 132 |
|     Treatment for AN | 133 |
| ETHICS: Safety and Eating Disorders | 135 |
| Bulimia Nervosa | 136 |
|     Neuroimaging and BN | 137 |

| | |
|---|---:|
| Cognition and BN | 137 |
| Treatment for BN | 138 |
| Binge Eating Disorder | 139 |
| SPECIAL TOPICS: Obesity and Therapist Role | 140 |
| Neurochemistry and BED | 141 |
| Neuroimaging and BED | 141 |
| Cognition and BED | 142 |
| Treatment for BED | 142 |
| Conclusions | 143 |
| Summary Points | 143 |
| **10. Sleep Disorders** | **145** |
| Purpose of Sleep | 145 |
| Normal Sleep | 146 |
| Stage 0: Wakefulness | 146 |
| Stage I Sleep | 147 |
| Stage II Sleep | 147 |
| Stage III Sleep | 148 |
| Stage IV Sleep | 148 |
| REM Sleep | 148 |
| Cycle of Stages | 149 |
| Dreaming | 149 |
| Circadian Rhythms | 150 |
| Life Span and Sleep | 150 |
| Sleep Deprivation | 151 |
| Cognition and Sleep Deprivation | 151 |
| Neuroimaging and Sleep Deprivation | 152 |
| Clinical Sleep Disorders | 152 |
| Insomnia | 152 |
| ETHICS: Are You Getting Good Sleep? | 153 |
| Hypersomnia | 154 |
| Parasomnia | 156 |
| SPECIAL TOPICS: Nocturnal Enuresis: A Behavioral Strategy | 157 |
| Conclusions | 158 |
| Summary Points | 158 |
| **11. Substance Disorders** | **159** |
| Background | 159 |
| SPECIAL TOPICS: Medical Marijuana—Helpful or Hurtful? | 160 |
| Behavioral and Neurobiological Response to Substance Exposure | 161 |
| Comorbidity of Substance Abuse and Psychiatric Disorders | 163 |

| | |
|---|---|
| Genetics and Substance Disorders | 163 |
|     Prenatal Exposure to Substances | 164 |
| Examination of Common Substances of Abuse | 164 |
|     Alcohol Abuse | 164 |
|     Cannabis Abuse | 167 |
|     Methamphetamine Abuse | 167 |
|     Cocaine Abuse | 168 |
|     Hallucinogen Abuse | 169 |
|     Opiate Abuse | 169 |
|     Caffeine and Nicotine Abuse | 170 |
| Conclusions | 170 |
| ETHICS: Use of Amphetamines as Performance Enhancers | 171 |
| Summary Points | 172 |
| | |
| **12. Medical Disorders** | **173** |
| SPECIAL TOPICS: Driving: When Is It No Longer Safe? | 174 |
| Delirium | 174 |
| The Dementias | 176 |
|     Alzheimer's Disease | 176 |
|     Vascular Dementia | 180 |
|     Frontal Lobe Dementia | 180 |
|     Dementia of Lewy Body | 181 |
| Seizure Disorders | 182 |
|     Seizure Subtypes | 183 |
|     Genetics and Seizure Disorder | 185 |
|     Neurochemistry and Seizure Disorder | 185 |
|     Neuroimaging and Seizure Disorder | 185 |
|     Cognition and Seizure Disorder | 185 |
|     Treatment for Seizure Disorder | 186 |
| Movement Disorders | 186 |
|     Huntington's Disease | 186 |
| ETHICS: Genetic Testing: To Test or Not to Test? | 187 |
|     Parkinson's Disease | 188 |
| Other Medical Events | 189 |
|     Tumors | 189 |
|     Anoxia and Hypoxia | 190 |
| Conclusions | 191 |
| Summary Points | 191 |
| | |
| **13. Traumatic Brain Injury** | **193** |
| TBI Risk Factors | 194 |
| TBI Defined | 194 |

| | |
|---|---:|
| Neurobiology of TBI | 195 |
| Neuroimaging and TBI | 197 |
| TBI: Severity Classification | 198 |
| Mild TBI: Definition and Symptoms | 199 |
| SPECIAL TOPICS: Sports-Related TBI | 201 |
| Malingering Issues in TBI | 201 |
| Outcome | 202 |
| Application of Treatment to Mental Health Providers | 202 |
|     Cognitive Rehabilitation | 202 |
|     Psychotherapy | 203 |
|     Neuropsychological Assessment | 203 |
|     Speech and Language | 203 |
|     Sensory Evaluations | 204 |
|     Headache | 204 |
|     Medication Effects | 204 |
| ETHICS: Confidentiality | 205 |
|     Diet/Nutrition/Exercise | 205 |
| Conclusions | 206 |
| Summary Points | 206 |
| | |
| **14. Personality Disorders** | **207** |
| SPECIAL TOPICS: Psychopathy, Neuroscience, and the Law | 208 |
| Antisocial Personality Disorder | 210 |
|     Genetics and ASPD | 211 |
|     Neuroimaging and ASPD | 211 |
|     Cognition and ASPD | 212 |
|     Treatment for ASPD | 212 |
| Borderline Personality Disorder | 213 |
| ETHICS: Dual-Role and Relationships with BPD | 214 |
|     Genetics and BPD | 215 |
|     Neuroimaging and BPD | 215 |
|     Neurocognition and BPD | 216 |
|     Treatment for BPD | 217 |
| Conclusions | 218 |
| Summary Points | 218 |
| | |
| *Glossary* | 219 |
| *References* | 243 |
| *Index* | 277 |

# Preface

It has been my experience that the academic study of neurobiology often focuses on a fundamental understanding of brain anatomy and functioning. Having this basic knowledge of brain functioning is necessary to being able to describe how the brain works. This understanding is valued in psychology to the extent that a biopsychology or neurophysiology course is offered in most universities' psychology departments, whether at an undergraduate, master's, or doctoral level. There are several clinical doctoral-level programs that even offer a neuropsychology track, to the extent that the field of neuropsychology is one of the largest and fastest growing fields within psychology.

However, not all psychology students are innately interested in the neuroscientific understanding of human behavior. In fact, it would come as no surprise to learn that most students who major in psychology or enroll in graduate-level programs in psychology have limited interest in, are intimidated by, or do not understand the necessity of learning about human behavior from a neurobiological perspective. They have a hard time seeing how learning this information will influence the psychotherapeutic work that is the long-term goal of most students in this field. I have spoken to many students who have become anxious when learning that they are *required* to take a biopsychology course. It is my perspective that it is the responsibility of the professors to help change this mindset. Although our understanding of brain–behavior connections has recently advanced tremendously through advancements in technology and scientific methodology, the information that has been gained, if not made practical, is limited in meaning and utility. Teaching the foundation of neurobiology is therefore necessary, but not sufficient, in the advancement of this field; rather, integrating our understanding of this information within other areas of psychology will increase interest, learning, and application of the material for students.

The goal of this book is not only to help students learn about fundamental brain functioning and to apply the information with various clinical populations with whom they may help to serve. This book is also intended to help the professor advance beyond the typical mindset of teaching only the basics in brain functioning. This text is intended to help the professor make neurobiology practical for all students, no matter at what level they are learning the material. Such an approach requires a slight pedagogical shift in teaching. Rather than separating courses by only teaching about their specific area of psychology and hoping that students are able to connect the dots of how their different coursework links together, this approach hinges on the integration of psychobiology within the context of psychopathology to aid understanding of how specific psychotherapeutic techniques are effective. This is often a challenging endeavor for even the most advanced professors and teachers. The goal of this text is to bridge the gap between what has been learned in the basic neurosciences, what is taught in coursework, and what is used by mental health professionals in therapeutic settings.

In Section I of the text, a foundational framework of neuroscience is provided, including important historical events, patients, and neuroscientists as well as an explanation of all the different techniques used in understanding human behavior. The first part of the text also focuses on core foundations of brain functioning, with an emphasis on the important neural systems often found dysregulated in psychopathology. The second section of the text explores many areas of psychopathology from a behavioral, cognitive, and neurobiological perspective before describing typical effective strategies used to treat the various disorders. An ethical topic is included in each section to help remind students and teachers alike that this area of psychology is not black and white, but rather is filled with challenges that occur daily and that should always be kept in mind.

An Instructor's Manual, test bank, and PowerPoint slides are also available to supplement the text. **To obtain an electronic copy of these materials, faculty should contact Springer Publishing Company at textbook@springerpub.com.**

Keeping up with the pace of what is learned in neuroscience is difficult for any professional or student. There will continue to be important advancements in this field for decades to come. I hope this book begins to bridge the gap and connect the dots for the next generation of mental health professionals to not only recognize the relevance and integrate this material within their own work, but also continue to advance the field through consideration of this area of psychology in whichever role they choose to explore.

# Acknowledgments

This book would not have been possible but for the countless neuroscientists striving to unravel the mystery of human functioning. It is because of their ideas, research, and writings that this field has advanced rapidly over the past few decades, inspiring this book. I have been fortunate to have been trained by some of the best, and I would like to thank them all for their teaching and guidance. Specifically, Stephen Strakowski and Michael Franzen have provided me with unwavering support and numerous opportunities. I would also like to thank Mary Beth Mannarino, who has trusted me and provided me the chance to teach this material to graduate students at a level that I believe is applicable to their field. I would also like to thank all the students whom I have taught, supervised, and trained during my career. Their enthusiasm inspires me daily. I would particularly like to thank all the students who helped with various aspects of this book, including Megan Yudin, Ben Edner, Alexis DeRiggi, Marisa Floro, and other students who provided valuable assistance near the end of the project.

I would like to thank the staff at Springer Publishing Company for helping make this book possible, particularly Nancy S. Hale, for believing in this project, and Kathryn Moyer, for her careful editorial assistance.

Finally, I am very fortunate to have extremely supportive and understanding family, friends, and coworkers. In particular, I would like to thank Tony and Chris for providing me with quiet space to write, my brothers for their support, and John F. Boyle for his levity throughout this process. Most important, I would like to say a very special thank-you to the love of my life, Dana, for believing in me and encouraging me while I wrote this book and in all my ambitions, as well as to my truly incredible children, Anna and Maia. Their patience, understanding, love, and support while I wrote this book and on a daily basis are truly remarkable. I could not have done this without them. Thank you!

CHAPTER 1

# History of Neurobiology

The understanding of human behavior as a result of brain-based functioning can be clearly traced back to at least the ancient Greek philosophers, and most likely even prior to that point, to the end of the Stone Age and ancient Egyptian times. Many authors and introductory chapters of neuroscience textbooks have detailed the origins and historical lineage of understanding the brain's role in terms of behavior. For example, there is clear documentation that philosophical debates discussing the role of the brain versus the heart occurred throughout history in an attempt to explain human behavior. Although the relationship has not been formally recognized, neuroscience and the study of human behavior have been integrated throughout time. The roots of understanding the importance of the brain date back centuries; however, it wasn't until the 1990s that the Decade of the Brain was declared by the Library of Congress and National Institute of Mental Health. Recognizing the recency of this declaration can help one to understand that applying our understanding of neurobiology for clinical purposes is still in its infancy. In order to have an ability to utilize knowledge gained from research for the purposes of clinical application, the fundamental knowledge of behavior, emotion, and cognition needs to be well established. A brief historical overview can help to explain how neuroscience has evolved to what we know it as today.

Although appreciating history is important for our understanding of how humanity initially connected the brain with human behavior, it is not often thought to be relevant for the purposes of the general mental health practitioner. For example, the historical information is not often utilized within clinical settings or for treatment purposes. Mental health clinicians do not typically discuss historical events from within their field with their patients, clients, or colleagues beyond an academic environment. As such, the history of mental health treatment and neurobiological understanding is often considered mundane, unnecessary for treatment purposes, or of little practical utility. On the other hand, it is important to learn from history in any field, whether government, business, or medicine. For the applied neurobiology student, a brief review of a few individual clinical cases, significant scientific milestones, and the introduction of specific treatments can help to explain how we have reached the current state of our understanding of behavior, cognition, and emotion and subsequent treatment for problematic areas.

By the time this book is published, more history and scientists will help further our understanding, as technology is advancing at a very rapid pace. This chapter focuses on the modern history of some of the most relevant aspects of neuroscience. Although this chapter examines specific, important clinical cases, significant neuroscience milestones, and important treatment modalities that have been implemented and that have led us to this point in our history, it is not meant to be an exhaustive review of the history of neuroscience. For a more thorough review of this area, the reader is encouraged to consult introductory neuroscience texts. Modern neuroscientists have strongly influenced the collective understanding of brain-based functioning and have guided the field to where it stands today. In this chapter and throughout the text, there is discussion regarding important ethical considerations in the neurosciences, as well as special areas of focus that generate an increased level of consideration, from the public and often from the media. The special topics sections have an increased level of application for the neurosciences. Having a more thorough understanding of the neurosciences can provide a unique perspective regarding these special topics and elicit increased interest from students, the media, the public, and others.

## CLINICAL CASE REPORTS

There are thousands of individual cases that have contributed to neuroscientific understanding throughout history. Examination of behavior and follow-up analysis of brain tissue have helped neuroscientists understand brain-based functioning. Clinical cases often lead to better understanding of specific illnesses or of functional implications of brain regions. Three clinical cases that stand out as having generated a wealth of knowledge for scientists, and which are deemed important for the general clinician to be aware of, are the cases of Phineas Gage, Mr. Tan, and Henry Molaison, also known as H.M. A synopsis of each of these cases and their historical relevance are provided.

### The Case of Phineas Gage

One of the best-known and most well-studied patients who have helped neuroscientists understand the connection between the brain and behavior was Phineas Gage, who lived in the mid-19th century. He was, by all historical accounts, a responsible, hard-working foreman, whose unfortunate accident fascinated scientists and people of all sorts for years, and continues to fascinate them even to this day. His story is so intriguing that he has become one of the most famous neuroscientific cases, and his skull can still be found in the Warren Anatomical Museum at Harvard University. Websites dedicated to him are easily found, and documentaries regarding his life and accident are utilized for teaching introductory-level classes through advanced graduate school course work. Gage's accident advanced brain understanding beyond what could have occurred in a laboratory setting.

In September 1848, Phineas was working as a foreman on a railroad project when a 43-inch-long tamping iron penetrated the front part of his skull and went through the back of his head at a very high rate of speed. Incredibly, Phineas survived the accident and was followed by doctors for a period of time thereafter. These doctors, among other things, documented his change in personality and behavior. Once a responsible, appropriate, and hard-working man, Phineas became irresponsible, socially inappropriate, and unable to work again. On the other hand, he reportedly did not demonstrate any changes in intelligence, memory, motor skills, or speech. The idea that the brain was responsible for personality, social behavior, and emotion processing was a novel concept at that time.

The examination of Phineas's skull and of the damage to his frontal lobe gave rise to the awareness of the frontal lobe's role in the areas of social functioning, personality, and decision-making abilities. As we document throughout the text, the role of the frontal lobe in these areas is very clear, in both nonclinical and clinical populations. The injury that Gage experienced was the beginning of opening our collective understanding of the frontal lobes. Modern-day injuries to the frontal lobes are known to cause changes in personality, mood, and social abilities. It was the tragic accident of Phineas Gage that allowed scientists to expect changes in these characteristics after injury to this area of the brain. The episode also formed the basis for further exploration of how the frontal lobes are important in a variety of areas, leading to future treatment strategies.

## The Case of Mr. Tan

Around the same time as Phineas Gage, Pierre Paul Broca, a French physician and neurosurgeon, was studying the brain and attempting to determine whether localization—the theory that certain parts of the brain are solely responsible for certain cognitive and emotional aspects of brain functions—was a legitimate scientific idea. Prior to that point, Franz Joseph Gall argued for the concept of phrenology, the study of brain functioning as it relates to bumps and indentations on the head. Certain portions of the skull, it was argued, were responsible for various behaviors. (See Figure 1.1.) There was a quest after that point to determine if the concept of phrenology could be proven or disproven.

In 1861, Dr. Broca examined Mr. Lebourgne, a 21-year-old male who was progressively losing his speech. This patient is referred to as "Mr. Tan," as all he could say was "Tan." Although he was losing his speech, he was not losing his comprehension of what others were telling him. Mr. Tan died in a hospital soon after Dr. Broca evaluated him, and Dr. Broca performed

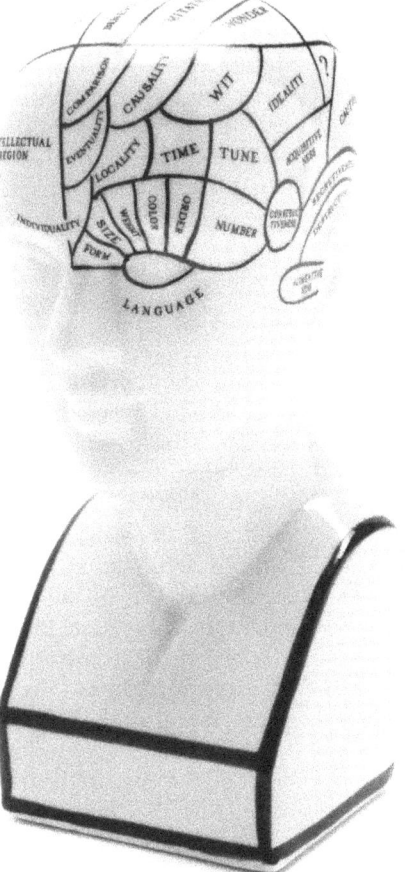

**FIGURE 1.1 Phrenology head.**
Source: © AMC Photography/Shutterstock

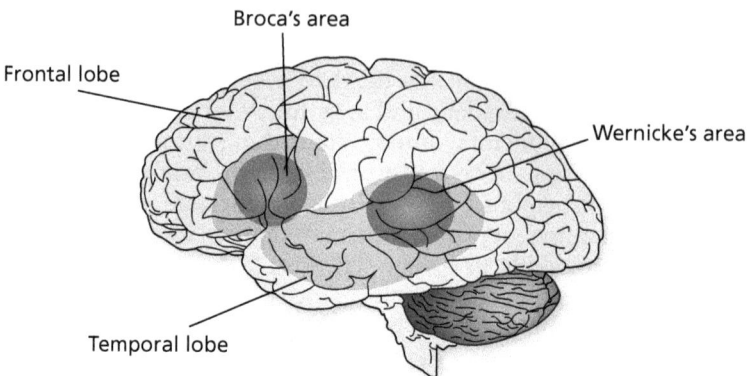

**FIGURE 1.2 Wernicke's area.**
Adapted from Holtz (2011).

an autopsy shortly thereafter. It was discovered that Mr. Tan had a lesion, an abnormal change in the tissue, in the left frontal lobe of his brain. Dr. Broca followed up with at least a dozen other cases of individuals with expressive aphasia, the inability to speak. He discovered that they were all found to have lesions in the same region of the brain. This region of the brain has since been labeled Broca's area. Dr. Broca's discovery helped to advance the notion that certain parts of the brain were in fact important or necessary for controlling certain cognitive functions. Since that time, other aspects of speech have been demonstrated to be controlled by other regions of the brain near Broca's area. For example, Wernicke's area, which is posterior to, or behind, Broca's area in the temporal lobe, is important for language comprehension. Wernicke's area will be discussed when describing alcohol-related diseases. The evaluation of Mr. Tan's brain led to an increased appreciation of the importance of the concept of localization. (See Figure 1.2.)

## The Case of H.M.

A more contemporary patient, Henry Molaison (better known as "H.M."), has been crucial in our understanding of memory. In 1953, H.M. underwent brain surgery due to debilitating seizures. It was thought that if his seizures weren't controlled more effectively, he would die from the events. A neurosurgeon removed the hippocampus in an attempt to stop his seizures from occurring. Although the surgery helped to control the seizures more effectively, H.M. experienced severe amnesia. He was able to recall the events that had occurred earlier in his life, but he was unable to form new memories. This condition is called *anterograde amnesia*. Interestingly, H.M. was also unable to recall events for the 11 years prior to the surgery, a condition known as *retrograde amnesia*. As a result of these outcomes, the neurosurgeon refused to complete any more of these surgeries, and H.M. became the only living case of this type of procedure.

H.M. was evaluated for 50 years by neuropsychologist Dr. Brenda Milner and her colleagues, including Dr. Suzanne Corkin. These scientists demonstrated that H.M.'s other cognitive abilities, such as intelligence, problem-solving, visuospatial skills, and language were all intact. H.M. was frequently evaluated, and, given his memory problems, he was an ideal candidate for studying various types of learning. In fact, during his life, he was able to demonstrate some learning. For example, he seemed to learn a motor task in which he was required

> **SPECIAL TOPICS**
>
> **Decision-Making Capacity**
>
> As you may have already learned or will soon discover throughout this text, evaluation of individuals who have experienced unusual or atypical behavior is often key to discovering and advancing neuroscience. When considering that these individuals play a vital role in advancing the science, it is important to consider how it is decided that it is acceptable for them to be examined. H.M., for example, was unable to form any new memories. How could somebody who does not have memory for events agree to participate in a research study, which ultimately may lead to his brain being dissected after death? In fact, it was not H.M., but rather a guardian had made the decision.
>
> A main consideration when determining whether an individual has the capacity to make decisions for herself or himself is whether the individual is able to understand what is being asked, has an appreciation of the risks and rewards, and is aware of the available options. For example, children, regardless of whether there is compromise to cognition, do not have the capacity to make their own decisions. Assenting to treatment is necessary, meaning that they must have an understanding of what is being asked. However, parents or legal guardians ultimately make the decision for the child. Similarly, guardianship is provided for individuals with impaired decision-making capability. Although it may be more clear-cut in the case of H.M. that he was unable to demonstrate decision-making capability, given his severe amnesia, memory problems in and of themselves do not render an individual incapable of making his or her own decisions. It is not uncommon for physicians who are questioning the capacity of patients to misunderstand this concept, as patients may have problems with memory but may be able to consistently indicate an awareness of the clinical situation and provide consistent rational reasons for their choices.
>
> Decision-making capacity will be discussed in greater detail when we examine schizophrenia, but consideration of each individual's ability to make decisions is important in science and treatment. It is also important to remember that patients are allowed to make bad decisions. People make poor decisions frequently, such as deciding to smoke, eat unhealthily, and make poor purchases, but if they are able to understand the possible ramifications of their decisions, then they have the capability to make those decisions. For mental health providers, it is difficult to allow patients to make poor decisions, but it is important to remember that, even with all the neuroscientific knowledge that one has, the patient is ultimately left to his or her own devices to make most decisions without interruption (with the exception of suicidal, homicidal, or child abuse situations, in which the provider is morally, legally, and ethically obligated to act appropriately). A clinician's goal is to help guide behavior and assist the individual in making the best decisions that he or she is capable of making.

to trace a star while looking in a mirror with a curtain blocking the direct view of the star. He was also able to learn some of the researchers' names near the end of his life. These discoveries revolutionized the understanding of the temporal lobes' role in memory and the different systems in place for various types of learning.

Unfortunately, H.M. passed away in 2008, when he was 82 years old. After he passed away, his brain was cut into 2,000 slices, live on video stream on the Internet. The dissection can still be seen online, and careful evaluation of his brain tissue is currently occurring. The case of H.M. will be as famous

100 years from now as that of Phineas Gage 100 years after his death. Video and audio recordings demonstrated that H.M. was glad that he was able to help neuroscience advance the understanding of brain-based behavior. The case of H.M. should lead a critical thinker to question whether he was able to properly consent to repeated examination, given his memory problems. The Special Topics box discusses the use of guardians to make decisions in treatment.

## Current Case Examinations

To this day, clinical case reports fascinate neuroscientists. Understanding healthy human brain functioning usually occurs through examination of unhealthy or atypical behavior. Various studies examining the brains of unique and often successful individuals or groups of people, such as the brains of athletes, have all played an important role in advancing scientists' knowledge of brain-based behavior. Some scientists have written about, spoken about, or been portrayed in movies as talking about their own experiences. This group includes Kim Peak, better known as "Rain Man," who was a savant and was demonstrated to have no corpus callosum. (The *corpus callosum* is the part of the brain that connects the left and right hemispheres.) Temple Grandin has spoken about her experiences with having an autism spectrum disorder, and how that affects her work with animals. The book *A Stroke of Genius*, by the neurologist Dr. Sandy Simon, is a first-hand account of the experience of stroke by an expert in the area. When severe violent behaviors occur, the media will often quickly attempt to provide a psychiatric explanation for the cause of the behavior and try to "diagnose" the individual. Attorneys have utilized neurobiological information to reduce the sentences of criminals. Individuals who have demonstrated extreme, violent behaviors and who have had their brains examined have shown atypical results. These contemporary cases, in an era with advanced technology, have helped us better understand brain functioning. The information can be used to assist the general clinician and the public to understand these neurobiological events. It is the mental health clinician's role to assist their clients and others in the psychoeducation of neurobiology.

Individual case studies are not the only, or even the best, way to help understand brain functioning. Groups of people who have been exposed to similar events, injuries, trauma, or other experiences are being collectively examined to understand brain-based changes. For example, this textbook will discuss traumatic brain injury in football players, the effects of trauma on soldiers' brains, and the brains of individuals who have been involved in violent crimes. With the advancement of technology, examination of brains within living individuals is continuing to help elucidate the heterogeneity found within groups of people. In vivo examination has led to a better understanding of how people with certain strengths or weaknesses process the world.

## IMPORTANT METHODOLOGICAL ADVANCEMENTS

In all sciences, there are "aha" moments, or moments of great relevance that transform a field. In neuroscience, there are likely to be countless numbers of these moments. Advancements in procedures, technology, and treatments have occurred rapidly over the past century. Being aware of certain events, the timing of the events, and, in some instances, what led to these events, not only can assist in a better appreciation of how we have achieved what we have achieved, but can also provide reasons for what has stopped us from advancing even further.

The following events were chosen not as the most important milestones in neuroscience, nor as "aha" moments, but rather because they are events that practitioners should know have a direct influence on their patients to the current day. These major scientific advancements have helped us to reach the level of understanding that we have today.

## Development of Neuropsychology Testing

The attempt to formally understand cognitive capabilities dates back nearly a century to the introduction of neuropsychological tests to help inform decisions regarding military placement. The first half of the 20th century was focused on understanding war-damaged brains. Objective cognitive testing was initially used to assist in determining cognitive capabilities in soldiers. The introduction of intelligence tests by Binet in the early 1900s allowed for the screening of recruits for military purposes. During World War I, there was a need to examine whether individuals were capable of fighting in a war. Results of this testing also helped to guide whether the soldiers were in need of rehabilitation (Lezak, Howieson, & Loring, 2004). During World War II, neuropsychologists continued to expand and develop techniques to examine brain functioning. Neuropsychological testing has been further utilized as a means to assist in clarifying diagnoses in unhealthy individuals. Having the ability to document cognitive strengths and weaknesses and to explore what happens to behavior with specific injuries or illnesses is fascinating and has strong functional purposes.

## Creation of Computed Tomography

Prior to the creation of computed tomography (CT), there was no way to safely examine an individual's brain until after the person had died. Having the capacity to look inside an individual's head to see pictures of the brain was thought to be particularly exciting for scientists. The invention of CT led to the first picture of a live brain. CT was invented in 1972 by a British engineer, Godfrey Hounsfield, and a South African physicist, Allan Cormack. Their Nobel Prize invention was created only to examine the head, and it initially required several hours to obtain one slice of a picture of the brain and several days to process scan results. With further refinement, CT scans are now very rapidly completed and analyzed. By 1980, CT scanners were widely available and were being used for a number of clinical and research purposes. CT scans have proven to be particularly effective in a variety of ways that will be discussed throughout the book. CT scans are found in community and academic hospitals, as well as in other imaging centers throughout industrialized countries. However, it wasn't until the development of magnetic resonance imaging (MRI) that less potentially harmful methods and a better clarity of brain images were available.

## Creation of MRI and Beyond

MRI was first developed in 1952 by a Harvard student completing his dissertation. It was slow to advance, as it was costly and difficult to re-create. It was 20 years later that the first MRI machine was created by an American scientist in New York. Though numerous scientists further developed the technology, it wasn't until 2003 that scientists Raymond Damadian, Paul Lauterbur, and

Peter Mansfield were awarded a Nobel Prize for their work in advancing MRI technology, dating as far back as the 1970s. Their safe way to repeat brain scans has helped us to explore healthy brain development, psychopathology, and the effects of treatment. These studies have in turn led to further techniques in the MRI machine that have allowed for evaluation of blood flow during cognitive tasks, and for evaluation of white matter tracts. Without the invention of MRI, we would still be significantly limited in our understanding of the brain.

The next chapter discusses in more detail neuropsychology testing, CT scans, and MRI scans. Having a better understanding of these methodologies will assist in informing the clinician about how to know when to make referrals for these specialized procedures, will tell him or her how to interpret some of the reports obtained from specialists, and, most importantly, will help the patient in terms of treatment or what to expect from undergoing these procedures and the results such procedures generate. Being aware that these techniques were all introduced no earlier than the past century can help one to appreciate why diagnosis and accurate treatment can at times be challenging. Other inventions and milestones, such as the mapping of the human genome, the ability to complete genetic testing, and the use of blood work to examine chemical functioning have helped to achieve a more thorough understanding of the brain. Throughout the book, examination of these different techniques will occur.

## THE HISTORY OF TREATMENT TECHNIQUE

Historically, individuals with bizarre behavior, delusions, cognitive or intellectual limitations, or social impairments were ostracized in various societies. If family members did not care for them or guard them from the public, these psychiatrically ill individuals would, in best-case scenarios, be institutionalized over time. At other less humane times, they were viewed as witches, as evil, or as having other negative connotations that led to lack of treatment, segregation, social isolation, and sometimes even torture and death. Therefore, initial treatments were not well funded, and ethical standards were not in place to ensure patient safety. In contrast, ethical standards regarding treatment are now well implemented (see the Ethics box).

However, although some early techniques would not have met modern-day ethical guidelines, the scientists who developed the treatments did not necessarily intend to harm patients, but were rather attempting to utilize the technology and knowledge that they had at that time to assist in improving the patients' functioning. These early inventions also contributed to the development of refined techniques that are safer and more effective, which we will discuss throughout the text.

### Lobotomy and Shock Treatment

Although depicted in movies and media as being gruesome, cruel, and painful, the development of invasive approaches such as lobotomy and shock treatment therapy were attempts to improve upon behavior and emotions during a time when it was clear that the brain played a role in these areas, but the science of treatment was still crude. During a lobotomy, a medical doctor removes a portion of the brain through a variety of means, including drilling holes into the head or using an ice pick through the upper eyelid. Although the procedure of lobotomy was already being questioned worldwide in the 1930s to 1950s for its relatively harsh approach and sometimes negative effects of cognitive dulling or even death,

> **ETHICS**
>
> **Safety of Patients**
>
> The Golden Rule of any practitioner, ranging from medical doctors, to researchers, to mental health providers, is "First do no harm." Ethical guidelines and principles are established with this simple goal as a central tenet. The Hippocratic Oath that medical practitioners swear to is intended to aim for ethical and safe treatment for patients. Underlying all clinical decisions for a mental health practitioner should be the thought of what will provide the best form of treatment and what needs to be done to ensure that there is no harm caused to the patient. Although this rule seems simple enough to follow, it can become particularly challenging to discern in situations of confidentiality, end-of-life decisions, and situations in which there is no clear ideal outcome. Yet, if the mental health practitioner has a solid explanation for his or her decisions guided by this principal, he or she is less likely to incur problems ranging from legal suits to ethical mistakes.
>
> In the American Psychological Association Code of Ethics, General Principles lay out a foundation for behaviors that psychologists strive to achieve. The first sentence of Principle A states that "Psychologists strive to benefit those for whom they work, and take care to do no harm" (American Psychological Association, 2010). It goes on to state that, when conflict occurs, attempting to minimize harm is the ideal goal of the provider. Whether new to the field or an experienced therapist, a mental health provider should remain mindful of this rule on a daily basis and allow it to guide treatment, research, professional behaviors, and personal relationships.
>
> Despite this seemingly straightforward core philosophy in ethics, this rule has at times been compromised in neuroscience as well. For example, concerns regarding falsified data that led to the theory that vaccines cause autism is an example in which researchers' misinformation led to children not being vaccinated, which in turn contributed to a recent rise of mumps, measles, and rubella. Although this theory of autism has been debunked, the notion is still publicly thought to be a possibility for the rise of autism. Similarly, researchers have been paid large sums of money by pharmaceutical companies, causing the release of some inaccurate information, as well as a distrust from others for very helpful medications. While it is ideal for mental health practitioners to be familiar with best available treatment strategies to help guide patient care, it is *necessary* for the mental health practitioner to do no harm to their clients while rendering their services.

by 1970 over 40,000 people had received lobotomies in the United States (Tranøy & Bloomberg, 2005). The Portuguese neurologist Antonio Egas Moniz, who developed the initial techniques for the procedures, even won a Nobel Prize for his innovation. It wasn't until 1977 that the United States determined that the procedure was not acceptable. However, to this day brain surgeries are used to control certain behaviors, such as seizures, with more controlled psychosurgery techniques.

Around the same time that lobotomy was being introduced in the 1930s, electroconvulsive shock treatment (ECT) was also being developed. Shock treatments were intended to improve upon various psychiatric treatments. Unfortunately, ECT often resulted in memory problems and confusion. Further, media portrayal resulted in a negative perception of ECT prior to the use of anesthesia to decrease patient fears and pain. As a result, although ECT is widely used to this day and is an effective FDA-approved technique, its use carries a continuing stigma. ECT has been used less following the introduction of antidepressant medication in the 1970s.

## Medication

The American Psychiatric Association was organized in 1844 in an attempt to advance the care of psychiatric patients (Hall, 1944). It was during and immediately after this time frame that medications initially developed for other purposes began to be used for mental health management. For example, medications aimed at improving seizures were found to have positive effects on sleep disorders as early as the late 1800s. Amphetamines were developed in 1935 to treat depression, and lithium was introduced in 1949 for mania. In fact, amphetamines and lithium were unusual, as they treated conditions beyond sleep, whereas most psychiatric treatment medications had been sedatives up to that point. The first antipsychotic medication, thorazine, was developed in the 1950s. Although thorazine was initially thought to be a chemically induced lobotomy, it was soon discovered that the medication was in fact working on dopamine receptors in the brain and was particularly effective for patients with psychosis. The first antidepressant medication, Tofranil, was created in 1956. In the 1980s and 1990s, psychotropic medications were substantially advanced in an attempt to target specific neurotransmitters that were thought to be dysregulated within certain illnesses. For example, Prozac was FDA approved in 1987 as a selective serotonin reuptake inhibitor (SSRI) prescribed to treat depression. Psychotropic development has flourished since that time, with several different types of medications being developed, including second-generation antipsychotic medications, extended-release medications, different forms of antidepressants, and various treatments for anxiety, to name a few. Today, medications are often used to manage specific symptoms and alleviate the degree of severity or eliminate symptoms from various disorders.

## Psychotherapy

Dr. Sigmund Freud, the father of psychoanalysis, was trained as a medical doctor in the 1880s and began his work thereafter. He believed that his methods of understanding the human psyche were associated with underlying neurobiological roots. Following Freud's development of various theories regarding the human mind, many other scientists postulated their own explanations of brain-based functioning and how talk therapy would assist in improving neurosis. Although his theories of the mind and the arrangement of the brain have advanced to more scientific explanations, Freud's idea that talk therapy could help an individual improve his or her emotional state and overcome neurosis has led to the creation of evidence-based treatments in modern psychology. One main area of advancement is cognitive behavioral therapy.

## Cognitive Behavioral Therapy

Psychotherapy has evolved into evidence-based treatment often rooted in cognitive behavioral therapy (CBT) methods to address a variety of disorders. Although part of the underlying CBT methodology is associated with the necessity for clinicians to be able to document improvement for insurance-based reasons, it has also often been found to be the best methodology for improving functioning in psychiatric patients. Research has demonstrated the effectiveness of CBT in a variety of patient populations, including those presenting with mood, anxiety, eating, and personality disorders, among others. Use of CBT techniques in conjunction with imaging techniques or other strategies such as biofeedback have demonstrated the effect of behavioral strategies on brain and other body regulations.

Over time, various forms of treatment have been created to assist in improving the mental health of individuals. Throughout this book, we will discuss more progressive forms of treatment that are proving to be invaluable in the care of patients. The use of shock treatment, medication, and psychotherapy have all been demonstrated to alter behavior due to brain-based changes. These hallmark advancements have led to other forms of treatment, including deep brain stimulation, transcranial magnetic stimulation, and cognitive rehabilitation techniques, which will all be explained in later chapters.

## NEUROSCIENTISTS

In addition to individual patient cases, there are countless neuroscientists who have played significant roles in our understanding of neurobiology. Contemporary scientists who have made important advancements include Camillo Golgie, Jean Marcot Charcot, David Wechsler, Antonio DeMasio, Alexander Luria, and Oliver Sacks, among others. These individuals have helped us to understand brain-based functioning and the neurological underpinnings of behavior and emotions, and to develop tools and strategies to assess cognitive and emotional functioning accurately. Their writings are popular with both scientists and the general public. Careful observation, study, and experimentation on the part of these and other scientists have led us to the point in history where we have a solid understanding of both the complexity and the homogeneity of the human brain. Although it is not important for the clinician to be familiar with each individual's contribution, having a general awareness that these neuroscientists are from the 20th and 21st centuries should help to develop an appreciation of why there remains so much that is not known about the brain. In fact, as recently as 2012, Oliver Sacks, a neurologist at NYU School of Medicine, published a book entitled *Hallucinations*, which continues to speculate on human brain functioning. This author continues to provide lectures regarding his experiences and observations, which have helped shape our understanding of the brain and mental illness.

The reader is encouraged to recognize, though, that there is still plenty that is not known regarding the brain, despite the great leap forward that has occurred over the past 50 years—a leap unrivaled in other sciences. Throughout the following text, several contradictory findings will be presented to demonstrate that there continues to be uncertainty in various areas. There is also the challenge of integrating the different areas of science that play a role in understanding the brain and its functioning. Neuroscientists specialize in particular fields, including genetics, neurochemistry, neurology, psychiatry, and neuropsychology, to name a few. However, it is necessary to have a holistic appreciation of the complexity of the brain to understand behavior, emotion, and cognition. The role of the mental health provider is to assist in developing strategies to improve functioning, and an awareness of these areas is vital for developing appropriate treatment strategies.

## CONCLUSIONS

The field of psychology is relatively young, and our attempt to understand our behavior as it relates to brain functioning is in its infancy. Several individual cases, along with certain neuroscientists, have helped us to advance to our current level of knowledge. It is certain that by the time you are reading this book, more neuroscientific information will have been published, further advancing our knowledge. The integration of this knowledge is vital for progressing our understanding of human behavior and disorders.

## SUMMARY POINTS

- The history of neuroscience dates back thousands of years, with evidence indicating that Greek philosophers debated the merits of the heart versus the brain in determining the importance of cognition.
- Clinical case examples are an important method that has been used to examine brain functioning. Case examples of Phineas Gage, Mr. Tan, and H.M., as well as modern examples of Temple Grandin and professional football players, have all advanced our understanding of the brain's functioning.
- Although treatment in mental health has not been perfected, an examination of the history of treatment clearly demonstrates advancements in psychotherapeutic contexts, as well as in medical interventions and medications.
- There have been several scientists who have been important in advancing the understanding of neurobiology. As scientists are able to integrate information from various sources, the complex biological aspects of psychology are becoming more clearly understood.
- Although knowledge of the history of neuroscience is not likely to influence the clinician on a daily basis, having an awareness and appreciation of the major advancements should result in a better understanding of how far neurobiology has advanced and what is currently available to patients as a treatment option.

CHAPTER 2

# Research and Clinical Methods

Although the previous chapter highlights humans' long-standing interest in understanding behavior, it has been only relatively recently that technological advances have allowed for understanding the brain beyond clinical observation or through unethical studies, some of which were described in Chapter 1. There has been a rapid expansion of scientific techniques that have allowed neuroscientists to better understand the role of neurobiology in explaining behavior. As a matter of fact, by the time that you are reading this chapter, it is likely that there has been newer technology developed to help us better understand brain functioning.

It is important to understand the techniques that are utilized in both research and clinical cases in order to assess the accuracy and clinical utility of research. Such understanding allows the clinician to be critical of the research while gaining knowledge of various disorders and diseases. Having an understanding of the techniques that are utilized can also assist the clinician in patient care, not only in terms of psychoeducating the patient but also for making recommendations or evaluations that may be beneficial, answering patients' questions about typical procedures, and helping to conceptualize diseases and disorders from a neurobiological perspective. Recent advances in technology have shed light on pathological processes that may be utilized to develop treatment strategies to improve behavior, emotions, and cognitions. This chapter aims to assist the clinician in understanding the various techniques, beyond clinical case examples, that have helped advance neuroscience research. Each technique, as well as its benefits and limitations, is explained. A discussion of how some of these original research methods are now used for clinical purposes is provided as well.

## MICROSCOPIC EVALUATIONS

Examining brain tissue under a microscope is one of the oldest assessment techniques in neuroscience. Examining pathological lesions dates back to the mid-1800s, around the time that Emil Kraeplin hired a number of neuroscientists to examine brain tissue in an attempt to understand the brain and behavior. The goal of these early microscopic studies was to understand the nervous system, while attempting to examine pathology associated with behavior through staining processes that allowed certain features of the nerve cell to be observed.

Neuroscientists still utilize microscopic techniques to examine postmortem brain pathology. Human brains that are donated can be utilized for clinical, teaching, and research purposes. It is through these studies that we have learned more about pathology in recent years, as well as the effects of environmental influences on behavior. Yet the techniques used to store, freeze, and dissect the brain vary according to the site in which the brains are donated. Vonsattel, Keller, and Amaya (2008) provide a detailed explanation of techniques utilized to process postmortem human brains for research purposes at the New York Brain Bank, including brain preservation techniques; nevertheless, they acknowledge that there is a lack of generally accepted methods across other sites. In general, the procedures are to process the brain as soon as possible after death. This process includes harvesting cerebral spinal fluid, submerging the brain tissue in a solution to maintain integrity of the existing brain, and freezing the brain for further evaluation. Microscopic evaluation is conducted on tissue beyond that point.

An obvious drawback of microscopic evaluation is that it can be completed only postmortem. Therefore, understanding the pathology of behavior will occur only after the behavior, instead of while the behavior is occurring. Postmortem study does not assist in treatment after the findings are obtained. As it is at a microscopic level and is only a sample of the brain tissue, it is also likely that pathology within the brain is not widespread, and portions of the brain may not be altered due to pathology. Therefore, negative findings do not rule out organic-related issues. Also, the challenges of proper storage and slicing of brains are difficult. Such systematic and highly detailed evaluations can typically occur only at larger medical centers.

## ABLATION

Another early method of scientific discovery in neuroscience is *ablation*, which is a surgical technique that involves removing or destroying a portion of neural tissue or brain region without killing the animal. This technique dates back to the early 1800s, and it was initially conducted in an attempt to assist in determining the location of various cognitive, motor, and behavioral abilities in animals. For example, destroying the cerebellum would result in loss of balance, whereas ablating the medulla would result in loss of respiration.

Ablation studies can still assist in determining general estimates of localization of brain-based functions, and they still occur in animal populations. It is a relatively inexpensive technique that can be conducted in a variety of scientific settings. However, it is well accepted that most, if not all, functions of human abilities involve a complex web of neurotransmitters and neuropathways. This phenomenon will be discussed in future chapters. Other concerns regarding ablation studies include use of animals in research (see the Ethics box), missing the target, and the fact that such research cannot be conducted in humans. Observations and assessment of patients with lesions provide the nearest example of a naturalistic ablation in humans, but, even so, the affected area needs to be identified by

## ETHICS

### Animal Research—Necessity in Neuroscience?

Science has a long-standing history of utilizing animals for research purposes. Known animal studies examining behavior have been on record at least since the time of the Roman Empire. Neuroscientific information, including emotional, behavioral, and biological information, has been gathered through animal studies. However, as technology has advanced and animal rights have expanded, it is reasonable to question the necessity of continued animal research. In fact, Greek and Greek (2000) argue that because animals and humans are different, research utilizing animal models is in fact dangerous to humans. It has also been argued that no species can be considered a biological model for another species, regardless of how closely they are related in terms of evolution, because responses to interventions or various stimuli are specific to individual species. Yet, history has time and again proven that animal modeling has been applicable to human behavior. Similarly, it has been found that animal brains overlap in terms of neuroanatomical structures and subsequent function. Therefore, animal research is relatively commonplace in the neurosciences.

The American Psychological Association Ethics Code, Section 8.09, discusses the humane care and use of animals in research (American Psychological Association, 2010). This section of the ethics code emphasizes that psychologists should follow laws in order to care for and dispose of an animal in an appropriate manner. It is suggested that animal research involving pain, stress, or death should occur only when no alternative is possible and when there is clear scientific value. This is relevant particularly as it relates to psychologists in neuroscience research. Efforts to understand brain pathology following injury, or alterations in neurochemicals following medication, are important. It also recommends that researchers attempt to minimize animal discomfort when conducting such research.

Although advances in imaging techniques, some of which have been described in this chapter, have been extremely useful in understanding brain-based functioning, studies utilizing animals are numerous, and they cannot merely be replaced with advanced technology. Yet psychology trainees and students may have strong reservations regarding animal research, even after understanding the APA Ethics Code on this topic. Regardless of personal beliefs, it is helpful to be familiar with the ethical guidelines protecting animals in these circumstances, if only to have a clearer understanding of how attempts are made to alleviate negative consequences of study procedures.

more advanced scientific techniques, such as magnetic resonance imaging. It is also possible that the natural lesion results in neurodegenerative difficulty in other areas of the brain.

## ELECTROPHYSIOLOGY RECORDINGS

### Single-Cell Technique

A single-cell electrode technique is used to describe when an electrode is inserted inside a brain, adjacent to the cell, in order to record electrical activity. The goal is to learn about the functioning of one neuron at a time. This process helps us to differentiate between the neurons and is relatively inexpensive. However, the drawback is that the procedure allows for examining *only* one neuron at a time. This is very time consuming and limits the understanding to one single neuron, which would not be expected to control behavior by itself. Further, it is conducted only in animals. Given that the human brain has up to 100 billion neurons, even if this technique were implemented in humans, examining one neuron at a time would be futile. Finally, such techniques as this one are invasive and not always successful.

## Electroencephalogram

An electroencephalogram (EEG) records electrical activity of the brain in a noninvasive manner by placing electrodes—small flat metal discs—on the scalp in order to record extracellular current flow of neurons. Essentially, this painless procedure monitors the electrical activity of the brain. Figure 2.1 is an illustration of the EEG procedure and Figure 2.2 shows the resultant clinical data.

**FIGURE 2.1 EEG electrode placements.**
*Source:* © Oleg Senkov/Shutterstock

EEGs can be used to examine whether a patient has epilepsy, sleep disorders, or delirium. The information obtained from an EEG includes whether the abnormal neuronal discharges that occur in epilepsy are present. Accompanied by video monitoring of the patient, EEGs in seizure patients can assist in determining psychogenetic seizures. A psychogenetic seizure, previously known as pseudoseizure, is behavior somewhat consistent with an epileptic seizure, but which is typically the result of emotional reaction to stress. Alterations in brain waves in a pseudoseizure are different from an epileptic seizure, and they can be documented on an EEG. Further details will be described in a future chapter. EEGs

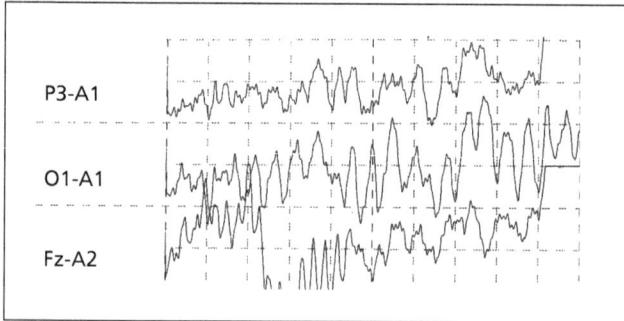

**FIGURE 2.2 EEG output.**
*Source:* © xpixel/Shutterstock

also examine whether there is electrical slowing of the brain, which is typical in patients with delirium, a confused state caused by organic factors. Although EEGs are not typically utilized to diagnose mass lesions, asymmetrical activity observed on an EEG can suggest such possibilities and can result in recommendation for further imaging. More commonly, EEGs are used in sleep clinics to examine brain wave length abnormalities during different stages of sleep. In this capacity, EEGs can determine whether sleep disorders such as sleep apnea, parasomnias, or narcolepsy are present. An EEG does not provide clinical information, such as structural integrity or blood flow during cognitive activity.

## Event-Related Potentials

An event-related potential (ERP) records electrical activity by using an EEG in conjunction with auditory, visual, somatosensory, or cognitive stimuli. This noninvasive technique is similar to an EEG, but it attempts to look at activity during specific cognitive processing. For example, an individual is connected to an EEG and is subsequently presented with a repetitive sound in order to determine how the brain responds to the sound. Measuring brain wave reaction to repetitive stimuli can assist in understanding how different information is processed. Results do not localize where the information is generated, but rather where the electrical activity is the strongest during a cognitive process. Both EEGs and ERPs are relatively inexpensive and are conducted by trained technicians, with the results analyzed by neurologists.

## NEUROIMAGING TECHNIQUES

## Structural Neuroimaging

### Computed Tomographic Imaging

Computed tomography (CT) scanning is a static brain-imaging technique that measures brain tissue density. CT scans are essentially X-rays of the brain, and they work like other X-ray machines. The scan requires a small, clinically insignificant amount of radiation to pass through the brain. Specifically, the CT is acquired when a beam of photons passes through the head, and portions of that beam are absorbed by brain tissue. Multiple detectors encompass the head of the patient, and those located opposite the beam source measure X-ray photons. As a result, tissue density is measured by the photons. Figure 2.3 is an example of a CT scan.

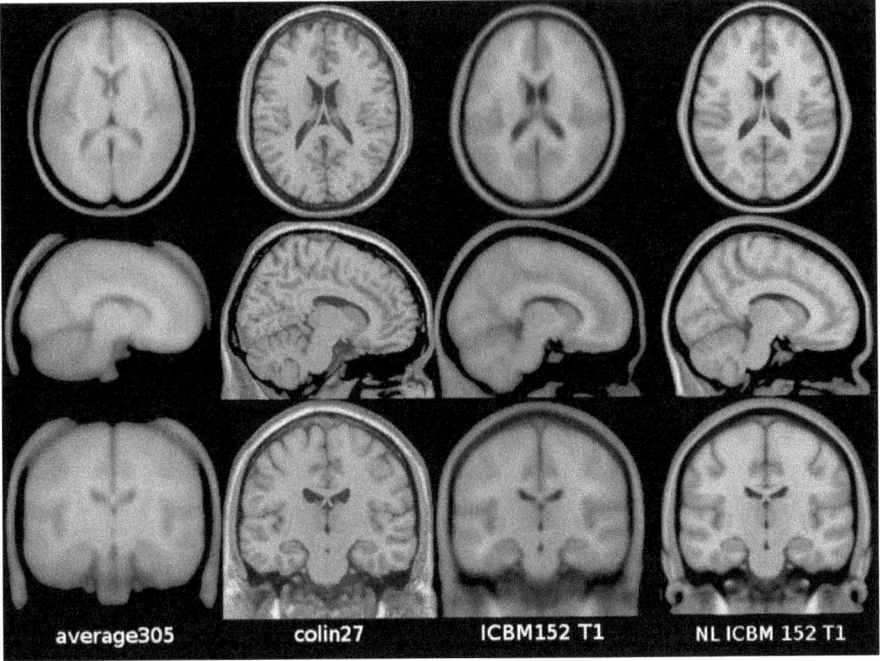

**FIGURE 2.3** CT scan.

The patient is asked to lie on a table while the CT scans are obtained. CT scanners are efficient, and several slices can be acquired rapidly. It is a relatively inexpensive procedure, and it is available in most community hospitals. The CT scan can be completed rapidly and, as a result, it is more comfortable than other imaging techniques for patients. It is effectively used to determine whether significant brain conditions, such as brain tumors or skull fractures, are present. Limitations of CT scanning include exposure to radiation, which restricts the frequency in which CT scans can be conducted. This problem can limit the ability of the physician to observe clinical changes over time. Further, though the acuity of CT scanning has improved with time, the scans are not very detailed in terms of brain structures. Specifically, deep-brain structures such as those found in the limbic system are not well represented on CT scans. CT scans are also limited in their ability to examine differences between gray and white matter of the brain.

*Magnetic Resonance Imaging*

Magnetic resonance imaging (MRI) utilizes nuclear magnetic resonance in order to obtain highly detailed images of the brain. MRI does not involve radiation, and it is therefore safe to repeat over the course of time. Although the physics of how an MRI machine works are complicated, a basic understanding of the mechanism by which the technique works is important in order to dispel patient concerns and to critically evaluate research.

Essentially, the human brain has atoms with hydrogen nuclei that are constantly spinning. A magnet aligns these spinning nuclei. When the magnet pulses, the protons return to their original positions and release energy that is detectable through radiofrequency. The response of hydrogen atoms to the magnet results in images that distinguish among different kinds of body tissues. Several images throughout the brain are obtained and these are called *slices*. Using slices, a computer creates three-dimensional images to represent a fairly high-resolution image of the brain. Figure 2.4 shows an MRI image that is created by these physics, and Figure 2.5 shows an example of an MRI machine.

The quality of the image is determined by the strength of the magnet. Therefore, it is not uncommon to find research magnets to be higher in strength and result in better resolution when compared to the results using clinical magnets. The reason behind this is that clinical magnets are used to visualize lesions, tumors, ischemic

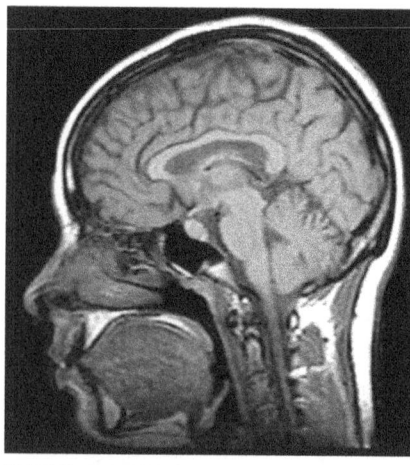

**FIGURE 2.4  MRI image.**

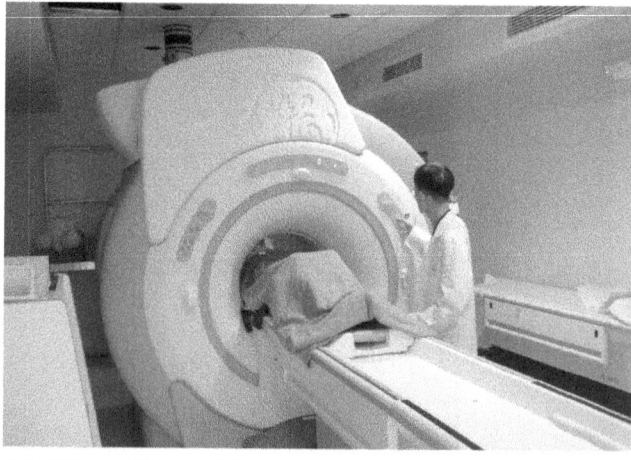

**FIGURE 2.5  MRI machine.**
*Source:* National Institute of Mental Health, National Institutes of Health, Department of Health and Human Services

alterations, and other structural abnormalities. Stronger research magnets are necessary to be able to visualize boundary differences among various neuropathways and anatomical structures in an attempt to differentiate the pathology of various neuropsychiatric disorders. Both types of magnets allow for examination of deep-brain structures, gray versus white matter differences, and volumetric analysis of various brain regions. Also, given the three-dimensional ability of MRI, images can be examined from an axial (from the top of the head and downward), coronal (from the face to the back of the head), or sagittal (from the left to the right side of the head) view. MRIs can be conducted on repeat occasions without concern for the exposure causing any known health risks.

Though MRI resolution is of high quality and can result in a remarkable image of the brain, there are some challenges involved with MRI that a clinician should be aware of and should inform patients about. For example, individuals with any sort of ferrous (iron-containing) metal or pacemaker in their body are not able to submit to MRI scanning (see the Special Topics box). Also, the narrow cylinder in which an individual is placed during an MRI procedure is relatively tight, and spending time in it can result in feelings of anxiety or claustrophobia. Although open MRI machines are available, they are typically not as strong, and their use results in a weaker resolution image. Also, the time to acquire an MRI scan of the brain can last between 20 and 45 minutes, requiring a person to remain still while the images are being obtained. In comparison, CT scans of the brain typically require 5 to 10 minutes. The cost of an MRI procedure is approximately double the cost of a CT. Therefore, the clinical provider needs to examine the clinical benefits of obtaining a more costly, higher-resolution MRI scan as compared to the less costly, lower-resolution CT scan.

Numerous research studies have attempted to differentiate structural differences among various diseases and disorders. The results of several of these studies will be addressed throughout the remainder of this text. It is clear that MRI is particularly useful in diagnosing traditional neurological disorders, such as strokes and trauma. The images obtained through MRI provide clear distinction of any structural brain abnormalities. There is also possible utility for clinical implication within psychiatric populations. MRI can be valuable in determining whether a medical condition is causing a psychiatric problem (Erhart, Young, Marder, & Mintz, 2005). This is particularly useful for late-onset psychosis. Also, MRI is used to assist in determining different types of dementia, such as frontal dementias. Any significant behavioral or emotional changes that occur in adults who are over 50 years of age can have a clear organic etiology to the changes, and an MRI scan should be taken. However, at the present time it is important to note that, though there are numerous MRI studies examining various psychiatric disorders, MRI is not used as a clinical technique to diagnose or evaluate effectiveness of treatment for psychiatric disorders.

## *Magnetic Resonance Spectroscopy*

Magnetic resonance spectroscopy (MRS) is based on the same physics and principles as MRI, but it detects specific neurochemicals in certain regions of the brain. For example, MRS focuses on lithium, carbon, phosphorus, and protium, rather than on hydrogen protons. As a result, this technique allows for measurement of certain chemicals including creatinine, choline, and *N*-acetyl. MRS provides information regarding neuronal damage through examination of cellular integrity and function. Although this technique has not been utilized in a clinical capacity, it has assisted researchers in understanding the neurochemicals involved in various psychiatric disorders. It has also assisted in better understanding neurobiological response to psychotropic medications.

## SPECIAL TOPICS

### Safety and the MRI Scanner

Approximately 20 million MRI scans of the brain are conducted annually worldwide, with about half of them occurring in the United States. MRI has proven to be an effective tool in diagnosing medical conditions such as stroke, traumatic brain injury (TBI), and some forms of dementia. Research utilizing MRI has helped provide a better understanding of the neuropathology of both psychiatric disorders and normal aging. MRI has numerous advantages over CT scanning, one of which is the lack of exposure to X-rays. Thus, MRI is thought to be a safe procedure, with replication being without risk for long-term health effects. However, a safe procedure does not mean that safety procedures are not necessary.

As described above, an MRI is essentially a giant magnet. It is important that the individual who is undergoing an MRI scan does not have any metal objects inside of his or her body. For example, an individual with a pacemaker or shunt could experience excessive heat and suffer burn-related accidents. Any ferrous metal objects near the magnet can also become hazardous because they are attracted to the machine. Therefore, any ferrous metal objects, including pens, piercings, metal cleaning equipment, chairs, gurneys, and other everyday objects are dangerous. In 2001, a fatal accident occurred when an oxygen canister struck a child in the head while the boy was in the machine. Similarly, a service technician was killed in 2010 when he was working on a portion of the machine.

In 2008, the Food and Drug Administration received nearly 400 reports of MRI-related accidents. This was a 270% increase in incidents over the previous 4 years. In that same year, the Pennsylvania Patient Safety Authority identified 148 reported cases of inadequate screening for ferrous metal in that state alone. In 2010, the ECRI Institute, a nonprofit organization whose mission is to research the best approaches to safety, quality, and cost-effectiveness, identified projectiles in an MRI as among the top unintended medical hazards. Assuming that there is no injury involved, the cost of repairing an MRI machine that has experienced a projectile accident has been estimated at an average of $43,172.

Effective screening of patients, visitors, and staff is important prior to allowing an individual near an MRI scanner, especially given that an MRI scanner is never turned off. The American College of Radiology has recommended and the Joint Commission has advised that health care institutions create safety zones, restrict access, and use trained screeners to double-check for metal objects. It might be useful for a clinician to review a safety checklist prior to suggesting a possible referral for such a scan. For a care provider, assisting in the education of this safety issue in patients is critical when scanning might be warranted.

### Diffusion Tensor Imaging

Diffusion tensor imaging (DTI; also known as *diffusion MRI*) utilizes MRI as well, but it maps the white matter fiber tracts of the brain by measuring the diffusion of water in brain tissues. The brain is made of approximately 80% water molecules. DTI measures the speed of the water in the white matter of the brain in order to identify existing white matter tracts. Although the speed of water is consistent in gray matter, it is not consistent in white matter. Rather, the motion of the water molecules is faster in the axons that run parallel with other axons due to

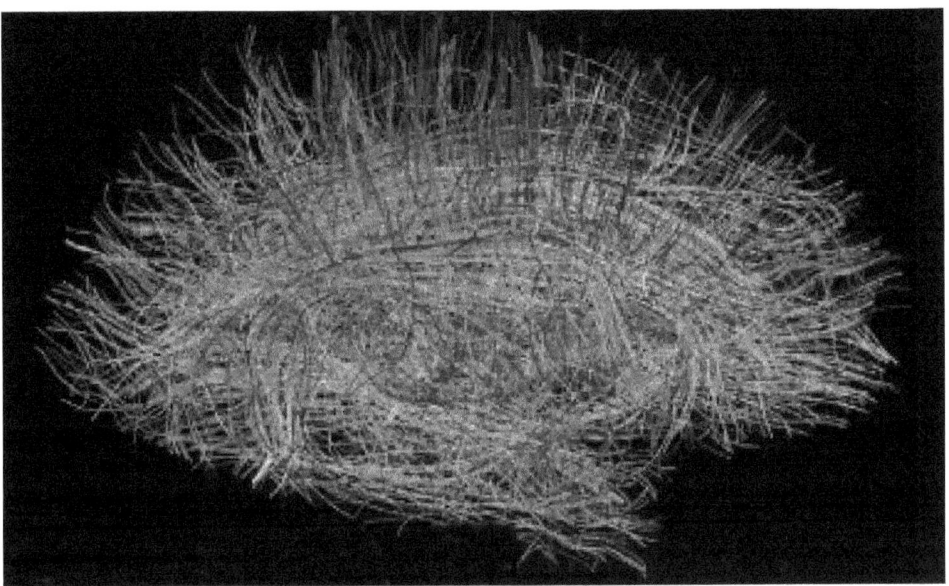

**FIGURE 2.6 Diffuse Tensor Imaging (DTI).**
*Source:* National Institute of Mental Health, National Institutes of Health, Department of Health and Human Services

the myelin sheath. Alterations in the diffusion of water help to identify damage of various forms in the brain. An example of an image produced by DTI is located in Figure 2.6.

DTI is a relatively new technology that has been shown to be effective in identifying medical injuries such as stroke, brain injuries, and multiple sclerosis. DTI is able to distinguish various types of brain tumors and helps to map out areas of infiltration in order to assist in preoperative planning (Lu et al., 2004). In multiple sclerosis, DTI is more sensitive to white matter abnormality, which may be more effective in documenting changes over time (Commowick, Fillard, Clatz, & Warfield, 2008). DTI has demonstrated white matter changes in chronic ketamine use in bilateral frontal and temporal areas of the brain, with more abnormalities associated with total amount used (Liao et al., 2010).

DTI has been researched in conjunction with psychiatric conditions as well. A literature review in 2008 indicated that over 50 studies utilizing DTI in schizophrenia demonstrated positive findings in the corpus callosum, cingulate bundle, and frontal and temporal regions (White, Nelson, & Lim, 2008).

DTI allows for improved objectivity and sensitivity in the detection of subtle developmental changes. Although DTI remains an investigational tool presently, this technology is likely to increase in utilization and scope in the near future, perhaps ultimately proving to be more useful than the relatively subjective evaluation of conventional MRI (Bammer et al., 2000).

## Functional Neuroimaging

### Single-Photon Emission Computed Tomography

Single-photon emission computed tomography (SPECT) utilizes CT technology to examine brain activity by providing scanned images of cerebral blood flow through utilization of a special injectable drug that has a radioactive ingredient

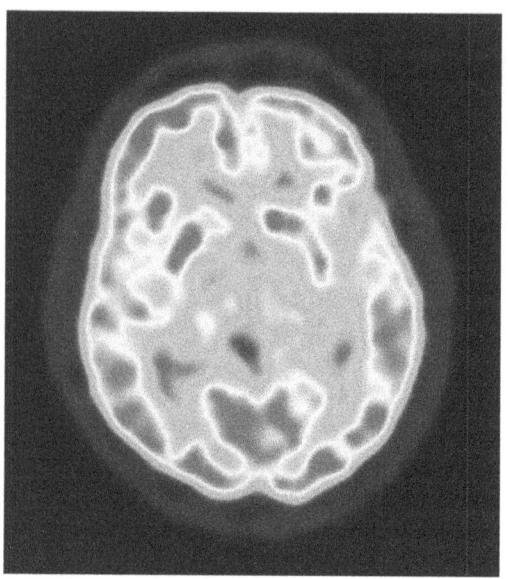

**FIGURE 2.7** PET scan.

(radioactive tracer), which passes into the neurons of the brain. After reaching a neuron, the drug releases high-energy photo emissions that are measured by a SPECT detector. From this information, a computer creates visual images that allow for visualization of cerebral blood. Brain structures in the subcortical areas appear bright on SPECT, whereas white matter appears dark. Although it has limited utility, SPECT shows how blood flows through veins and arteries.

*Positron Emission Tomography*

Positron emission tomography (PET) examines the glucose metabolism in the brain by using a radioactive tracer that enters neurons and emits positrons, colliding with electrons within tissue. These collisions result in high-energy protons that can be detected by a PET scan. PET scans utilize CT technology to examine regional blood flow, blood volume, and oxygen usage. The patient is asked to sit still for a time to allow the tracer to be absorbed into the blood stream.

PET can be used to assess brain activity during cognitive tests or at rest. The cost of PET is relatively inexpensive, and the resolution of the images is more defined than with SPECT. However, the resolution of the images is not as defined as with fMRI (discussed in the next paragraph). PET also provides limited specificity of brain structures and, therefore, determining the exact location of activity is more challenging. PET is used for both clinical and research purposes because of its relatively inexpensive nature, and it has assisted investigators in understanding various aspects of behavioral disorders. Figure 2.7 is an example of a PET scan.

*Functional Magnetic Resonance Imaging*

The fundamental physics associated with MRI apply to functional magnetic resonance imaging (fMRI) as well. However, the goal of fMRI is to measure the changes of oxygen level in brain tissue to map the neuroanatomical activation patterns associated with different cognitive tasks, as seen in Figure 2.8. There are several different techniques that are utilized to measure brain activation with fMRI. Essentially, each approach examines the neuronal activity in terms of cerebral blood flow in response to an activity, either cognitive or sensorimotor, that the patient is asked to do while in an MRI scanner. For example, brain activity is measured with oxygen saturation in the blood and is altered by the cognitive or sensorimotor activity. At the same time, MRI scans provide structural images of the brain, which allows for mapping of where the blood flow changes are occurring in the brain.

The benefits and drawbacks of utilizing fMRI are similar to those of MRI. fMRI provides for a three-dimensional, high-resolution view of the brain in a relatively safe environment. No exposure to potentially harmful radiation or injections occurs during an fMRI procedure. However, the patient is asked to perform a task in a relatively small space with limited motion. This provides challenges in creating fMRI-friendly cognitive and motor tasks. This also leads to decreased standardization of possible tasks across various institutions. In addition, it has been argued that random false positive results can occur using fMRI. An extreme

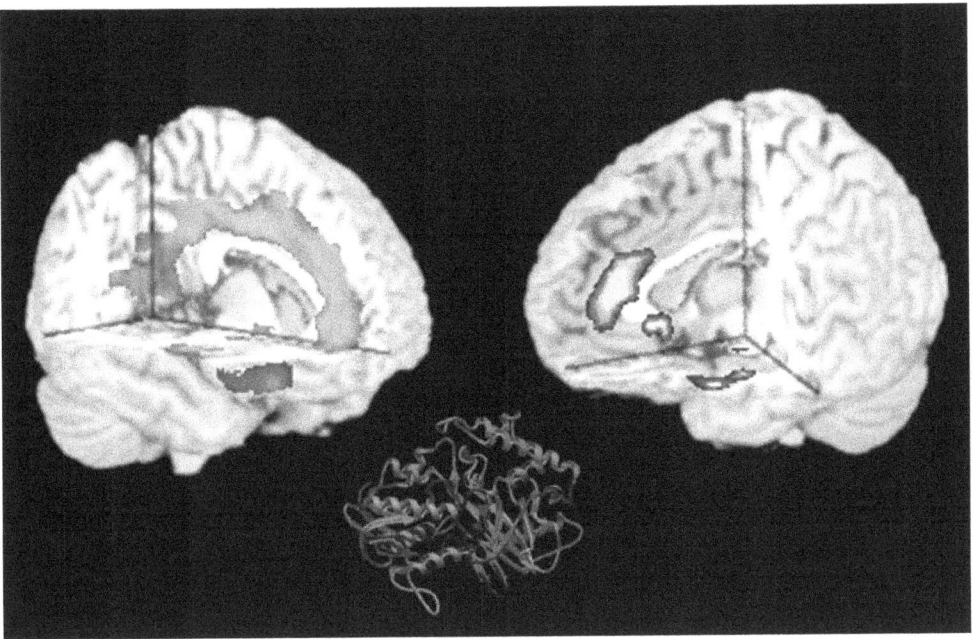

**FIGURE 2.8 fMRI.**
*Source:* National Institute of Mental Health, National Institutes of Health, Department of Health and Human Services

example of this occurred in a study that resulted in an fMRI scan indicating brain activation of a deceased fish (Bennett, Baird, Miller, & Wolford, 2010), which will be discussed in the next chapter. This is also part of the reason why fMRI is still used only for research purposes in an attempt to better understand the neurocircuitry implicated in various psychiatric and behavioral disorders.

Another limiting factor is the inadequate availability of fMRI. Although the availability of MRI is still somewhat limited, it has substantially increased over the past decade. In comparison, fMRI technology and the personnel to utilize and analyze fMRI information are scarcer and unavailable at many sites that have MRI technology. This is due partly to the increased cost of implementing fMRI as well. Regardless, the potential for fMRI to play a role in understanding neurobehavioral disorders is strong.

## *Magnetoencephalography*

Magnetoencephalography (MEG) utilizes EEG technology to examine small intracellular electrical currents of neurons in the brain. These electrical currents generate magnetic currents over the surface of the brain. MEG is able to provide an image of these surface-level currents. MEG allows for the identification of the specific location of cell activity. By doing so, one can identify brain regions that emit unusual electric currents. As demonstrated in Figure 2.9, MEG allows for cognitive activity during the procedure. Using MEG, one can record electrical activity during various cognitive activities, including motor, language, memory, and executive functioning tasks. This process allows for identification of abnormal electrical brain activity during specific tasks in order to assist in clarifying diagnosis, including within seizure patients prior to brain surgery. MEG examines only electrical activity, without consideration of structural or chemical abnormalities. Regardless,

MEG can be useful for determining functional documentation prior to surgery for seizures, tracking recovery from treatment for seizures and traumatic brain injury, as well as understanding normal activity within a healthy brain.

## NEUROPSYCHOLOGICAL EVALUATIONS

Neuropsychological evaluations are conducted to determine objective cognitive and behavioral functioning in patients diagnosed with neurobehavioral difficulties. The goal of neuropsychological testing is to examine cognitive abilities in areas of attention, memory, language, visual–spatial and perceptual skills, motor skills, and executive functioning. These are examined through a variety of paper–pencil, computer-based, and oral questions and game-like activities. These assessments utilize normative data in order to compare performance to other people of the same age, race, gender, and level of education, and they can assist in determining whether there are any organic or psychological reasons for changes in behavior. There are several textbooks dedicated to neuropsychological assessment. See Holtz (2010) for a detailed explanation of applied neuropsychology purposes. A brief overview of the utility of neuropsychological testing is presented here to assist the clinician in understanding the purposes and utilization of neuropsychology, as well as helping him or her know when a referral to neuropsychology might be warranted.

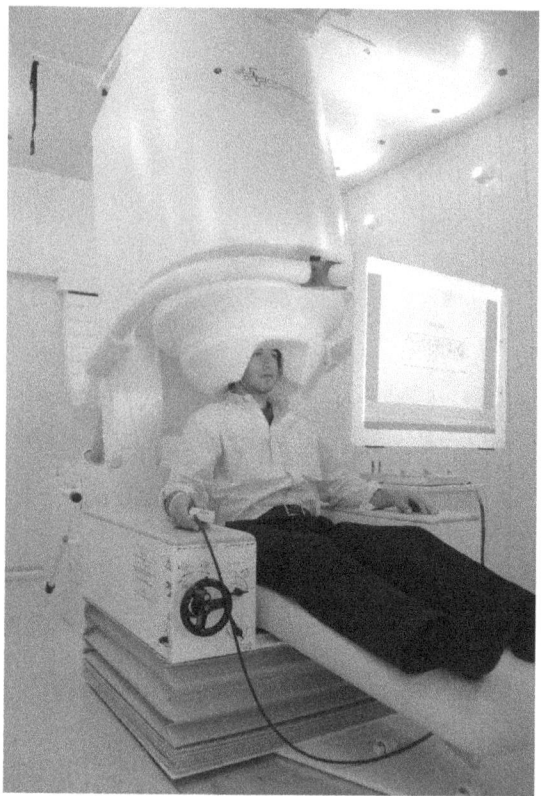

**FIGURE 2.9 MEG machine.**
*Source:* National Institute of Mental Health, National Institutes of Health, Department of Health and Human Services

The goal of a neuropsychological evaluation is often to document the current level of cognitive functioning, to clarify diagnosis, and to assist in treatment planning. By examining areas of cognitive strengths and weaknesses, a clinician is able to better understand how the patient is functioning. Whereas other types of brain-based examinations such as MRI, CT, and EEG provide information regarding structural integrity, blood flow, and electrical activity, neuropsychological evaluations comprise the only type of brain-based evaluation that examines functional abilities. These evaluations are conducted by neuropsychologists, clinical psychologists who have extensive training in brain–behavior relationships. Neuropsychological evaluations also take into consideration psychological factors that can play a role in cognitive performance and clinical presentation. For example, an individual might have nonepileptic seizures, and a neuropsychological evaluation would be able to examine not only whether cognitive functioning is intact, but also whether the person demonstrates personality characteristics in which he or she internalizes stress, which could result in the physical manifestation of stress in seizure-like behavior. This type of somatic-like behavior is considered psychogenetic seizures. The personality assessment component of the neuropsychological evaluation would be able to determine whether the individual is prone to such behavior.

Neuropsychological evaluations are conducted for both research and clinical purposes. Clinical evaluations can help determine whether an individual is capable of functioning independently in a variety of settings. For example, following a closed head injury, a neuropsychological evaluation can assist in determining whether the patient is capable of returning to his or her previous level of activity, including school, work, driving, living independently, and so on. Most times, imaging techniques are not sensitive to the neurobiological changes that occur in closed head injuries because they are occurring at a microscopic level. Similarly, neuropsychological evaluations assist in determining whether an individual meets the criteria for mental retardation, dementia, or mild cognitive impairment. None of these diagnoses can be accurately made without a formal neuropsychological evaluation. Neuropsychological evaluations are conducted for research purposes for a variety of reasons, including attempting to assist in efficacy studies and to determine whether different cognitive profiles exist for various disorders.

Neuropsychological testing is particularly useful when there is baseline testing, in which an individual had an evaluation prior to sustaining an injury. Baseline evaluations, though rare, are growing in popularity, particularly in populations where there is a higher risk of injury, such as with football players. However, most people do not have baseline evaluations. In the absence of a prior baseline evaluation, comparing performance on measures used to assess cognitive functioning to other people with similar characteristics (amount of education, gender, age, etc.) will occur. Although normative-based testing is typical, it can lead to misinterpretation of whether there is cognitive change if other characteristics are not taken into consideration. For example, occupation and performance in school are two features that normative data are not typically based on, though they play a very important role in performance. Further, motivation during the testing session is a central factor in performance. If an individual is not motivated to perform well or is fatigued from a poor night of sleep, test scores are not likely to be an accurate representation of typical cognitive functioning.

## BEHAVIOR GENETIC TESTING

Genetic testing is a relatively new science in the study of understanding behaviors. Although the field of genetics has been around for over 100 years, it has traditionally been used to examine physical hereditary traits, such as eye color and hair color. However, epidemiological research has consistently demonstrated increased incidence of psychiatric disorders in families. Therefore, hereditary studies showing the increased prevalence of such disorders are typically found in psychology textbooks.

With the completion of the Human Genome Project, an understanding has come to light of how behaviors and ultimately psychiatric conditions are linked to genetics. Although still new, behavior genetics rests on the theory that no single gene causes behavior. Yet information regarding chromosomes is also important; for example, an extra chromosome 21 is known to cause mental retardation. Behaviors are the result of genetic interaction, along with other biological processes and experiences. Understanding which genetic links are associated with or increase the risk for certain behaviors is particularly useful in the neurosciences, and this fact is often overlooked by clinicians. While they are not expected to be geneticists, mental health practitioners should certainly be familiar with the role of genetics in mental health disorders.

## CONCLUSIONS

Several techniques are useful in advancing neuroscience. For example, clinical case examples and observations of animals in controlled experiments continue to advance a better understanding of the complexity of brain–behavior relationships. Similarly, brain imaging technology and systematic assessments of humans are advancing the understanding of brain–behavior relationships as well. Recent advances in technology and science will significantly alter our knowledge of mental disorders and how to predict, prevent, and treat them (McFall, 2006). Clinicians need to maintain awareness and general understanding of these advances in order to better serve their patients as well as to assist in advancing the practice of psychotherapy.

## SUMMARY POINTS

- There are several ways to study the brain, including microscopic evaluation, electrical activity analysis, structural imaging studies, functional imaging studies, cognitive studies, and genetics.
- The clinician should be familiar with what is obtained in the various ways behavior is studied, as it can help with treatment planning, appropriate referrals, and critical evaluation of existing literature.
- Combining our knowledge in these separate areas of neuroscience provides a better understanding of brain-based behavior.

CHAPTER 3

# Nervous System and Brain Structure

The brain is a complex organ that sets humans apart from all other species. It has led people to great achievements, such as traveling in space and being able to create and use highly technical electronic devices such as smartphones and computers. The brain assists us in composing music, achieving athletic feats, and developing medications for deadly diseases, among other accomplishments. The brain even allows us to have insight into understanding our own brain function and dysfunctions that occur. In order to understand brain dysfunction or pathology, a description of normal brain processes, including brain structures and the functions of those structures, is necessary.

The brain controls all bodily activity, including heart rate, learning, motor functioning, and sexual behaviors. It plays a role in influencing the immune system, controlling our sleep, and developing our personality. Essentially, the brain is what makes us human. Despite the fact that the adult human brain weighs only approximately three pounds and it is clearly an essential aspect of being human, there is still much we do not know regarding the brain. Many mysteries and uncertainties remain due to the limitations in scientific tools to help us understand this intricate organ. It is almost guaranteed that from the time that this chapter is written to the time this book is published, more information regarding brain functioning will be discovered and/or disputed. This chapter will attempt to provide a straightforward overview of our current knowledge and understanding of normal brain functioning. Multiple books have been written on that topic: therefore, this chapter is not intended to provide an exhaustive understanding of normal brain functioning. Rather, it is focused on aspects of brain functioning that are necessary for the clinician to understand from an applied perspective. As one better understands normal brain functioning, the connection among behavioral, emotional, and cognitive abnormalities found in psychopathology and neurological disorders is elucidated. It could also be argued that, given the ongoing understanding of the role of the brain in behavior and emotion, a therapist must have a working knowledge of such matters in order to hold the competency to be a clinician (see the Ethics box). The aim of this chapter is to provide a basis for a working understanding of brain functioning. In

# ETHICS

## Competency and Neurobiology

The American Psychological Association and American Counseling Association provide clear ethical principles and codes of conduct describing multiple standards of competence. An overarching theme is that providers who are offering therapeutic services should have education about and understanding of the population that they will serve. Population factors to take into account include race, gender, ethnicity, age, sexual orientation, language, socioeconomic status, disability, and religion. Further, it is clearly indicated that therapists need to take reasonable steps to gain competence in emerging areas. Given the role of neurobiological factors in the presentation of human behavior and emotions, neurobiology is one area that providers should focus on.

Many graduate programs require a biological basis of human behavior as part of their core requirements. However, several Master's-level programs in psychology either do not offer the topic or offer it only as an elective. The lack of education in this area can negatively impact a client if the therapist is unfamiliar with how many medical disorders may present, co-occur, or begin by appearing as psychiatric disorders. Lacking knowledge in this area could not only prevent the therapist from making appropriate referrals in a timely manner, but it could also lead to significant harm to the patient. Further, it is not uncommon for students who complete formal coursework to become clinicians to never attend further training in the area. Given that the field of neuroscience is ever expanding, it is difficult enough for a specialist to maintain up-to-date knowledge; however, a general clinician who does not receive formal, continuing education in this area might also cause harm, or at least not provide the most effective treatment for his or her patients. According to Shiles (2009), competency is not a fixed construct. Rather, competency is a continuum in which there are many components. It is clearly indicated that the professional should be familiar with scientific knowledge for the population that he or she serves.

The reader is encouraged to consider understanding neurobiology at a fundamental level as a prerequisite to practicing psychology at any level. Lack of an understanding of neurobiology means the clinician should recognize significant limitations in his or her ability to provide thorough treatment, and, in most cases, it should result in referral for care to a specialist who can ensure that the psychological treatment being offered is appropriate. Neuropsychologists often play the role of consultants to determine appropriate diagnosis and treatment, but typically not until after treatment has resulted in limited or no positive outcome. Continuing education in neurobiology is strongly encouraged in order to assist the professional in ensuring that the patient receives the best care within an optimal timeframe. Given the clear link between brain and behavior, students learning the material will be able to enhance their own treatment skills and will ultimately provide better, more competent services.

turn, this is a relatively simplistic presentation of normal brain functioning, and the reader is referred to other texts discussing clinical neuroanatomy for more exhaustive knowledge.

## OVERVIEW OF THE NERVOUS SYSTEM

The nervous system is divided into the central nervous system and peripheral nervous system. The majority of this text and of neuroscience in general focuses on understanding the central nervous system and the complicated brain–behavior relationships that exist. The peripheral nervous system communicates with the central nervous system to allow for interaction with the environment. A brief overview of the peripheral nervous system is important to helping one recognize the complexity of being aware of our environment and subsequently reacting to that environment (see Figure 3.1).

## Peripheral and Central Nervous Systems

### Peripheral Nervous System

The peripheral nervous system (PNS) is located outside of the brain and spinal cord. Its major responsibilities are to send to, and receive information from, the central nervous system. The PNS consists of the somatic nervous system and the autonomic nervous system. The somatic nervous system is responsible for responding to environmental stimuli by connecting the voluntary skeletal muscles with cells that are responsive to sensations, such as touch, vision, and hearing. It comprises afferent nerve cells that connect the eyes, ears, skin, and skeletal muscles to the central nervous system, allowing sensory information to be transmitted to the brain. In turn, efferent nerve cells carry information from the central nervous system to the skeletal muscles, which allows for movement. Essentially, all somatosensory information runs through the somatic nervous system.

The autonomic nervous system consists of neurons connecting the central nervous system to internal organs (see Figure 3.2). The role of the autonomic nervous system is to regulate the response of the body. This regulation allows for involuntary control of the heart, endocrine system, exocrine glands, and smooth muscles. The autonomic nervous system is subdivided into the sympathetic and

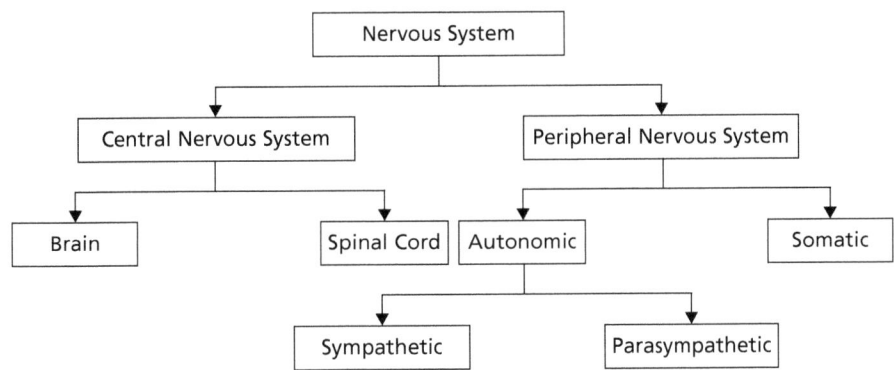

**FIGURE 3.1 The nervous system.**

parasympathetic nervous systems. The sympathetic nervous system regulates arousal by mobilizing energy during times of stress and high arousal. It releases adrenaline in response to fear, danger, and excitement, increasing activity and metabolic rate. It is the activation of the sympathetic nervous system that encourages humans to watch scary movies, read a thrilling book, and race into action when an emergency occurs. The parasympathetic nervous system, on the other hand, conserves energy during relaxed states. It lowers the metabolic rate and is important for sleep, digesting food, and restoring blood pressure and heart rate. The sympathetic and parasympathetic nervous systems are complimentary, and they need to work harmoniously for optimal functioning. The dysregulation of the autonomic nervous system occurs in anxiety disorders, posttraumatic stress disorder, and some other psychiatric disorders.

The PNS is vital for human functioning, and it is often dysregulated in psychiatric disorders. The interconnection between the PNS and CNS is important in symptom presentation, internal feelings, and physical response to the external world. The role of the PNS often gets overlooked by the clinician when he or she is working with patients, and treatment—in terms of medicine, therapy, and other newer treatment approaches—is typically focused on the central nervous system.

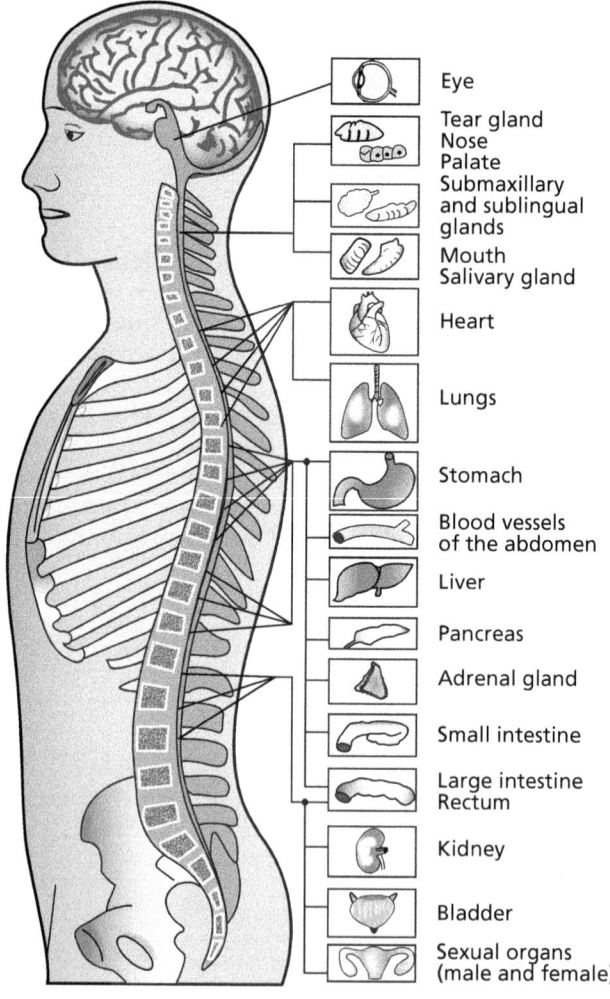

**FIGURE 3.2 Body functions and the autonomic nervous system.**

Despite this tendency, the clinician is encouraged to consider the role of the PNS when working with and educating patients, as well as when developing effective treatment strategies. Deep-breathing strategies and other techniques can improve upon PNS functioning.

## Central Nervous System (CNS)

*The Neuron* The human brain is composed of several billion neurons, with studies showing counts ranging up to 180 billion (Weiss, 2000). It has been estimated that, at birth, a baby might have as many as 100 billion neurons (Fair & Schlagger, 2008). It has also been estimated that brains in other species range from one billion to one trillion brain cells (Society for Neuroscience, 2002). The exact number of neurons is unclear and disputed. Newer research examining a small sample of deceased middle-aged men has found 14 billion neurons, which may be a more accurate depiction of the quantity of neurons found in the human brain (Herculano-Huzell, 2009). Although the amount of neurons found in the human brain remains a mystery, the structure and function of neurons are much better understood.

Generally speaking, a neuron is a specialized cell found within the brain that transmits electrical and chemical signals to other nerve cells, muscles, or gland cells. It is essentially the basic working unit within the brain. The brain is able to operate based upon the functional and structural properties of neurons. While each neuron has a unique role, the structural property of the neuron is essentially identical among them.

*Properties of the Neuron* The neuron consists of several structures that have unique functional properties, as demonstrated in Figure 3.3. As mentioned earlier, the scope of this chapter is not to provide an exhaustive description of the neuron at the microscopic level, but to provide a review of the essential structures. The cell body, also known as the soma, contains the nucleus, cytoplasm, and axon. The soma is the location where neurons assemble proteins and maintain metabolism. In the middle of the cell body resides the nucleus. The nucleus of the neuron contains deoxyribonucleic acid (DNA). DNA is the genetic code of human beings, and it is within the nucleus of the neuron that certain genetic characteristics reside. The cell is also made of cytoplasm, which is the clear, internal fluid of the cell that helps to give the cell shape. The cytoplasm holds all of the organelles beyond DNA, including the mitochondria and lysosome, in place. This allows the mitochondria, the site of energy production, to reside within the cell body. It is here that fats, sugars, and proteins react with oxygen to produce adenosine triphosphate (ATP). ATP is thought to be the fundamental source of energy for neurons and all other cells. Also found within the cell is lysosome, which breaks down bacteria and viruses into proteins that can be reused by the cell to build new structures and organelles. It is a cellular recycling system that contains digestive enzymes.

The axon of the neuron is an electrically excitable output fiber that gives rise to many smaller branches before ending at a nerve terminal. Neurons signal by transmitting electrical impulses along the axon. Axons can range in length from a small fraction of an inch to up to 3 or more feet. Upon reaching the end of an axon, voltage changes trigger the release of neurotransmitters, the brain's chemical messengers, at a synapse. The synapse is the communication point with other neurons. A synapse is also commonly referred to as a *synaptic cleft*, *synaptic gap*, and *synaptic junction*. Dendrites extend from the cell body and receive messages from other neurons. The dendrites and cell body are covered with synapses from the axons of other connecting neurons. The communication of two neurons occurs at the end of the axon, also known as the axon terminal.

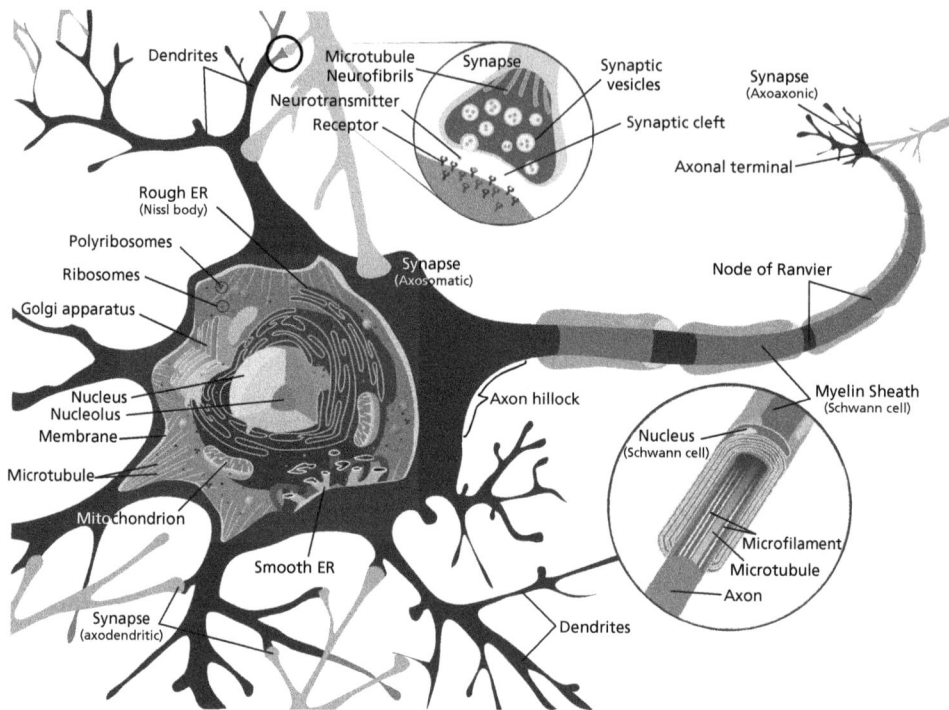

**FIGURE 3.3 The neuron.**

Also, many axons within the brain are covered with specialized cells called *oligodendrocytes*. These specialized cells coil around the axons, forming a layered, insulating myelin sheath, which in turn assists with the speed and transmission of electrical signals along the axon.

*The Firing of a Neuron* The ability of a neuron to activate depends on a small difference in electrical charge between the inside and outside of the cell. As nerve impulses begin, an action potential occurs at one point on the cells' membranes. An action potential is the dramatic change that takes place when the electrical signal passes along the membrane of the axon, which can occur at speeds up to several hundred miles per hour. Nerve impulses involve the opening and closing of ion channels, which are water-filled molecular tunnels that pass through the cell membrane and allow ions to enter or leave the cell. The flow of ions creates an electrical current that results in voltage changes across the membrane. In the resting state, the neuron is polarized, and the concentration of sodium ions and chloride ions are greater outside the neuron than inside the neuron. At the same time, there are more potassium ions on the inside than on the outside of the neuron. With activation, sodium ions are transported out of the neuron, and potassium ions are pumped into the neuron through a mechanism called a sodium–potassium pump (see Figure 3.4).

These voltage changes trigger the release of neurotransmitters, chemical messengers of the brain. Neurotransmitters are subsequently released at nerve ending terminals, spread across the intrasynaptic space, and bind to receptors on target neurons. The neurons at the adjoining cell are switched on and off by these binding receptors, which in turn selectively recognize specific chemical messengers. When a transmitter is in place, the neurons' outer membrane potential is altered and triggers a change, including increased activity of an enzyme. This mechanism is similar to how a key fits into a lock (Society for Neuroscience, 2012). At the time

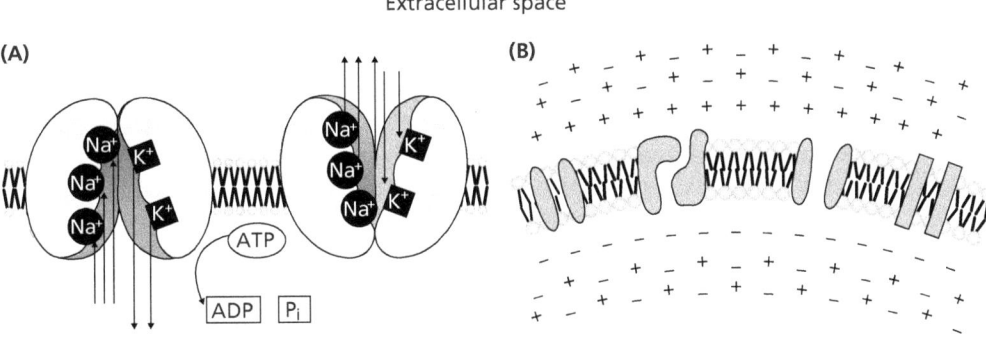

**FIGURE 3.4 Sodium–potassium pump.**
Adapted from Holtz (2011).

of the release of neurotransmitters, also known as *exocytosis*, the electrical process of the cell firing transforms into an electrochemical process. The dendrites at the adjacent cells receive the chemical information, passing it through the cell bodies and in turn firing an electrical signal through the axon to the synapse, releasing a neurotransmitter.

## NEUROTRANSMITTERS AND PSYCHIATRY

At the molecular level, the function of neurotransmitters is to either excite or inhibit the postsynaptic cell. Neurotransmitters are the brain's chemical messengers that allow for communication among neurons. They are packed and stored by Golgi bodies, which are flat membranes that otherwise sort and ship cellular substances. When the appropriate signal is received, they release their contents. Neurotransmitters are synthesized in the cell body and are transported to nerve terminals.

Neurotransmitters are either excitatory or inhibitory. Excitatory neurotransmitters increase the likelihood that the neuron will fire an action potential. Inhibitory neurotransmitters decrease the likelihood that a neuron will fire an action potential. It has been known for nearly 100 years that neurotransmitters play a role in the chemical transmission of nerve impulses. Currently, over 100 neurotransmitters have been identified (Hyman, 2005). Each neurotransmitter has a unique function. Many neurotransmitters combine, which results in such important human abilities as emotions and thoughts. Understanding this complex process is helpful when conceptualizing the reason for the effectiveness of psychotropic medications. It has been long established that neurotransmitters play a central part in emotional and behavioral functioning. Several receptors of these neurotransmitters play specific roles, but each neurotransmitter has general functional abilities. An awareness of the role of a handful of key neurotransmitters will help to develop a basis for understanding numerous psychiatric disorders.

### Acetylcholine

Nearly 100 years ago, acetylcholine (Ach) became the first identified neurotransmitter. As such, it is the most studied and perhaps best understood neurotransmitter. Most of the acquired information about Ach concerns its action outside of the brain. Specifically, Ach is released by neurons that are connected to the voluntary muscles, which causes them to contract. It is also released by neurons that control heartbeat. Ach is created at axon terminals after an action potential arrives, and the electrically charged calcium ions are introduced. Ach then releases into the synapse and attaches to Ach receptors. This process, in turn, opens sodium channels and subsequently causes muscles to contract. Afterward, Ach breaks down and is resynthesized in the nerve terminal. Antibodies that block the receptor for Ach cause muscle weakness and fatigue, and they can lead to myasthenia gravis, a chronic autoimmune neuromuscular disease. Ach links motor neurons to muscles, and it is found throughout the CNS and PNS.

It is well-established that Ach is also found in many brain regions; however its role in brain regions is less clear. It has been thought to play an important role in attention, memory, and sleep. As we will learn in a later chapter, Ach-releasing neurons have been found to be decreased in patients diagnosed with Alzheimer's disease. In fact, Ach will be discussed throughout the second section of the book, as the clinical implications of Ach dysregulation are clear.

## Serotonin

Serotonin is present in many tissues, and it was initially hypothesized to be involved in high blood pressure because of the role it has in inducing strong contractions of smooth muscles. It is also present in blood. Serotonin has, in fact, been found to play an important role in the transmission of other neurochemicals, including the catecholamines. Serotonin acts as the switch affecting various mood states and plays an important role in sleep.

## Catecholamines

Dopamine, epinephrine, and norepiniphrine are catecholamines that are found throughout the brain and PNS. The catecholamines are a group of neurotransmitters that arise in sequence from tyrosine (Society for Neuroscience, 2012). Tyrosine is an amino acid used by the body to synthesize proteins. In general, catecholamines help the body prepare for fight or flight reactions, as they help the body respond to stress. The adrenal glands, which are found on top of the kidneys, make a large amount of catecholamine as a reaction to stress.

### *Dopamine*

Dopamine is important in controlling movements, regulating hormonal response, and causing psychotic symptoms. Dopamine is inhibitory, and it is involved in control of voluntary movement. It has also been found to play an important role in the brain reward system. Dopamine is produced in the substantia nigra. Dopamine has been shown to be depleted or nearly depleted in individuals with muscle tremors and rigidity and difficulty moving due to Parkinson's disease. As a result, medication aimed to regulate dopamine through synthesis has improved upon those symptoms. Dopamine is thought to be over-released in schizophrenia, as it is known that medications that block dopamine receptors in the brain diminish psychotic symptoms. Unfortunately, Parkinsonism often occurs as a consequence. The endocrine system has also been found to be regulated by dopamine, as the hypothalamus is directed by dopamine to create hormones and release them into the bloodstream or to release hormones found in the pituitary gland.

*Norepinephrine*

Norepinephrine is synthesized from dopamine. Norepinephrine is released into the blood as a hormone and into the nervous system by neuroadrenergic neurons. It is secreted by the sympathetic nervous system in the periphery to regulate heart rate and blood pressure. Nerve fibers containing norepinephrine are found throughout the brain. Norepinephrine is known to increase arousal. In a portion of the brain stem called the *locus coeruleus*, activation increases production of norepinephrine, which subsequently projects to many neuropathways within the brain's limbic system and throughout the cortex. This activation affects alertness and arousal, and it influences the reward system as well. Norepinephrine is known to play a role in psychiatric disorders, including attention deficit disorder and depression.

*Epinephrine*

Epinephrine, also known as adrenaline, is a catecholamine that acts on nearly all body tissue. It is released into the bloodstream as a response to any type of stressful situation. Epinephrine rapidly prepares the body for action when necessary. It quickly increases the blood and oxygen supply to the brain and muscles, while diminishing other bodily processes, such as digestion and the immune system. For example, epinephrine dilates the pupils in the eyes, increases heart rate, and constricts the gastrointestinal tracts.

*Glutamate*

Glutamate, an excitatory neurotransmitter, regulates cortical and subcortical functioning. Glutamate is particularly important for cognitive functioning. Although stimulation of glutamate can be beneficial for cognition, overstimulation can result in cell damage and death that is found in brain injury and stroke.

*Gamma Amino Butric Acid*

Gamma amino butric acid (GABA) is an inhibitory neurotransmitter that decreases activity level and lowers arousal. There is a high concentration of GABA in the cortex. GABA regulates seizure activity, and is augmented by alcohol. Medications aimed to decrease anxiety, such as benzodiazepines, aim to increase GABA receptor activity.

## CENTRAL NERVOUS SYSTEM DEVELOPMENT

The development of the central nervous system begins shortly after conception. A process of cell division occurs after a sperm unites with the egg. After the first week, a blastocyst—a mass of cells with a fluid center—is formed. By the second week, the blastocyst, which will become the embryo, implants in the uterine wall. At this point, the growth of the cells becomes rapid, and shortly afterward the nervous system develops. At three weeks, the neural plate is formed and alters the cells so that they no longer have the ability to become any type of cell, but instead become specific to the nervous system. By the fourth week, the neural tube is developed, which subsequently becomes the central nervous system. The inside of the neural tube becomes the spinal cord and cerebral ventricles, and the outer layer becomes various portions of the brain. Within the first two months the embryo resembles a human being, and by three months the brain is clearly being developed.

During embryonic growth, different layers of cells occur: ectoderm, mesoderm, and endoderm. These layers subsequently develop into portions of the body. The ectoderm is the outermost layer of the embryo, and it eventually develops into the nervous system, sense organs, and skin. The mesoderm is the middle layer, and it develops into the excretory system, muscles, and blood. The endoderm is the innermost layer: it develops into the internal organs, including lungs and digestive system. Several processes occur along the way for this to happen, including induction, cell proliferation, migration, and aggregation as well as maturation. These terms are used to describe various stages of cell development, and the clinician should be aware that disruption of normal cell process through environmental factors, stress, or trauma can severely affect normal cell developmental during prenatal stages. Otherwise, these processes, if uninterrupted, lead to subsequent brain division and differentiation of cell responsibility for a healthy neurodevelopment progress.

## BRAIN DIVISION

Understanding the brain requires appreciation of the intricate neurodevelopment and subsequent growth. Many texts describe the differences among the forebrain, midbrain, and hindbrain along with differences found within the telencephalon, diencephalon, mesencephalon, metencephalon, and mylencephalon. This explanation and terminology often confuses or intimidates general clinical students, and it frequently complicates basic brain understanding beyond what is necessary for the general clinician. Further, though it can help one to understand neurodevelopmental division, from an applied perspective for the general clinician it is rarely used. Rather, understanding the various lobes of the brain, neuropathways connecting the lobes involved in various functions, and the main neuroanatomical structures found within the pathways is a useful way to understand pathology involved in various psychiatric disorders, and that is what is presented in this chapter. First, however, it is necessary to briefly describe other important aspects of the brain that we will encounter when learning about brain structure and division.

### Gray Versus White Matter

The brain includes gray and white matter, which is discernable in brain MRIs. Gray matter of the brain comprises neurons, cell bodies, dendrites, capillaries, and glial cells. Gray matter does not contain myelinated axons, and thus the gray color remains. The gray matter is necessary for cognitive abilities, including memory and language abilities, emotional responses, and sensory abilities. Gray matter has been associated with cognitive performance in areas of executive functioning, memory, and attention in various clinical populations.

In comparison, white matter is the color white because it is the area of the brain that contains many long, myelinated axons between the brain and cerebrum. The role of the white matter is to send signals to and from various areas of the cerebrum. There has been more recent interest in white matter, as it was once thought to be passive. White matter controls the distribution of information across the brain. It has been found to be effected in brain trauma, and the volume of white matter is correlated with learning, as new cells in the white matter are generated with learning. The diffusion imaging techniques discussed in Chapter 2 have shed light on the role of white matter in healthy brain development, as well as in injury and impairment.

Recognizing that gray matter and white matter serve different roles and comprise different portions of the cell will help us to understand behavioral symptoms involved in various disorders. Understanding the roles of white matter and gray matter will elucidate our understanding of psychiatric problems.

## Left and Right Hemispheres

Another important distinction within the brain is the role of the right and left hemispheres. They were once thought to play very specific roles, but recent neuroscience research has suggested that the interaction of these hemispheres is complex. Historically, however, the left hemisphere of the brain was thought to play more of a role in language abilities and logical and analytic abilities. The right hemisphere has been traditionally thought to be more creative, intuitive, emotional, and involved with musical abilities. Although it is clear that the 21st-century understanding of brain-based functioning is much more complex than that simple theory, the concept of lateralization still holds some relevance in psychiatry. Specifically, individuals with lateralized seizure activity display focal, or specific, limitations. Language abilities appear to be more prominent in the left hemisphere, whereas spatial abilities are more prevalent in the right hemisphere. It has also been shown that if an individual injures one side of the brain, there are more likely to be localized impairments. For example, the right hemisphere of the brain controls the left side of the body, and the left hemisphere controls the right side of the body. Though the neuropathways and structural aspects found within the different areas of the brain are more likely to be involved with cognitive abilities, lateralization is important to consider when conducting neuropsychiatric research.

## Cerebellum

The cerebellum is located behind the brain stem, and is sometimes referred to as *the little brain*. It is attached to the bottom of the brain, underneath the cerebrum. Small lobes comprise the cerebellum. The cerebellum is important for sensory and motor skills and also balance. It is particularly important for coordination and fine motor skills. Recent research has also demonstrated that it is important for memory, learning, and emotional functioning. Though not considered part of the other lobes, it receives information from many neuropathways, and it is therefore involved in many functional, emotional, and behavioral functions.

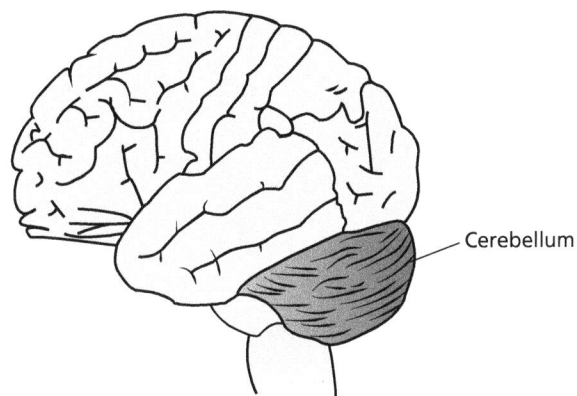

**FIGURE 3.5 Location of the cerebellum.**

## Corpus Callosum

The corpus callosum is a thick collection of white nerve fibers that connect the right and left hemispheres of the brain. This allows for communication between the two hemispheres, permitting increased processing speed. Agenesis, that is, partial or complete lack of development or disruption of the corpus callosum, can lead to a variety of behavioral and cognitive symptoms. Evaluation of split-brain patients, in which there is no corpus callosum connection, has supported lateralization concepts.

## Glial Cells

Glial cells play an important role in the nervous system. They are inherently different from neurons because they do not conduct nerve impulses. Rather, the role of glial cells is to assist the neurons by holding them in place, supplying nutrients and oxygen to them, assisting in repairing damage to the nervous system, and insulating the neurons from one another. Glial cells account for about half of the volume of the CNS, with neurons accounting for the other half.

## Meninges and Cerebral Spinal Fluid

The brain is contained inside of the skull, which is encapsulated by several layers of meninges for cushioning and protection. The dura mater, the toughest outermost layer of protection of the brain, is nearest the skull. Directly under the dura mater is the arachnoid layer. The layer closest to the brain is the pia mater.

In between the arachnoid layer and the pia mater lies the cerebral spinal fluid (CSF). The CSF is a clear fluid that fills the subarachnoid space, the canal of the spinal cord, and the cerebral ventricles. CSF is produced by the choroid plexuses, which are found inside the ventricles. There are four cerebral ventricles that are connected to one another, as well as the subarachnoid spaces (Holz, 2011). CSF serves a protective role by cushioning and supporting the brain. CSF flows through the ventricles and eventually drains into large veins found in the neck.

The brain is also protected by the blood–brain barrier. The blood–brain barrier is a collection of closely arranged cells that make up the walls of the blood vessels and that assist in keeping toxic chemicals from invading the brain. The blood–brain barrier allows nutrients, glucose, oxygen, and sex hormones to pass through, but it prevents other chemicals from breaking the barrier. It also communicates with cells and secretes material. Our understanding of the blood–brain barrier is still in its infancy with technological advancements, but it is known that traumatic brain injuries, neurologic disorders, and psychiatric disorders can result in a breakdown of the blood–brain barrier, contributing to functional problems.

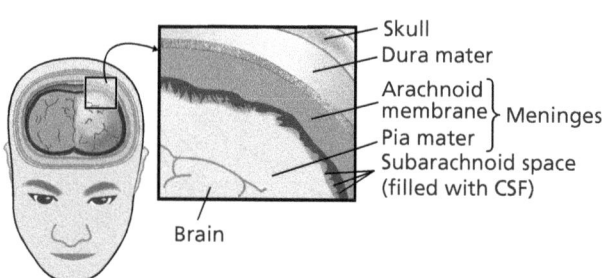

**FIGURE 3.6 Layers of meninges.**
Adapted from Holtz (2011).

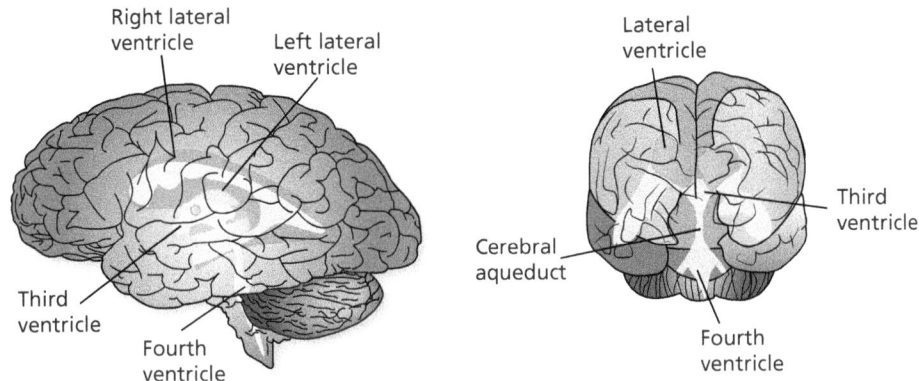

**FIGURE 3.7 Cerebral ventricles.**
Adapted from Holtz (2011).

## THE LOBES OF THE BRAIN

The brain, at a very basic level, can be divided into separate regions demarcated by specific gyri and sulci. These regions of the brain are responsible for specific functions and abilities. The four lobes of the brain are the occipital, parietal, temporal, and frontal lobes. In a healthy brain, there is constant communication among these lobes and neuropathways involving some or all of these lobes. These four lobes have symmetrical halves, but distinct functions within each hemisphere. Despite this technicality, the lobes will be viewed as integrated entities, and left and right hemispheres will be considered unitary. Upcoming chapters will differentiate the characteristics between the left and right hemispheres of these lobes. It is necessary for the clinician to have an understanding of the role of each lobe in order to better understand more complex pathways and neuroanatomy, discussed in the next chapter.

### The Occipital Lobe

Starting at the back of the brain, the occipital lobe is primarily responsible for vision. The occipital lobe is the most specialized lobe of the brain. The occipital lobe has pathways involved with visual processing. The optic chiasm is the point at which some axons of the optic nerve cross over to the opposite side of the brain, permitting each eye to view part of the left and right visual fields. As a result, injury to different portions of the optic chiasm can lead to visual field cuts in both eyes. This deficit can lead to difficulty in viewing the entire visual field from both eyes.

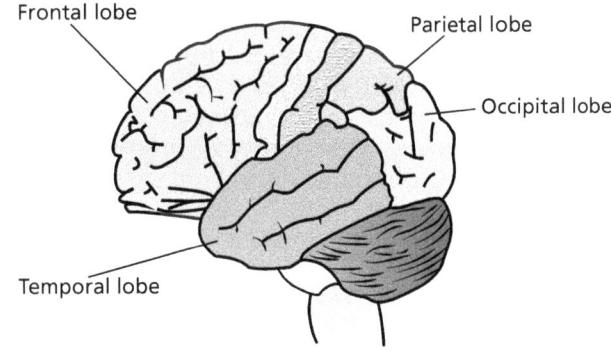

**FIGURE 3.8 Lobes of the brain.**

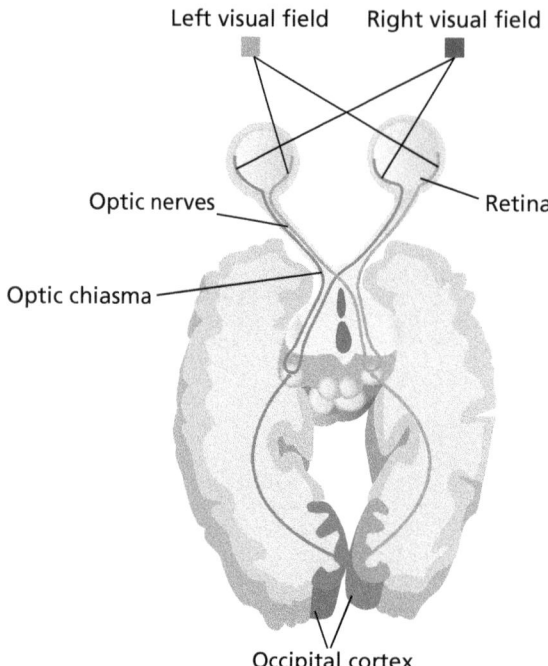

**FIGURE 3.9** Optic chiasm.

## The Parietal Lobe

Moving up from the occipital lobe to the top of the brain is the parietal lobe. The parietal lobe is responsible primarily for processing somatosensory information, such as speech, pain, and spatial orientation. The parietal lobe processes somatic sensations and integrates sensory information for movement control. It is therefore important for tasks such as note-taking, eating, and brushing our teeth. Injury or damage to the parietal lobe can result in agraphia, the loss of skilled movement not caused by general weakness, inability to move, poor muscle tone, or intellectual deterioration. Given the parietal lobe's role in integrating information from all the senses, damage to this area can yield a variety of complex behaviors and limitations.

## The Temporal Lobe

The temporal lobe lies on the sides of the cortex, where individuals consider their temples to be located. Subcortical temporal structures include the limbic cortex, amygdala, and hippocampal formation. Neuropathways that are connected throughout the rest of the brain include the temporal lobe, with its importance in hearing, processing auditory information, processing affective tone of nonverbal cues, memory, and recognition. The case of H.M. described in the last chapter occurred due to the removal of the hippocampus within the temporal lobe, resulting in significant memory difficulties. Lateralization is observed in the temporal lobe, as the left and right hemispheres play different roles and have different abilities, with the left hemisphere typically being more involved in verbal memory and skills, while the right hemisphere is more involved in nonverbal memory and skills.

## The Frontal Lobe

The frontal lobe of the human brain encompasses one-third of the cortex, and it is considered the part of the brain that makes humans unique from other species. Specifically, the frontal lobe is responsible for executive functioning, which includes higher-order cognitive processes such as problem-solving, planning, organization, and inhibition. The frontal lobe is located in the front part of the brain, and it can be subdivided into specific regions. The frontal lobe includes the premotor cortex, motor cortex, and prefrontal regions. Further demarcations of the various regions of the frontal lobe include the medial basal, dorsal lateral, and orbital frontal areas, which will be expanded upon in the next chapter. Each specific region is thought to be important for various behavioral, emotional, and cognitive controls. Damage to the frontal lobe can result in a variety of cognitive, emotional, and behavioral difficulties.

## SPECIAL TOPICS

### Errors in Analyzing MRI for Understanding the Brain

It is not uncommon for students who are unfamiliar with scientific data collection from complex methodology and high-level statistical analysis to accept the results of research without questioning the scientific methods. Though the neurosciences strive to provide objective data collection and analysis, human error and variability in data collection and analyses frequently occur. A common technique that is thought by the public to be foolproof and that is often considered extremely reliable by practitioners is MRI. However, MRI is not without its methodological concerns.

As you will discover throughout the course of this text and in research in general, there are often follow-up studies that do not confirm earlier results. Although confirmatory studies are difficult to publish in scientific literature, publishing them is necessary to ensure the validity of the initial findings. Scientific research relies on replication in order to confirm hypotheses. Yet different conclusions, often completely contradicting each other, are frequently discovered by various researchers examining the same scientific question. It is even common for the same researcher to find different results within his or her own laboratory, examining the same scientific question. The reasons for these inconsistent results are several. They are often hypothesized to be due to the patient population subsample. For example, as the scientific community understands the clinical presentation of various subtypes of schizophrenia, it has been suggested that different neurophysiological underpinnings are at least partially responsible for inconsistent results. Similarly, with the relative power of MRI scanners at different locations and even at different sites, there can be different strengths of magnets, which can alter the resolution of findings. It is reasonable to consider error in analyzing imaging as the explanation for various findings. In doing so, one must understand the methods used in obtaining the results.

In older, structural MRI studies, research assistants would often be responsible for measuring the different neuroanatomical structures of the brain obtained from MRI. Ideally, two or more people would become trained to determine different neuroanatomical boundaries in order to locate specific structures of the brain. This would allow for agreement as to where certain structures end and others begin. The different raters would establish high intrarater reliability when measuring these brain regions. Although these methods decrease the likelihood for inaccurate measurements, they are not foolproof. Semiautomatic measurements were created through computerized programs to identify the size of brain structures. This model is based on a model brain that the computer is able to utilize when comparing structures throughout the brain. Given that this standard is also human-generated and based on a model brain, error can still occur on the part of the computer when one is identifying the brain gyrus or other demarcations when measuring brain structure volume.

Functional MRI (fMRI) is also suspect. Beyond the already discussed concerns regarding the demarcation of various brain structures and perhaps region analysis that fMRI utilizes, fMRI also has a risk of false positives. A popular fMRI study conducted by Bennett, Baird, Miller, and Wolford (2010) analyzed results from a dead adult Atlantic salmon being presented with photographs of humans interacting. Figure 3.10 demonstrates that the results revealed voxel activation in an area of the salmon's brain, indicating that the fish appeared to be thinking about the pictures that were being presented. Although a dead fish clearly cannot think, the study aimed to remind scientists that

*(continued)*

*(continued)*

there can be random noise within sophisticated technology such as fMRI. Scientists are encouraged to correct for multiple comparisons as standard practice, but this advice is often ignored by some scientists.

Although understanding the physics behind complex technology is difficult for the typical practitioner, being aware of the possible error risks involved in the techniques and how the results are understood can help the practitioner assist his or her patients with understanding the overwhelming amount of neuroscientific information that is available. Be cognizant of how limitations in understanding the brain neuroanatomy, neurochemicals, and functional implications are valuable when describing brain functioning to clinical populations and their families.

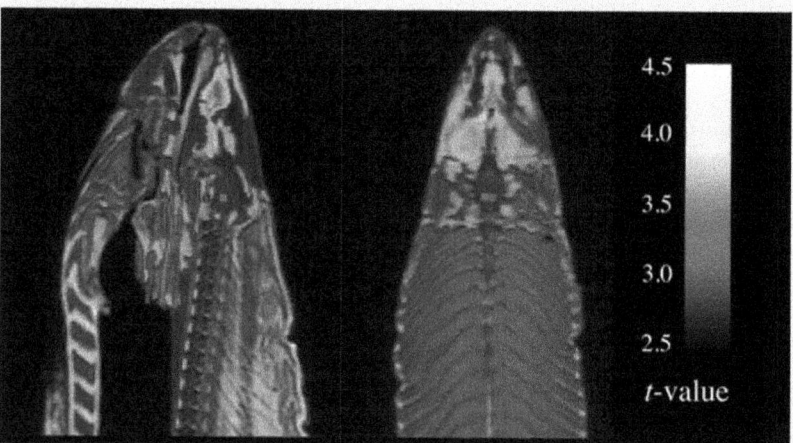

**FIGURE 3.10 Brain activity in dead salmon.**
*Source:* Bennett (2013).

Though defining the lobes of the brain is relatively easy, neuroimaging techniques used to examine these areas are not without their flaws. Clinicians are encouraged to critically analyze research, and be aware of the limitations that exist in our understanding of the brain (see the Special Topics box).

## CONCLUSIONS

The brain is a complex organism that can be understood in a variety of different ways. Healthy brain functioning requires a multitude of stable neurochemicals, structural anatomy, communication among different brain regions in different hemispheres, and an overall healthy nervous system. Disruption of the brain at a variety of different levels can cause mental health difficulties, which can disrupt behavior, cognition, and emotions.

## SUMMARY POINTS

- The nervous system consists of the peripheral and central nervous system.
- Within the central nervous system are neurons; a collection of similar neurons constitute neurotransmitters.
- Neurotransmitters that are released throughout the brain are vital for emotional and cognitive well-being.
- The brain is divided into four lobes, all of which interact and communicate with one another but provide unique abilities.

CHAPTER 4

# Major Neurobiological Brain Systems

The human brain is a very complex system of chemicals, electrical charges, and anatomical structures that interact with one another and with the environment to elicit behaviors, emotions, and cognitions. The previous chapters have examined our historical understanding of the brain–behavior connection, as well as how we are aware of brain-based function. In addition, the fundamentals of microanatomy were presented in the last chapter, along with basic brain knowledge and important neurochemicals as they relate to psychiatric disorders. Yet the brain is more than just a collection of chemicals, countless quantities of electrons, and conglomerate blocks of gray and white matter. The brain is capable of communicating incredibly fast and without conscious thought or control.

As previously described, various portions of the brain are responsible for different functions and abilities. The understanding of the brain has since been expanded, as it is more sophisticated than the notion that one part of the brain is important and solely responsible for a certain function. Rather, specific neuropathways that encompass various brain structures, lobes, regions, and highly concentrated neurochemicals interact with one another for even the most basic of human emotions and behaviors. In fact, though our knowledge has advanced substantially over the past two decades, it is likely that scientists will continue to discover further aspects of the brain that we cannot even fathom at this point. Many mental health clinicians become overwhelmed and intimidated by the vast intricacies of the brain. The clinician can overcome these difficulties by learning just a few specific brain systems (also referred to as neuropathways) that are relevant in the field of mental health, the major structures involved in those brain systems, and the functions in which these pathways are important.

This chapter aims to focus on the most relevant brain systems, or neuropathways, of the brain, as they relate to psychiatric disorders. Although there are several ways to understand the brain's microanatomy, the clinical- and counseling-level student and practitioner should have a clear understanding of those brain systems that play a central role in behavior, emotions, and cognitions. This chapter is not intended to provide an exhaustive review of all the known neurobiological systems found within the brain, although all systems play a role in psychiatric disorders in some capacity. For example, the visual pathways that extend from the visual cortex to the occipital lobe and back out, transcending throughout the rest of the brain, are clearly important for all aspects of behavior, and they are relevant to understanding how we perceive the world.

Similarly, the somatosensory and auditory pathways are well-defined and important for human behavior. Yet, from a mental health perspective, disruption of these pathways is not thought to play a central role in most mental health disorders. The clinician who is interested in understanding these systems further is encouraged to read the numerous biological or introductory psychology texts that are available, where these processes are typically described. Rather, this chapter will allow the general clinician to become familiar with common neuropathways that are disrupted in various psychiatric disorders. It is intentionally focused on only the most pertinent neurobiological systems in psychiatry, and these are described concisely. The aim of this chapter is not to overwhelm the general mental health trainee, but to direct him or her to the most applicable material for a selected field of study.

The reader might return to this chapter frequently throughout the remainder of this text as a reminder of the role that different brain structures and neuropathways play in different psychiatric disorders. The end of the chapter does briefly discuss some important neurobiological systems that are secondary in psychiatric disorders, but that are necessary for normal functioning, in order to provide the reader with a better understanding of some of the other neurobiological systems. It is important to note that, although somewhat arbitrarily labeled as *secondary* in this textbook, these systems, as well as the ones described here, are far from secondary in human functioning. A lesion or disruption of functioning in any of these

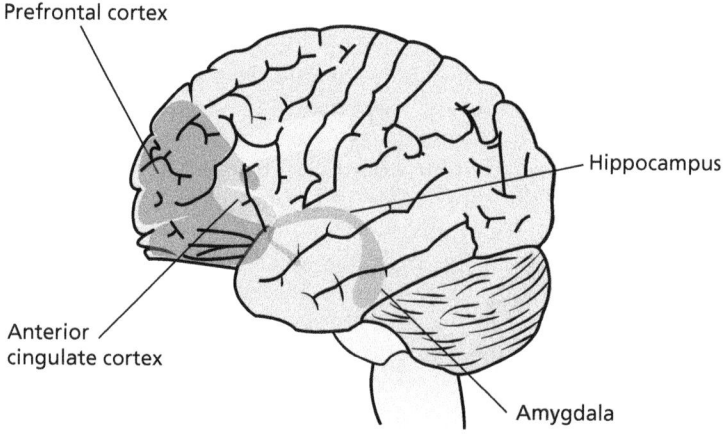

**FIGURE 4.1 Communication of various brain regions.**
Adapted from the National Institute of Mental Health, National Institutes of Health, Department of Health and Human Services.

or numerous other systems can lead to severe emotional, cognitive, and behavioral impairment. It is also important to note that the section is not intended to be an exhaustive compendium of all other systems, but rather a brief description of some vital systems that play a role in human functioning.

Another critical fact to keep in mind is that not all scientists agree as to which structures are involved with each system. In fact, a particular area of the brain might be considered a complete neurobiological system for some, but only a part of a system for other scientists. This lack of absolutism is further evidence of the complexity of the brain and of our limited understanding of it. For the purposes of this chapter, the neuronal structures that are discussed as part of the neural systems are those that are commonly thought to be involved within the system. Given the role of the cingulate gyrus, examination of it as its own separate system is warranted, although there are numerous neural connections with other systems.

The Ethics box discusses the importance of cooperating with other professionals. This is particularly relevant for the mental health professional who does not plan to specialize in a neurobiological understanding of human behavior, but who still desires to help people in therapy. The concept of neuroplasticity is described in the Special Topics box toward the end of the chapter. Trainees, clinicians, neuropsychologists, other providers, and clients find the concept of neuroplasticity

## ETHICS

### Collaboration and Cooperation

An ethical standard inherent in all medical fields is cooperation with other professionals. It is expected that the mental health clinician appropriately and professionally cooperates with other professionals to serve his or her clients in the best capacity possible. Working with other professionals in some capacity will be almost inevitable for the mental health clinician. Whether the clinician works in a hospital setting, in a school setting, in independent practice, on a mobile team, or in some other type of practice, the client, be he or she a child, adult, or geriatric in nature, will have had contact with another medical provider in most industrialized countries. While balancing the higher level of confidentiality requirements that mental health professionals have, which also varies depending on the degree obtained, the mental health clinician should ideally have communication with other providers, including primary care physicians, psychiatrists, family therapists, neurologists, and other medical professionals. Child mental health professionals often have contact with teachers or with other treatment teams as seen to be appropriate.

The perspective that a mental health professional has regarding behavior and cognition will be unique in comparison to that of other professionals. The conceptualization of brain-based dysfunction as it relates to behaviors and cognitions might serve in the client's best interest in some circumstances. Whether working with a hyperactive child who is causing disruption in school, a teenage boy who ignores his parents about curfew, the brain-injured adult who is prone to road rage, or an older adult who forgets appointments, the mental health clinician who understands neurobiological processes might help not only to explain the root causes of the behaviors but also to identify how the behaviors can be corrected, and then implement corrective actions. By cooperating with other professionals, mental health clinicians can often improve the emotional and functional aspects of their clients' lives.

hopeful and relevant in this field. In fact, the notion of neuroplasticity helps to explain how therapeutic interventions can be beneficial and is a crux in bridging our fundamental knowledge with application of the neurosciences.

## PRIMARY NEUROBIOLOGICAL SYSTEMS IN PSYCHIATRIC DISORDERS

### The Limbic System

The limbic system (LS) is considered to be the human emotion regulation system. Given its role in emotion processing, it is seemingly involved in most psychiatric disorders in some capacity, whether it is from a structural, functional, and/or neurochemical dysregulation. It is responsible for several functions, but it serves primarily as the emotion center of the brain. It maintains control of a broad range of behaviors, feelings, and cognitions. The LS provides affective tone to situations. It allows for people to understand other people's emotions through observation of facial affect and other nonverbal cues. The LS regulates how and why an individual becomes angry and upset, along with internal feelings of disappointment and sadness. The LS is also activated when a goal is achieved, success occurs, or other positive accomplishments result in internal states of well-being. It is located in the center of the brain and consists of subcortical structures (see Figure 4.2). It is approximately the size of a walnut and is considered to be the most primitive part of the brain.

The role of the LS extends beyond the expression and recognition of emotion. The LS is also responsible for determining whether events are internally important and it plays a valuable role in other areas of functioning outside of emotion. For example, within the LS, highly charged emotional memories are stored, and our LSs are activated when these memories are recalled. Given the role in emotion, it might surprise some people to discover that the LS plays a role in controlling the appetite, regulating the sleep cycle, and processing smells, and that it is involved in the libido. Further evaluation of this phenomenon helps the clinician to link

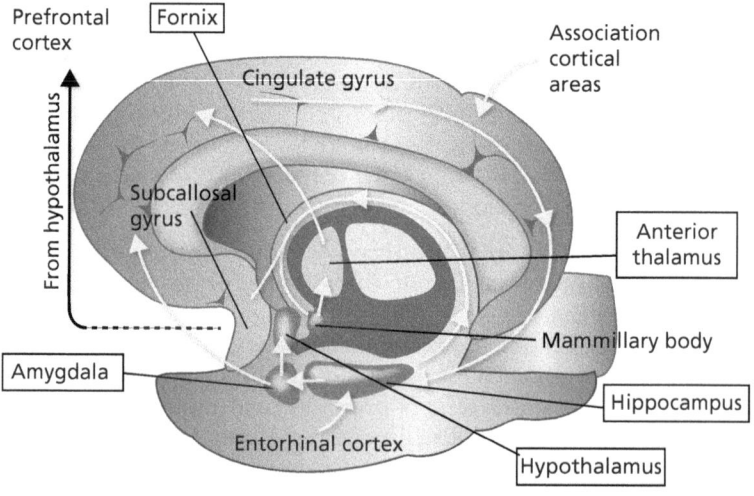

**FIGURE 4.2 Limbic system pathways.**
Adapted from Holtz (2011).

physical symptoms, such as decreased appetite, sleep dysregulation, and changes in libido that occur during the course of a mood episode, such as mania or depression. Further, motivation is modulated through the LS.

There are several important neuroanatomical structures that are a part of the LS (see Figure 4.2). The LS is thought to encompass the amygdala, hippocampus, hypothalamus, anterior thalamus, fornix, and portions of the cingulate gyrus. Table 4.1 provides a general overview of the major structures that are a part of the LS, the main role of these structures within the system, and clinical symptoms that could occur as part of dysregulation of the structure within the system. This table, and the tables that accompany the remaining systems, is a simplified summary and is in no way intended to offer a complete explanation of the actual complexity of each structure as it diminishes the role of the connections within the system. In fact, the clinician should be aware that dysregulation of any one area within a brain system can lead to disruption of the vital functions that the system controls. However, the schematics presented in this chapter can assist the mental health clinician in understanding general actions involved in the various brain structures and provide a basis of understanding.

The amygdala plays an important role in arousal and emotional response. The amygdala consists of a number of small nuclei that connect to the hypothalamus, olfactory system, and brainstem (see Figure 4.3). There are a number of connections to the frontal areas of the brain. These connections help to explain how humans are able to temper their immediate emotional reactions and keep calm when angered. The connections between the amygdala and the basal ganglia are part of a limbic feedback loop that is activated in numerous psychiatric disorders (Heimer, 2003).

As part of the LS, there are several links between the amygdala and the hippocampus, as well as the thalamus. The hippocampus, also referred to as the hippocampal formation, is important in consolidating and forming new memories. The hippocampus is divided into several different sections, which may be relevant in the expression of different disorders, but which are not pertinent to the general

**TABLE 4.1 The Limbic System**

| Major Structures | Vital Functions | Symptomatic Presentation |
| --- | --- | --- |
| Amygdala | Mood regulation; emotional response center; arousal | Emotional dysregulation: depression, mania, irritability, anger; negative thinking |
| Hippocampus (hippocampal formation) | Consolidate new memories; navigation; spatial orientation | Forgetfulness; inattention; amnesia; problems encoding new material |
| Thalamus | Motor control; receives and sends sensory information. | Abnormal movement; disrupted sleep; decreased ability to integrate sensory input; attention difficulty |
| Hypothalamus | Controls autonomic and endocrine functions; regulates sleep/wake libido cycle; homeostasis | Disrupted sleep; decreased appetite; increased appetite; weight changes; body heat dysregulation |
| Fornix | Connects hypothalamus to cerebrum and hippocampus | Disrupted sleep; appetite and memory problems |

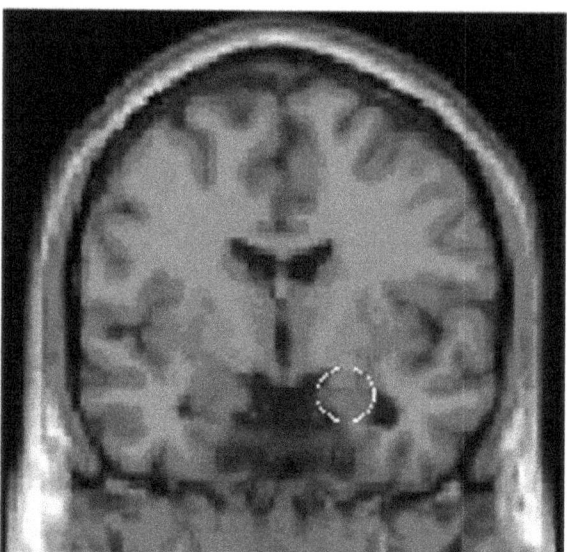

**FIGURE 4.3 Location of the amygdala.**
*Source:* National Institute of Mental Health, National Institutes of Health, Department of Health and Human Services.

clinician. The hippocampus performs many complex functions, and in fact it mediates various skills and abilities. Hippocampus lesion studies in animals have found that behavioral changes include decreased attention, problems with modulating arousal, and forgetting where they have placed food (Isaacson, 1982). Similarly, it is well-known that damage to the hippocampus in humans results in inability, or decreased ability, to form new memories. There are lateralized effects of hippocampal damage, as the dominant side (typically the left hemisphere) is important for the understanding and processing of language, long-term verbal memory, retrieval of words, and auditory processing. The nondominant hemisphere (typically the right hemisphere) is important for recognition of facial expression, decoding vocal intonation, musical abilities, and visual learning. Damage to the dominant hemisphere will often lead to aggression, violent or dark thoughts, mild paranoia, reading problems, and auditory processing difficulties. Damage to the nondominant hippocampus can result in problems decoding vocal intonations and recognizing nonverbal cues such as facial recognition, and issues with other social skill abilities. The previously discussed case study of H.M. is a prime example of symptoms resulting from bilateral hippocampal damage, or, in his case, removal!

The thalamus is known as the motor relay center, as it receives sensory information from the nervous system and subsequently relays the sensory signals to the proper areas of the cerebral cortex. The thalamus is also divided into different areas that are responsible for various functions. The integration of the visual, auditory, and somatosensory systems occurs at the thalamus. The thalamus has extensive feedback loops for motor, cognitive, and emotional systems that project to and from the cortex to the basal ganglia (Mendoza & Feundas, 2008). The thalamus is also known to play a role in wakefulness, directing attention, and determining how to proceed with sensory information. In the latter role, it filters, modulates, and decodes the sensory information that is relayed from the peripheral nervous system.

The hypothalamus can also be subdivided into various sections with numerous roles. In general, the hypothalamus controls autonomic and endocrine functions. As such, the hypothalamus is important for homoeostasis of the basic human behaviors of sleep, eating, sex, and body temperature. It assists the endocrine system in growth and reproduction. It is also responsible for controlling the autonomic nervous system. The hypothalamus plays a role in alertness as well.

The role of the fornix is a little less understood. The fornix is known to be a part of the LS, as it carries information from the hypothalamus to the hippocampus. Thus, it plays a vital role in the same processes as these other two structures.

The constant communication of these various brain structures within the LS allows humans to feel, eat, sleep, remember, and experience emotional reactions.

The disruption of the LS, by neurochemical means, trauma, or developmental deficiencies, leads to a variety of clinical problems. One subsection of the LS that the clinician should be aware of is the mesolimbic system.

## *The Mesolimbic System*

As suggested by its name, the mesolimbic system is found within the LS. The mesolimbic system is a specific dopaminergic pathway that is the brain's reward pathway. It is responsible for linking certain behaviors with feelings of pleasure, reward, desire, and enjoyment. It is found to be heavily implicated in addiction, compulsion, and motivation. The mesolimbic system is thought to play a role in mood disorders and schizophrenia as well. The mesolimbic system, as well as other portions of the LS, has connections with other important systems, particularly the basal ganglia, frontal system, and cingulate system.

## *The Hypothalamic–Pituitary–Adrenal Axis*

The hypothalamic–pituitary–adrenal (HPA) axis is a feedback loop found partly within the LS, linking the hypothalamus, the pituitary gland, and the adrenal glands. The HPA axis is a part of the endocrine system that regulates our reaction to stress. It regulates digestion, mood, the immune system, and other body functions. Within the HPA axis, the hypothalamus secretes chemicals that are responsible for cortisol production. The release of these neurochemicals is influenced by numerous factors, including stress, disease, exercise, sleep, and food consumption. Therefore, HPA activation is facilitated by other LS activity, including activity in the amygdala, thalamus, and hippocampus. For example, if a person is in a situation that was previously anxiety-provoking, neurochemicals in the amygdala and hippocampus will be activated, recalling the previous negative experience. Subsequently, the hypothalamus will activate stress hormones within the HPA axis. In general, within the LS the hypothalamus will either activate or not activate neurochemicals associated with stress hormones. The HPA is particularly activated in stress and posttraumatic stress disorder (PTSD). Further discussion of the HPA axis is found within the PTSD section of this text.

## The Basal Ganglia

The basal ganglia (BG) is a large brain system that serves several functions, but it is often thought to be particularly important for integrating feelings and movements. As demonstrated in Figure 4.4, it is the neural system that surrounds the LS, and it has numerous projections throughout the frontal lobe and LS. In fact, the amygdala is often also considered to fall within the BG as well as within the LS. Otherwise, the typical structures are listed in Table 4.2. It should be noted that, given the various projections within this system and connections with the other systems, the symptomatic presentations listed in the table are not meant to represent a comprehensive understanding of the role of the structures found in this system. Rather, the table's purpose is to help the mental health clinician understand general difficulties that can occur as a result of dysregulation.

The role of the BG extends beyond the integration of feelings and movements. It is also responsible for shifting and smoothing fine motor behavior and for

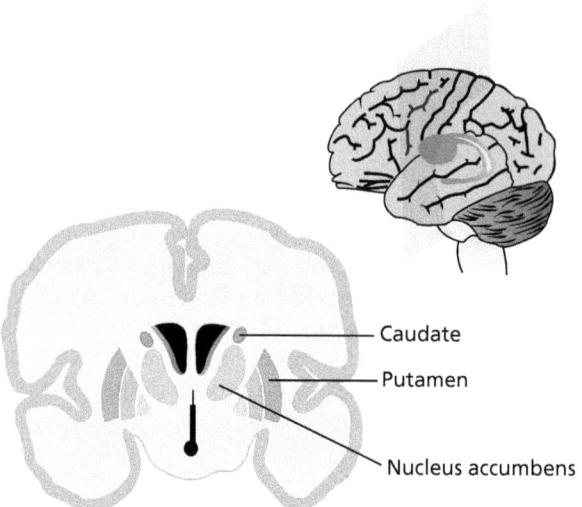

FIGURE 4.4 The basal ganglia.

suppressing unwanted behaviors. It facilitates transition of different motor activities. Given the physical symptoms of motor aspects, it is not surprising that it sets the body's anxiety level. The BG assists in motor memory of previously learned material. The BG has a series of feedback loops that allow for these activities. The feedback loops contain the caudate nucleus, putamen, and nucleus accumbens, which, combined, are commonly referred to as the *striatum* in scientific research articles. For the purposes of this text, the BG also includes the globus pallidus and substantia nigra, as they are all clearly connected and communicate with one another within the various loops as well.

The striatum is primarily responsible for movement and planning. It has the greatest number of connections to the frontal lobe and is therefore important for executive functioning components as well as feelings. Specifically, it plays a role in regulating all thoughts, feelings, and movements. The striatum has connections to the substantia nigra as well, and hence the interaction of the dopaminergic functioning on brain systems responsible for integration of movement and feelings.

The globus pallidus also assists in the modulation of the cerebral cortex. It relays information from the striatum to the thalamus through different pathways. The substantia nigra has a number of tracks connected to the caudate and putamen. GABA, an inhibitory neurotransmitter, is highly active in these tracks. The substantia nigra plays a vital role in the production of dopamine, which is found to be dysregulated in schizophrenia and Parkinson's disease, causing motor-based impairment.

TABLE 4.2 The Basal Ganglia System

| Major Structures | Vital Functions | Symptomatic Presentation |
| --- | --- | --- |
| Caudate | Regulates all information; filters, regulates, and gates all movements, thoughts, and feelings | Anxiety; nervousness; panic symptoms; tremors; chorea (rapid uncontrolled (movements); catastrophizing |
| Putamen | Acts with caudate to influence motor activities | Tics; fine motor problems; poor gait |
| Nucleus accumbens | Liaison with limbic system | Paranoia; depression; decreased motivation |
| Globus pallidus | Relays information from caudate and putamen to thalamus | Poor concentration |
| Substantia nigra | Produces dopamine | Tremors; dystonia; rigidity; bradykinesia |

## The Prefrontal System

The role of the frontal lobe was addressed in the previous chapter. The frontal lobe accounts for approximately one-third of the overall volume of the cortex. It is thought to play a role in most psychiatric disorders, and it will be discussed repeatedly throughout this text. In general, the frontal lobe is known to be responsible for executive functioning. Further, the frontal lobe includes the prefrontal area, as well as the motor cortex and premotor cortex. The premotor and motor sections of the frontal lobe are responsible for the planning and execution of a movement. The prefrontal area is more complex, and a better understanding is necessary to effectively navigate the brain systems and their relationship to behavior.

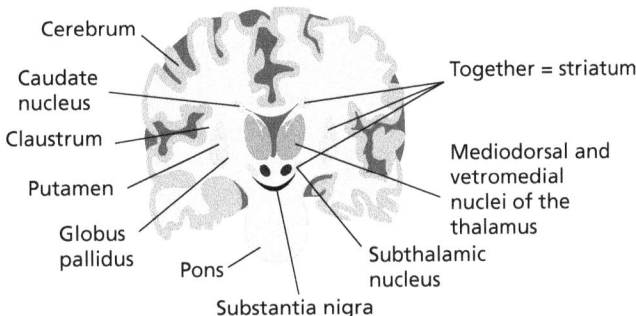

FIGURE 4.5 Coronal view of the basal ganglia.

Within the frontal lobe lie several subsections and, ultimately, brain systems that are responsible for various components of executive functioning and behavioral regulation. The three areas of the prefrontal cortex that are commonly implicated in psychiatric disorders are the dorsolateral, orbital frontal, and medial basal. Table 4.3 provides a brief description of the roles of these different prefrontal areas. It is important to remember, however, that these prefrontal areas are connected to each other, to other areas of the prefrontal cortex, and to yet other brain systems, particularly the limbic and BG systems. In this way, each contributes to other functional abilities as well.

The dorsolateral prefrontal cortex (DLPFC) is located in the back part of the prefrontal cortex. It is mostly responsible for higher-order cognitive abilities, including problem solving, planning, self-regulation, and sequencing. The DLPFC is involved in all complex mental activity. Damage to the DLPFC can lead to mild to severe executive impairments, which can be localized or broad. Disruption of normal DLPFC functioning is also thought to be involved in perseveration, procrastination, and poor time management. The DLPFC has been demonstrated to be involved in attention skills, and it is thought to play a role in attention deficit hyperactivity disorder.

**TABLE 4.3 The Frontal System**

| Major Areas | Vital Functions | Symptomatic Presentation |
| --- | --- | --- |
| Dorsal-lateral (back part of frontal lobe) | Problem solving; planning; self-regulation; sequencing; critical thinking; temporal ordering | Poor planning, perseveration; procrastination; trouble learning from feedback and experience; poor time management |
| Orbital frontal (front part of frontal lobe) | Inhibition; socially appropriate behavior | Disinhibition; impulsivity; poor judgment |
| Medial basal (middle part of frontal lobe) | Goal-directed behavior; ability to feel and express emotions; forward thinking | Bradykinesia; flat affect; limited response to reinforcement |

The orbital frontal prefrontal cortex (OFPC) is located in the front portion of the prefrontal cortex, above the orbital bone. The OFPC is responsible for guiding people to make socially appropriate behaviors. It is important for inhibition and good judgment. Dysregulation of the OFPC results in disinhibited and socially inappropriate behavior. The behavioral symptoms associated with Pick's disease, described in the medical chapter of this text (Chapter 12), provide an example of disruptions of the OFPC. The disruption of the OFPC might be best exemplified by a legal case of medical malpractice in which an obstetrician carved his initials into a woman's abdomen after a successful caesarean section. He was referred to as Dr. Zorro, and a settlement occurred in which news articles reported that his attorney's explanation for his behaviors was that he had Pick's disease. Assuming that diagnosis was accurate, his behavior is a good example of what could happen when there is disruption of the OFPC.

The role of the medial basal prefrontal cortex, which is located in the middle portion of the prefrontal cortex, is more closely associated with goal-directed behaviors and emotional tone. Damage to the medial basal prefrontal cortex leads to bradykinesia, which is the lack of goal-directed behavior. It has also been discovered that damage to this area of the prefrontal cortex leads to flat affect, mutism, and apathy.

## The Cingulate System

The cingulate system is a long brain structure or system that is found in the deep portions of the medial cerebral cortex, near the corpus callosum. It encompasses almost every lobe, and it has several connections with the LS. In fact, many theorists surmise that the cingulate gyrus, which is another name often used to describe the cingulate system, is in fact a portion of the LS. Regardless, the cingulate system plays a unique role in human functioning and behavior (see Table 4.4). It is responsible for several functional abilities, but it is known primarily as being important for cognitive flexibility, shifting attention, and adaptability. It can be broken down into anterior and posterior sections. Often referred to as the anterior cingulate cortex (ACC), the front section of the cingulate system has been found to play a role in a number of psychiatric disorders, including eating disorders, obsessive-compulsive behaviors, and anxiety. It is linked to road rage, poor cooperation, and inability to get past periods of being stuck. People who have perseverative tendencies—that is, they try the same solution to the same problem despite not being successful—have been found to have ACC dysfunction. The role of the posterior cingulate cortex is less clear. While it likely contributes to the same cognitive and emotional components as does the ACC, it is also thought to play a role in memory.

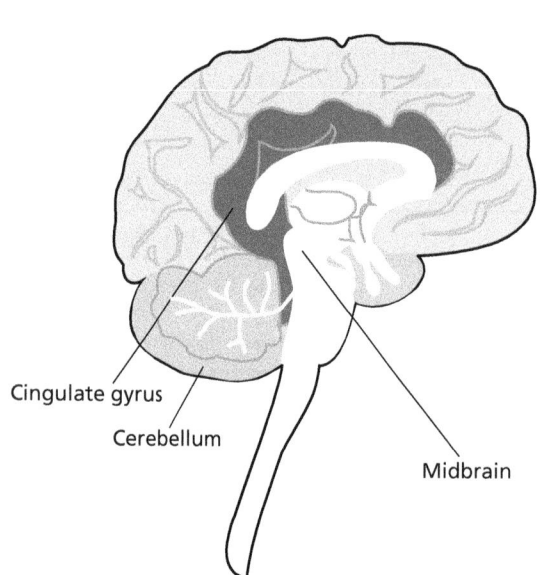

FIGURE 4.6 The cingulate system.

## The Fusiform Gyrus

The fusiform gyrus is located in the temporal and occipital lobes, with projections to both areas. Although not technically

**TABLE 4.4  The Cingulate System**

| Major Areas | Vital Functions | Symptomatic Presentation |
|---|---|---|
| Anterior cingulate cortex (front portion) | Impulse control; decision making; reward anticipation; cognitive flexibility; adaptability; cooperation | Cognitive inflexibility; obsessing; compulsive behaviors; addictiveness; uncooperative behaviors; worrying; road rage |
| Posterior cingulate cortex (back portion) | Memory | Holding onto negative past events |

considered its own separate brain system, it is important for the mental health practitioner to be aware of the fusiform gyrus. It is known to be important for facial recognition. A person with damage to the fusiform gyrus might have prosopagnosia, the inability to recognize familiar faces. The fusiform gyrus also plays a role in word recognition and processing of colors. Damage to the fusiform gyrus can occur in dementias as well as in childhood disorders, which will be discussed later in this text.

## SECONDARY NEUROBIOLOGICAL SYSTEMS IN PSYCHIATRIC DISORDERS

There are several secondary systems that are necessary for understanding normal human behavior. It is unlikely that the mental health clinician will have much day-to-day need to understand how these systems relate to psychiatric illness in as much depth as the previously mentioned systems; however, having a basic foundation in some of these systems can be helpful for more thoroughly understanding the complex interactions among the various neuropathways. A few of these systems are briefly described later. As mentioned at the beginning of this chapter, a detailed presentation of the sensory and motor systems can be found in Kolb and Whishaw (2008), as well as in many other texts.

### Cranial Nerves

As described in the previous chapter, a nerve is a bundle of axons that are classified based on the neurons found within them. Neurons are classified as either sensory neurons, motor neurons, or mixed neurons. Outside of the spinal nerves, there are 12 cranial nerves that emerge directly from the brain. The first two cranial nerves originate in the cerebrum, whereas the remaining 10 originate from the brainstem. The cranial nerves are a part of the peripheral nervous system, with the exception of cranial nerve II, which originates in the central nervous system. Table 4.5 provides the names of the different cranial nerves as well as their functions. Overall, the cranial nerves are necessary for the human brain to obtain and process sensory and motor information. Disruption of these nerves can cause speech, language, eating, smell, facial, and other abnormalities that can contribute to functional problems.

TABLE 4.5 The Cranial Nerves

| Number | Name | Type | Function |
| --- | --- | --- | --- |
| I | Olfactory | Sensory | Smell |
| II | Optic | Sensory | Vision |
| III | Oculomotor | Motor | Eye movement |
| IV | Trochlear | Motor | Eye movement |
| V | Trigeminal | Motor and sensory | Masticatory movements and face sensations |
| VI | Abducens | Motor | Eye movement |
| VII | Facial | Motor and sensory | Tongue sensitivity |
| VIII | Auditory vestibular | Sensory | Balance and hearing |
| IX | Glossopharyngeal | Motor and sensory | Movement and taste of tongue |
| X | Vagus | Motor and sensory | Movement of lungs, heart, muscles for voice |
| XI | Accessory | Motor | Movement of neck |
| XII | Hypoglossal | Motor | Swallow and speech articulation |

## Mirror System

Mirror neurons are neurons that are activated when a person observes another person's behaviors. These neurons are found in the frontal and parietal lobe cortex of the brain. Recent research on the mirror system suggests that this system is important in understanding and learning from other people's behaviors. It might serve to explain why imitation is an important means of behavioral alteration. These sets of neurons may be important for empathy as well. It has been demonstrated that portions of the LS that are activated when a person experiences sadness are also activated when a person observes other people feeling sadness. There is speculation that mirror system dysregulation occurs in autism, which helps to explain the lack of social reciprocity that occurs in the disorder. It is likely that in the next few years research will continue to bring light to the mirror system and its role in other psychiatric disorders, including mood disorders and possibly personality disorders, given the difficulties in social functioning found in these disorders.

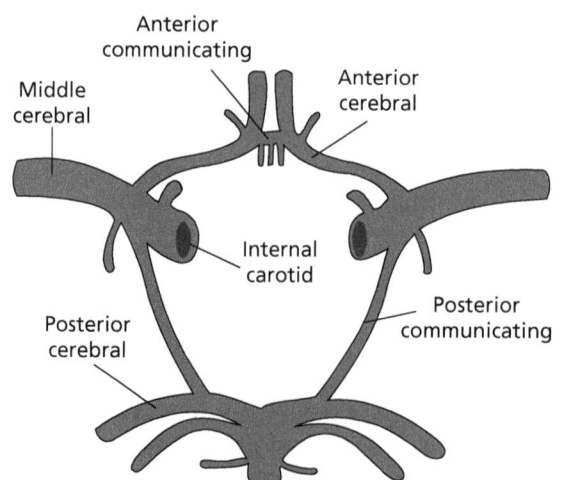

FIGURE 4.7 The Circle of Willis.

## Circle of Willis

The Circle of Willis is a connection of blood vessels that is located at the base of the brain and supplies blood to the entire brain. Although it is not specifically responsible for cognition, emotions, and behaviors, with disruption of the Circle of Willis a person will lose either efficiency or functioning of brain abilities. It is arranged in a looping pattern, and it comprises the anterior cerebral artery, anterior communicating artery, internal carotid artery, posterior cerebral artery, and posterior communicating artery. This pattern is important, as it helps other arteries assist malfunctioning arteries to maintain blood flow pressure in the brain. That said, a severe disruption can lead to a hemorrhage, stroke, or other vascular abnormality, some of which are described in the medical chapter (Chapter 12) of this textbook.

## APPLICATION OF THE NEURAL SYSTEMS

Although they can be overwhelming to learn about, it is recommended that the mental health clinician become familiar with the LS, the BG, the prefrontal systems, the cingulate cortex, and the fusiform gyrus. Being aware of the structures and various roles of these brain systems will help guide the clinician to a better understanding of pathology and the role of these brain systems and structures. Although useful in understanding brain functions, an in-depth awareness of the cranial nerves, mirror neurons, the Circle of Willis, and other systems is not typically utilized in the mental health clinician's practice, but is important in understanding healthy human functioning. For many students, studying these secondary systems can be overwhelming and a distraction from learning the systems that are consistently implicated in mental health disorders, but for others, it can help piece together some of the interactions between behaviors and brain functioning.

Throughout the remainder of the text, an understanding of the role of the systems in behaviors, emotions, and cognitions will be demonstrated. This knowledge can help the clinician conceptualize problems, guide treatment, and assist in making recommendations. It can provide a useful perspective on which medications may or may not be effective in certain circumstances. Though a fundamental understanding of biological processes is required in many undergraduate- and graduate-level programs, the clinician should be able to apply the material in a way that goes beyond mere conceptualization and more toward using our technology, knowledge, and understanding to further advance treatment and resolution of the difficulty.

## CONCLUSIONS

There are several brain systems that are responsible for human behavior, emotions, and cognitions. Although the ones presented in this chapter are not the only systems or structures found within the brain, they have been found to account for or play an important role in numerous mental health disorders. For example, dysregulation of the LS accounts for emotional turmoil in some capacity. Along with the LS, the HPA axis is largely involved in stress reactions. The BG is responsible for anxiety level and movement disorders. Fortunately for us, particularly for counselors, the brain is ever-evolving. The notion of neuroplasticity allows for therapeutic interventions to be effective, as people can learn with time, experience, and repetition.

The second section of this book attempts to describe neurobiology as it is known at the time of the writing of this book. Though the underlying neurobiological

## SPECIAL TOPICS

### The Application of the Concept of Neuroplasticity

*Neuroplasticity* is the concept that the brain is plastic. It is a dynamic, ever-changing organ that is adaptive and can be trained to function in certain ways—particularly during development, but also after injury. A quick Google search for neuroplasticity will generate hundreds of thousands of hits that range from advertisements selling techniques to slow the progression of aging, to magazine articles on its role in the utility of medications, and other stories about successful treatment of debilitating disorders. There have been TED Talks that have focused on neuroplasticity. In fact, the concept of neuroplasticity is the reason why therapy can be useful. If the brain could not adapt, repeated exposure to positive cognitions, behaviors, and environment would be ineffective. It could be argued that therapy is, in and of itself, a form of attempting to utilize neuroplasticity for improved abilities.

Neuroplasticity can be described as the brain's ability to grow or develop neuronal connections through repetitive behaviors and routines. It is the brain's way of reorganizing neuronal pathways to be efficient and capable of learning new material. It can occur throughout a lifetime, but it is a part of normal brain development. Neurons, the myelin sheath, glia, and other brain cells are involved with neuroplasticity. Research has shown that enriched environment, repetition of certain behaviors, and repeated exposure can alter the brain in volume, cortical connection, and brain activation.

It has long been known that neuroplasticity occurs during childhood. There is a case study of a woman named Barbara Guerra, who at the age of 2 years sustained an electrical shock, causing her to lose both of her arms (Thorpe, 2002). However, she was trained to perform all vital functions, including bathing, eating, using her phone, and driving, with her feet and legs. To this day, she is capable of performing these tasks for herself and for others with her feet. The brain reorganized itself to perform these vital options when she was required to learn to do them through repeated exposure and practice.

In adults, although it might take longer and be more frustrating, it is possible to learn new tasks with repeated exposure and practice. The old adage "you can't teach an old dog new tricks" is not correct. Rather, it just takes longer and requires more practice. This explains why you might find it more frustrating to learn a new language as an adult, but you can accomplish it in time. In comparison, a child can learn nonnative languages relatively quickly. The same is true for physical and motor skills, including learning to play an instrument, driving a car, walking, and learning a new sport. The fact that neuroplasticity can occur in adults should give the mental health clinician a feeling of hope and happiness. Even working with very difficult clients, recognize that, with repetition, practice, and consistent intervention, most people are capable of changing behaviors and thought processes.

---

connections are much better understood now as compared to 20 years ago, it is without doubt that the complexities explaining human behavior, cognition, and emotions will continue to be elucidated over the next 20 years. Although the rest of this text is intended to provide the mental health practitioner with an understanding of the neurobiology involved in each section, it does not suggest that any of these disorders is completely understood.

Rather, the discussion is presented to help the mental health professional understand how far neuroscience has come in breaking the mystery of each of these disorders. The astute reader will find that many of the disorders have similar neurobiological processes, including involvement of various brain structures, regions, neurotransmitters, and neural systems, with some still very different neurobehavioral and cognitive outcomes. In fact, mental health disorders that fall under the same general psychiatric condition, such as an anxiety disorder, have by definition similar but at times very different clinical presentations. Factors including genetic contribution, the environment, and as-yet-unknown variables are likely to be responsible for this heterogeneity. The Special Topics box provides an example of how our understanding of the brain can be effective in treatment.

## SUMMARY POINTS

- The brain is made up of several systems that are interconnected as well as intraconnected, resulting in emotions, behaviors, and cognitions.
- The main systems, from a mental health perspective, are the LS, BG, prefrontal system, cingulate system, and fusiform gyrus.
- Some other important brain systems that are vital for human functioning include the cranial nerves, the mirror system, and the Circle of Willis.
- Each of these systems has specific neuroanatomical structures, or regions, that perform, or assist in performing, specific tasks and abilities.
- The theory of neuroplasticity essentially underlies the reason for the effectiveness of therapy.

CHAPTER 5

# Childhood Disorders

Emotional and behavioral problems in children have become substantially more prevalent over the past three decades. The rates of attention deficit hyperactivity disorder (ADHD), autism, learning disorders, and mood disorders have doubled and in some cases tripled over the past 10 years. With the advancement of genetics research, more genetic disorders are being identified in children, even during pregnancy. Researchers have examined the effects of trauma on children, have disagreed about the clinical diagnosis of bipolarity in youth, and continue to learn about the short- and long-term effects of environmental toxins on the developing human brain. Throughout these endeavors, it has become increasingly clear that many psychiatric disorders are developed over time and result from a combination of genetic predisposition, environmental factors, and neurobiological expression. As a result, behavioral, cognitive, and emotional dysregulation often occur in children.

A complete review of the neurobiology of every childhood disorder would require a whole other textbook. In fact, several texts have examined autism, ADHD, and other childhood disorders separately. The mental health practitioner who will be working exclusively with children should be familiar with other literature examining the connection between neurobiology and the behavior of the pathology. The mental health practitioner who does not plan to work with children at all should be cautious about believing that he or she does not need to learn the neurobiology of these disorders, particularly because children eventually mature into adults. In fact, the *DSM-5* now characterizes these disorders not as *Childhood Disorders*, but rather as *Developmental Disorders*, which likely better indicates that these disorders do not stop as children grow into adolescence and adulthood. Yet characterizing only these disorders as developmental likely diminishes the conceptualization that only these disorders have a continual progression, when in fact most psychiatric disorders discussed in the remainder of this book have a developmental

## SPECIAL TOPICS

### Psychotropic Medication in Children

Psychotropic medications are often prescribed to children who are experiencing significant behavioral and mood dysregulation. In fact, it is estimated that 6.9% of children are prescribed at least one form of psychotropic medication (Zito, 2008). This prescribing results in over 8 million children in the United States receiving at least one psychotropic medication. The population of children in foster care is prescribed at an even greater percentage. This high prevalence is suggestive of medications' effectiveness at controlling behavioral problems, improving mood, and enhancing quality of life. In 2008, American children were 3 times as likely to be prescribed medications as were their European counterparts. The reasons for the higher amount of usage in the United States might be associated with a greater availability of psychiatrists and a different classification of disorders in the United States, which together increase the likelihood of more severe diagnoses being identified in this country. Regardless of the reason, the neurodevelopmental effects of psychotropic exposure remain uncertain.

Parents and students in mental health–focused programs are often opposed to the use of psychotropics in children. Public perception of the negative effects of medication is fueled by governmental black box warnings and negative media coverage. It is reported that selective serotonin reuptake inhibitors (SSRIs), common antidepressants used to treat adults with depression, might increase the likelihood of suicidal behaviors in depressed children. Indeed, antidepressants now carry a black box warning describing that risk for children. Less severe but important consequences accompany other medications as well. For example, commonly used psychostimulants, when prescribed to control attention symptoms in children who have anxiety or autism spectrum disorder, often increase anxiety, irritability, and agitation, while also potentially disrupting sleep.

On the other hand, it is possible that improved functioning in children with behavioral and emotional difficulties is associated with positive neurobiological changes within the brain. It has been suggested that regulation of neurochemicals and other neurophysiological factors increasing compliant behavior can result in more effective neuronal communication and neurodevelopment. Some theorize that if the behaviors can be improved upon with the implementation of medications, one can assume that the central nervous system is itself self-regulating in a positive capacity.

It is clear that there might be both positive and negative consequences of psychotropic medications for children. The mental health practitioner should be aware of his or her own bias toward whether medications are appropriate, but should also be well versed in the utility and risk factors of such practice. Although the mental health practitioner for whom this text is intended may not be prescribing psychotropic medications to children, he or she

**FIGURE 5.1 Medications.**
*Source:* © zhekoss/Shutterstock.

*(continued)*

> *(continued)*
> will have clients who do use medications. Being able to provide a balanced perspective while describing the feasible neurobiological implications of the introduction of psychotropic medications may help parents make these important decisions for their children.

course, even if that development occurs after childhood. In fact, the mental health practitioner is encouraged to consider a life span approach for all psychiatric disorders.

The goals of this chapter are to provide a brief overview of healthy neurodevelopment, as well as to examine the prevalent literature regarding the most common childhood disorders, including ADHD, autism spectrum disorders, intellectual disability, and learning disorders, as well as the effects of trauma. Other childhood disorders, including those thought of as motor disorders and communication disorders, are not presented in this text, but are the subject of extensive neurobiological literature. The Special Topics box discusses psychopharmacological treatment in children. The Ethics box later in this chapter explores the clinical implication of how inaccurate research can be particularly devastating to the public by examining the reported connection between early childhood vaccinations and autism. Although this chapter does not provide an exhaustive review of the literature, it allows the reader to understand and evaluate the complexities of diagnosing childhood disorders in an attempt to assist in providing the best clinical care.

## NEURODEVELOPMENT OVERVIEW

As described in Chapter 3, the central nervous system begins to develop shortly after conception. Throughout the stages of mitosis, embryogenesis, cell migration, and myelination synaptic pruning, the brain becomes an efficient, highly specialized, amazing organ that allows us to perform, understand, learn, speak, and feel. Disruption of any of these processes by genetic variables, stress, environmental components, or other neurobiological processes can result in functional difficulty.

Brain development occurs through a series of stages, sometimes simultaneous, that are affected by our genetic makeup and experience. Intrinsic, delicate processes occur daily that, though they might seem routine and "normal," are very complex and specialized. The brain initially undergoes mitosis, in which cells multiply and divide in the neural tube in the part of the brain that will become the ventricular surface. Over time, mitotic division occurs during a process called *embryogenesis*, and neuroblasts are laid out. Neuroblasts are the immature nerve cells that are produced after final mitotic division. At this stage, the brain produces two to three times the amount of neurons that will be found in the adult brain. Although neurogenesis stops in most regions of the brain by birth, stem cells continue to grow in the hippocampus throughout life (Gould et al., 1999).

Next, neurons migrate to their final destination within the brain. Neuronal migration is the complex process wherein the neurons establish connections with appropriate targets and other neurons, aided by glial cells. Neurons

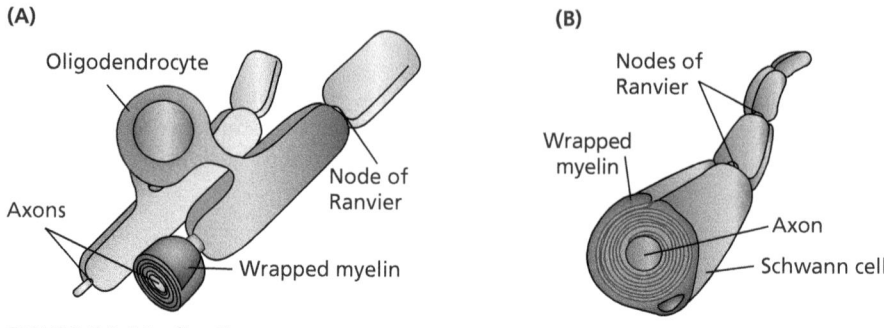

**FIGURE 5.2 Myelination.**
Adapted from Holtz (2011).

typically reach their destination approximately four months post-conception. At this point, cell death occurs in about 50% of these neurons as a result of inactivity and glial activity, among other factors. Synaptic density decreases dramatically during this time frame in the cerebellum, although it continues to occur throughout childhood in the cortex. Thereafter, myelination occurs. Myelination is the process by which fatty layers accumulate around nerve cells, speeding neurotransmission. Myelin forms around the axon, allowing for more complex processes. Myelin is produced by oligodendrocytes in the central nervous system and by Schwann cells in the peripheral nervous system. Along the axons, where myelin does not reside, nodes of Ranvier are present that further aid neurotransmitter transmission. The development of this process has been shown to occur into adolescence (Snook Paulson, Roy, Phillips, & Beaulieu, 2005).

As the person grows, a process called *synaptic pruning* occurs. Synaptic pruning is the stage wherein unused or inefficient nerve cells are eliminated to increase the processing speed and effectiveness of more commonly used neuropathways and connections. During this stage, high synaptic density found in the brain, useful for new learning and skill acquisition, is replaced with lower synaptic density for faster processing of established connections. This process begins to occur as early as seven years of age (Giedd, 2004), but it lasts throughout late adolescence and into early adulthood. Disruption of this pruning process has been linked with many developmental disorders that we will discuss, including autism spectrum, psychosis, ADHD, and intellectual disability (formerly known as mental retardation).

Brain development relies on natural, complex processes that are influenced by environmental and genetic factors. During sensitive periods of development, experience might significantly affect development. Furthermore, there are critical periods in which certain experience must occur for proper development. Positive environmental factors and enriched experiences can increase neuronal development during sensitive periods (Meaney & Szyf, 2005). As a mental health clinician, you must therefore ensure that your child clients are receiving the maximum possible level of positive reinforcement and exposure to healthy experiences.

Multiple factors play a role in the expression of mental health disorders. Though the general clinician might not often be asked to explain the neurobiological processes of healthy neurodevelopment, it is important that you recognize that these processes are typically altered in childhood disorders. Genetic, neuroimaging, cognitive, and treatment studies help us better understand these disorders' underlying neurobiology.

## ADHD

ADHD is likely the most common psychiatric disorder in childhood, with between 3% and 18% of children documented as meeting the criteria (Rowland, Lesesne, & Abramowitz, 2002). According to the Centers for Disease Control, parents report that approximately 9.5% of children between the ages of 4 and 17 were diagnosed with ADHD in 2007. Since 1997, rates of diagnosis have increased by between 3% and 6% annually, with boys nearly three times as likely to be diagnosed. This phenomenon has been attributed to better psychoeducation and identification and greater acceptance, as well as to overdiagnosis, poor parenting, and environmental influences. Some have argued that immaturity is misdiagnosed as ADHD, as children with later birthdays than their classmates are 30% to 70% more likely to be diagnosed with ADHD (Morrow et al., 2012). Regardless of the explanation, the rates of ADHD are staggering, and neurobiological factors play a vital role.

Behaviors associated with ADHD include poor concentration, limited attention span and vigilance, impulsivity, disorganization, forgetfulness, disinhibition, hyperactivity, and procrastination. ADHD is linked with other psychiatric disorders, with research suggesting that anywhere between 40% and 70% of individuals with ADHD have a co-occurring diagnosis. The most common disorders include conduct disorders, learning disorders, and anxiety and mood disorders. ADHD has been linked with several risk factors, including low birth weight, malnutrition during the first year of life, fetal alcohol exposure, and lead exposure. What was once thought of as a disorder that children grew out of is now recognized as manifesting itself differently into adulthood. Though only 15% of children who have ADHD meet the full criteria at age 25, a staggering 65% of these will show signs of social impairment (Faraone, Biederman, & Mick, 2006). Adults with a childhood history of ADHD have increased rates of divorce, employment difficulties, lower educational achievement, and head injuries.

The mental health practitioner will undoubtedly encounter ADHD in his or her practice. The increased diagnosis of ADHD has contributed to a shortage of medications at times and has increased scrutiny by the FDA to ensure that those who are prescribed medications are receiving appropriate treatment and follow-up. Physicians are more likely to refer patients to psychologists and other mental health professionals for help confirming diagnosis, as well as to implement treatment modalities geared toward lessening symptoms.

## Genetics and ADHD

Genetic factors have been described as "playing a major role in the development of ADHD" (Teicher, Tomoda, & Andersen, 2010). It is well known that ADHD occurs in families, with twin studies suggesting a heritability factor as high as 0.91. Several studies have attempted to locate the genetic contributions, and research has indicated some likelihood that multiple polymorphisms could play a role. Duplication of a chromosome at 15q has been shown to be significant in the expression of ADHD (Williams, et al., 2012). Stergiakouli and colleagues (2012) revealed several genetic pathways that could contribute to the expression of ADHD. Polymorphisms on the dopamine D4 receptor gene, serotonin, and noradrenergic genes have all been identified as playing a potential role in the expression of ADHD (Comings et al., 2000).

## Neurotransmitters and ADHD

Several neurotransmitters are also thought to play a role in the expression of ADHD. A marked overproduction of dopamine in the striatum during childhood might account for hyperactive symptoms frequently observed in ADHD. In light of the role of the dopamine D4 receptor, it is not surprising to find that dopamine, which is important for movement, is consistently found to be overproduced in ADHD. Catecholamine dysregulation likely accounts for several of the symptoms, but not for the entirety of neurotransmitter dysregulation in ADHD. Serotonin, responsible for mood state as well as for transmission of the catecholemines, has been found to be decreased in ADHD. Furthermore, norepinephrine is thought to be dysregulated in ADHD as well. Dopamine and norepinephrine play important roles in attention systems within the brain. A dysregulation of these neurotransmitters helps elucidate the behavioral and cognitive difficulty.

## Neuroimaging and ADHD

ADHD is perhaps the most widely studied child developmental disorder in terms of imaging research. There is a high occurrence of ADHD: children with ADHD are typically able to tolerate receiving scans of their brain, and investigators are strongly interested in discovering underlying neurobiological factors that play a role in this common disorder.

Studies have shown a reduction in both cortical and subcortical volumes in children who have been diagnosed with ADHD. The basal ganglia, an area of the brain that plays an important role in the production of dopamine, and an area that is also sensitive to dopamine, has consistently been reduced in volume in children with ADHD (Qui et al., 2009). A reduction in the caudate and putamen as well as the cerebellum has been found in ADHD as well (Bush 2008; Durston et al., 2004). A reduction of the frontal lobes has consistently been found, which may help to explain the poor decisions, lack of planning, and other executive dysfunction that often occurs in children who have been diagnosed with ADHD. Structural imaging studies have implicated a reduction in the corpus callosum as well (Hill et al., 2003). Some studies have suggested that the volumetric differences occur primarily in the right hemisphere, which may help to explain the largely intact verbal abilities that are typically found in ADHD. Figure 5.3 provides a representative sample of areas of the brain that have been found to be affected in children with ADHD. Even as early as preschool, MRI studies have shown that children who have been diagnosed with ADHD have decreased volume of the basal ganglia, particularly in the caudate (Mahone et al., 2011).

Several functional imaging studies have been conducted for ADHD. A recent meta-analysis of 55 fMRI studies indicated that frontal parietal systems, ventral attention, and somatomotor networks have hyperactivation (Cortese et al, 2012). This extensive meta-analysis also suggested that the prefrontal and parietal networks demonstrate hypoactivation. Once

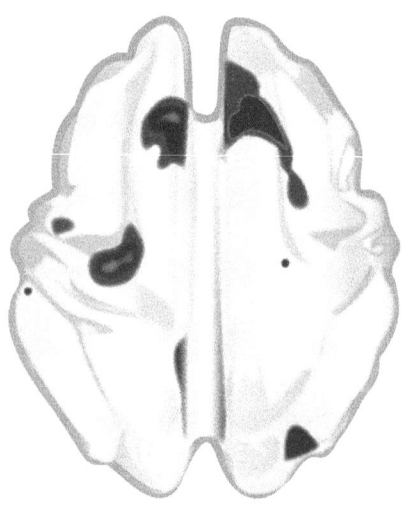

**FIGURE 5.3 Decreased brain volume area in ADHD.**
*Source:* National Institute of Mental Health, National Institutes of Health, Department of Health and Human Services.

again, this is consistent with the behavioral and cognitive symptoms of ADHD. They describe the frontal-parietal network as including the frontal pole, anterior cingulate cortex, anterior prefrontal cortex, cerebellum, and inferior parietal lobe. A meta-analysis examining single-photon emission computed tomography (SPECT) and positron emission tomography (PET) studies demonstrated that striatal dopamine transporter density is contingent upon psychostimulant exposure, with lower density occurring in drug-naive ADHD patients (Fusar-Poli, Rubia, Rossi, Sartori, Balottin, 2012).

## Cognition and ADHD

Neurocognitive studies have extensively examined ADHD. There are relatively consistent limitations of cognitive functioning in the areas of executive functioning, sustained attention, concentration, processing speed, and working memory. Although poorer cognitive performance is typically found, there is heterogeneity within children who have been diagnosed with ADHD. For example, it is not uncommon for children who have higher IQs but who demonstrate behavioral symptoms of ADHD to not demonstrate poorer performance on formal cognitive tests. Similarly, attention span is often found to not be problematic. Rather, inconsistency in performance is found on these tasks. Maintaining performance over time is particularly challenging for children who have been diagnosed with ADHD.

Similarly, concentration difficulties might account for poorer performance on formal measures of problem solving, planning, and organization. Barkley (1997) notes the wealth of evidence indicating that ADHD is an executive dysfunction disorder linked with difficulties in terms of inhibition of responses. It is possible that the dysregulation of prefrontal functioning contributes to executive impairment that occurs in ADHD, rather than being merely secondary to inattention. Cognitive limitations found in ADHD might also be contingent upon the subtype of the disorder. It is possible that individuals who have a more hyperactive presentation have more executive impairment than those who have the inattentive subtype.

## Treatment for ADHD

Psychostimulant medications (listed in Table 5.1) are the most common form of treatment for children who have been diagnosed with ADHD. These medications have slightly different neurophysiological reactions. Methylphenidate and dexmethylphenidate both increase dopamine, thereby improving attention and concentration. They include both short-acting and long-acting forms of the medication. Long-acting forms of these medications are typically preferred, with an eye to maintaining stability and preventing a rebound effect. Amphetamine, dextroamphetamine, and lisdexamfetamine work by blocking reabsorption of dopamine and norepinephrine. Strattera is a nonstimulant medication that has also been found to be effective for treating ADHD. Initial results from the multisite Multimodal Treatment of Attention Deficit Hyperactivity Disorder (MTA) Study indicated that although medication controlled the core symptoms more effectively than behavioral treatment did, behavioral therapy resulted in better long-term outcomes. Furthermore, although the MTA initially indicated that behavioral therapy did not initially contribute to improved functioning beyond the effects of medication, time revealed that the behavioral treatment was in fact more effective. It was

TABLE 5.1 Common Psychostimulants

| Methylphenidates | Dexmethylphenidates | Dextroamphetamine | Lisdexamfetamine |
|---|---|---|---|
| Ritalin | Focalin | Adderall | Vyvanse |
| Concerta | | | |
| Daytrana | | | |
| Metadate | | | |

also found that children who have a co-occurring diagnosis benefited most from the combination of medications and therapy. Of note, however, was the discovery that 30% of children did not respond to methylphenidate, the most commonly prescribed medication for ADHD. Neuroimaging studies that have been conducted to identify which children will respond to stimulant medication have shown promise as predictors of outcome.

Even when medication is able to improve attention and concentration, many parents continue to express concern about the behaviors that accompany executive dysfunction. Many behavioral treatment approaches have been developed to assist with improving children's choices, organization, and inhibition. Behavioral strategies often need to be practiced in therapy, and parent training can be particularly important for ADHD. Consistent routine, expectations, and encouragement are essential for successful treatment. Awareness of executive dysfunction, working memory problems, and attention difficulties should prepare the clinician for repetition of behavioral strategies during the course of treatment. Studies have indicated that the use of positive reinforcement focused on reducing impulsivity and encouraging positive work habits might be more effective than medications.

## AUTISM SPECTRUM DISORDERS

The diagnosis of autism spectrum disorder (ASD) is one of the most increasingly common diagnoses in the field of mental health. As recently as 1996, ASD was thought to occur in 4 to 5 children per 10,000 births (Fombonne, 1996), but incidence of ASD has been increasing. From 2006 to 2008, there was an increase of 20% in the number of those diagnosed with ASD. In fact, in the period from 2002 to 2012, there was a staggering 78% increase in ASD diagnosis (CDC, 2012). The increase of ASD has been attributed to a variety of factors, ranging from older age of fathers and genetic factors to better identification of symptoms and inaccurate diagnoses. Now debunked research even led to the popular notion that ASD was caused by early vaccine exposure (see the Ethics box). Pediatricians now consistently screen for ASD in young children, consistent with the American Academy of Pediatrics's 2009 recommendation that every child who goes to a pediatrician's office be screened for autism. Regardless of why, ASD prevalence rates are very high, with 1 in 88 children now being diagnosed with ASD by age 8. The numbers are even higher in boys, of whom approximately 1 in 54 have been estimated to have ASD (CDC, 2012). The care and treatment for a child with ASD is timely, challenging, and costly. Though specialized clinics exist for ASD, it is often the responsibility of the mental health provider to help identify ASD, provide resources to the family, and even sometimes treat ASD. There are very few resources for individuals who have been diagnosed with ASD as they enter into adulthood, particularly if they

> ## ETHICS
>
> ### The Harmful, Long-Lasting Effects of Falsifying Neurobiological Data
>
> A 1998 study published in the *Lancet*, examining a possible link between autism and vaccine, received a high degree of publicity and notoriety. In the now infamous study, a link was identified between the onset of autism and the administration of the measles, mumps, and rubella vaccine, commonly known as the MMR vaccine, which is given to children as part of normal inoculation. The study reported that eight of twelve children given the MMR vaccine started to demonstrate behavioral symptoms consistent with autism within days of the vaccination. The authors suggested that MMR not be administered, but rather that single vaccines be provided. After this study and the media presentation of the study, there was a decrease in use of MMR, with a subsequent rise of measles, mumps, and rubella. Although several dozen other studies concluded that there was no link between MMR and autism, many of the public continue to believe that MMR causes autism, and many parents opt not to have their children vaccinated.
>
> An investigation in 2004, 5 years after the original publication, indicated that the lead author of the study was provided a large sum of money by a lawyer to find evidence that the vaccine was dangerous so that the evidence could be used against the vaccine manufacturer in legal proceedings. It was also disclosed that the lead author had applied for patents on a vaccine that would compete with MMR. The results of the study were partially retracted in 2004, accompanied by statements that further studies needed to be conducted and that full disclosure should have occurred. Even further investigation revealed that data were fabricated and manipulated, the study's results were fully retracted in 2010.
>
> The disregard for ethical standards in this research was staggering. It is clear that conflict of interest had great potential to play a role in encouraging the falsification of data. Perhaps a better review of the original study by the journal's reviewers would have occurred had disclosure of financial gain been provided. Furthermore, every ethics board and every ethics guideline includes a provision for not falsifying or manipulating data. The effects of falsified research can be devastating to the public, as noted here, and can lead to problematic research and financial waste in other less popular research. In any case, the mental health practitioner needs to be honest in all aspects of his or her profession. Integrity is a necessary ingredient in all aspects of mental health care. Honesty about our knowledge, about our ability to help, and in all information that we provide is necessary to maintaining the confidence of the public and serving our clients as effectively as possible.

do not meet criteria for intellectual disability. Having a working understanding of ASD will be necessary for all providers, given the high rate of occurrence and co-occurring emotional, behavioral, and cognitive difficulty.

ASD is characterized by persistent deficits in communication, social functioning, and a restrictive pattern of behavior that presents early in development. These difficulties result in functional impairment in terms of relationships, education, occupation, and other domains. Although the *DSM-IV-TR* used the term *Pervasive Developmental Disorder* and distinguished among autism, Asperger's disorder, and pervasive developmental disorder, NOS, the revised *DSM-5* eliminated these distinctions, leading to a universal ASD diagnosis. The ramifications of this change are still unknown, but some have been concerned about the changes in diagnostic criteria, which will decrease identification of those who have been diagnosed with the condition formerly known as Asperger's disorder as also being diagnosed with ASD. As a result, many mental health professionals and families of children who have been diagnosed with Asperger's are fearful that the new definition will reduce the availability of services to many children. The clinician should be familiar with the difference between the old terminology and the new diagnostic criteria in order to best serve his or her patients. Awareness of the neurobiology of ASD may help to clarify the use of spectrum terminology in the new diagnostic manual.

### Genetics and ASD

ASD has long been known to have a genetic component. Twin studies have estimated the heritability of ASD to be as high as 90% (Freitag, 2007). Additionally, some genetic disorders result in ASD, including Fragile X and Rett syndrome. However, it has been concluded that no one genetic mutation accounts for the majority of ASD (NIMH, 2012). Though the chromosomal abnormalities remain unclear, it has been suggested that multiple gene interactions are the most likely genetic influence of ASD. A series of studies published in 2012 suggested that spontaneous mutations in the DNA sequence occur that are not present in the parents. These spontaneous mutations are referred to as *de novo* mutations. According to the *DSM-5* (APA, 2013), 15% of ASD cases feature de novo mutations. Otherwise, it has been estimated that DNA changes play an important role in the expression of ASD. Numerous candidate genes have been linked to ASD, and it has been estimated that over 100 minor genetic abnormalities may be present. Deletions and duplications have been documented in a variety of locations, including on chromosome 15 and the serotonin transporter gene, among other sites. Although there are likely genetic factors contributing to ASD, most people who have been diagnosed with ASD do not have an identifiable etiology.

### Neurotransmitters and ASD

The neurobiology of ASD is as complex as the genetic findings. The most consistent neurotransmitter dysfunction appears to be serotonin and endorphin elevations. Numerous neurotransmitters appear to play a role in the expression of ASD. High levels of serotonin (5-HT) have been found in ASD, as well as in first-degree relatives. Cerebral spinal fluid analysis has indicated that elevated endorphin levels also occur in ASD. This elevation likely accounts for the high pain tolerance and self-injurious tendencies of ASD patients. Mental health clinicians and parents are often surprised by a child with ASD who performs acts that seemingly should inflict serious pain, yet apparently without suffering such pain.

### Neuroimaging and ASD

Abnormal brain growth often precedes the emergence of ASD symptoms. In general, volumetric increases have been found in autism, suggesting neuronal overgrowth and lack of developmentally appropriate synaptic pruning. Multiple brain regions have been shown to demonstrate neuronal dysfunction in early ages of ASD. Some have focused on the lack of synaptic pruning of the prefrontal cortex, which is known to play a role in communication, social skills, and executive functioning. Furthermore, premature cessation of the cerebellum, cerebrum, and overall limbic system have been well documented (Courchesne, 2001; Zilbovicius, Meresse, & Boddaert, 2005).

Other areas of brain development remain unclear. Some studies have demonstrated an increase in amygdala volume, whereas others have suggested a decrease in amygdala volume. Similarly, imaging data have been inconsistent regarding hippocampal volume in ASD. Postmortem studies have found a reduced number of Purkinje cells, which likely play an important role in synaptic communication in the cerebellum. Yet neuroimaging studies do not consistently indicate either increased or decreased hippocampal volume.

Functional imaging research has clearly suggested that abnormal responses occur in ASD. Magnetoencephalography (MEG) studies have indicated that children with ASD have longer delays when responding to recorded vowels, beeps, and sentences (Robert et al., 2010). This may account for delays in language processing abilities. Functional MRI studies have consistently demonstrated decreased amygdala activation in children who have been diagnosed with ASD who are presented with fearful faces (Schultz, 2005; Schulkin, 2007). This decreased activation may contribute to the social difficulties and lack of social reciprocity in ASD. Difficulty with recognizing social–emotional cues likely contributes to limitations on social skill abilities and relationship development. Bilateral slowing in the prefrontal regions has also been documented in ASD.

## Cognition and ASD

Cognitive impairment is a hallmark feature of ASD. It has been estimated that up to 70% of individuals who were diagnosed with the condition formerly known as autism met the criteria for what was formerly known as mental retardation. However, with the inclusion of all Asperger's disorder patients as a part of ASD, this number is likely to drop significantly. Language impairments occur in ASD. Delays in reaching milestones, such as talking in single syllables, or putting together two or three words, are typical among many people who have been diagnosed with ASD. Limitations on verbal production, verbal problem-solving, and verbal abstract reasoning are common. The individual who has been diagnosed with ASD will often have problems with cognitive flexibility and will perseverate and remain focused on one topic. Cognitive limitations can also occur in areas of attention, memory, and working memory. These limitations result in the functional impairment found in ASD.

## Treatment for ASD

There is no medication specifically designed to treat ASD. However, numerous medications have been used to treat the behavioral and emotional problems that often accompany ASD. Selective serotonin reuptake inhibitors, atypical antipsychotics, and typical antipsychotic medications have been prescribed to control the anger outbursts, irritability, mood instability, and self-injurious behaviors that may accompany ASD. Long-term community-based behavioral interventions are effective therapeutic strategies. Early identification of ASD is particularly important to enable early facilitation of intervention services, which have been found to be particularly effective in the development of social skills and functioning.

Social scientist Temple Grandin has been very public regarding her own diagnosis of ASD. She has written books and presented at conferences (available online) to discuss her ASD, during which she has demonstrated diffuse tensor imaging (DTI) of her own brain. She concludes that children who have been diagnosed with ASD perceive the world in unique ways that can help society advance. Specifically, nonverbal strengths can be garnered to advance technology, and use of current technology can help an individual who has been diagnosed with ASD overcome his or her language impairments. Treatment of ASD is contingent upon the severity of symptoms and defined cognitive strengths and weaknesses. Use of cognitive testing to help develop specific treatment suggestions is encouraged.

## INTELLECTUAL DISABILITY

Intellectual disability (ID), formerly referred to as *mental retardation*, is characterized by cognitive impairment as reflected on measures of intelligence, executive functioning, memory, and learning that result in adaptive functioning impairments. Deficits in mental abilities affect everyday functioning because the individual is well below the level of those of his or her peers in areas of relationships, academic abilities, and life management skills, such as caring for oneself. The course of addressing ID needs to begin during the first 18 years life, and the earlier the better, but the condition is often identified before age 2. There are several causes of ID, including genetic inheritance, such as chromosomal disorders or extra copies of genes, seen in trisomy 21 (Down syndrome). Other potential causes of ID include in utero exposure to toxins such as alcohol or drugs, stress, brain malformations, seizure disorders, or environmental factors. Examining each of these causes is beyond the scope of this text. However, a general, broad understanding of neurobiological and cognitive expectations of ID is important for the clinician.

There are several levels of severity of ID, ranging from mild to profound, with various degrees of cognitive and functional impairment. Approximately 90% of individuals with ID fall under the mildly impaired range, with another 7% falling under the moderately impaired range. This suggests that fewer than 3% meet criteria for either severe or profound ID. Individuals who have been diagnosed with mild ID have an IQ between 55 and 70, can learn many adaptive skills, and can achieve approximately a sixth-grade education level. They can be trained to do many rudimentary jobs, and they can live in communities with assistance. They might even be able to handle some aspects of their own financial affairs. In comparison, individuals who have been diagnosed with moderate ID typically live in supervised settings, attain approximately a second-grade level of education, and need an increased level of support. People who have been diagnosed with severe and profound ID have severe limitations, demonstrate few adaptive skills, and need to be closely supervised for safety. Typically, the profoundly ID pass away by their early 20s because of complicating medical problems.

The hallmark feature of ID is an IQ of less than 70. This is one of a few disorders whose diagnosis relies heavily on objective cognitive results. Without a documented IQ of less than 70 before age 18, individuals will often be rejected for appropriate public services. Although imaging findings vary depending on the etiology of the ID, in general, smaller brain and hippocampus volumes, with specific areas of increased gray matter contingent upon etiology, have been reported. For example, Down syndrome has been shown to have larger parietal gray matter, whereas Fragile X has been linked to increased caudate gray matter. Typically, ventricle volume is increased in ID as well. Two other common and interesting syndromes leading to ID are Prader–Willi syndrome and Angelman syndrome. These two genetic disorders affect chromosome 15. Prader–Willi syndrome is characterized by a substantial increase in weight, with obesity commonly occurring. Though the behavioral and cognitive functioning associated with Prader–Willi varies, the individual typically meets the criteria for ID, is of reduced height, and has small hands and feet. Angelman syndrome is characterized by severe ID, ataxia, speech limitations, seizure, and inappropriate affect—often extreme happiness.

## Learning Disorders

Three different learning disorders are characterized in the *DSM-IV*: Reading disorder, mathematics disorder, and disorder of written expression. Although there is often a comorbidity of these disorders, reading disorders are the most prevalent learning disorders. Another learning disorder—not recognized by the *DSM-5* but nevertheless thought to occur—is nonverbal learning disorder (NVLD). NVLD has been studied largely by Byron Rourke, and controversy continues over whether this is indeed a separate learning disorder, or owing to another etiology such as ASD. The reader is encouraged to research Rourke's work for additional information if so inclined. For the purposes of the general clinician, the most prevalent learning disorder should be understood from a biological perspective.

### Reading Disorder

The most common learning disorder, and one best understood from a neurobiological perspective, is reading disorder. Specific types of reading disorders have also been commonly referred to as dyslexia, and research examining reading disorders has used this term frequently. The characteristics of dyslexia include letter reversals, word reversals, word omissions, and slow reading pace. Other symptoms of dyslexia include problematic auditory sensory processing, impaired perceptual motor skills, and difficulty differentiating left from right. Impairments in sound discrimination and phoneme discrimination are also seen.

Twin studies have indicated that dyslexia has a high heritability component, perhaps up to 91%. Although exact chromosomal abnormalities remain unclear, several genes have been linked to dyslexia. It is likely that candidate genes interact with the environment to produce dyslexia. One gene that has commonly been found to play a role is located on chromosome 6 (Knopick et al., 2002).

Although traditional school-based evaluations for learning disorders are predicated on an IQ/achievement discrepancy analysis of 1.5 to 2 standard deviations, new research suggests that such an approach is unsophisticated. Using this approach, those who test as having a lowered IQ would then have a more difficult time demonstrating a reading disorder. Functional neuroimaging research suggests that reduced left parietal lobe and fusiform gyrus occurs in individuals who have been diagnosed with reading disorders. Eckert (2004) provides a thorough review of the neurobiology underlying dyslexia. Brief disruptions of neural systems underlying phonological and orthographic information have been discovered. Both include frontal, parietal, and cerebellar regions, with various other neurobiological underpinnings. Children who have been diagnosed with reading disorders have been found to have decreased left hemisphere activation in fMRI studies, accompanied by reduced left hemisphere dorsolateral prefrontal cortex.

Treatment for dyslexia should keep in mind that the individual who has been diagnosed with dyslexia will need individualized education contingent upon his or her own strengths and weaknesses. Typically, individual tutoring, small classroom instruction, and use of strategic resources such as books on tape can be helpful in to the endeavoring reader. Use of community resources offering specialized training techniques for aiding those who have been diagnosed with reading impairments are strongly encouraged. There is no evidence to support the idea that medications can assist in the treatment of learning disorders, but if co-occurring ADHD is present, then consideration of the medication used for treating ADHD is suggested and might improve concentration ability and thereby increase reading tolerance.

## CHILDHOOD TRAUMA

The term *childhood trauma* is a broad term that can be used to describe a variety of different experiences encountered as a child. These can include the loss of a loved one; witnessing violence or tragedy; maltreatment—including physical, sexual, emotional, and verbal abuse; and being diagnosed with medical conditions, among other experiences. Often, childhood trauma does not receive treatment or does not receive appropriate treatment. People consider children to be resilient, and they may be, but people often forget to make sure that children who have experienced a traumatic event cope with the trauma effectively. If they do not, the effects of the trauma may degrade that child's emotional functioning, cognitive thought processes, relationships, and overall psychosocial functioning well into adulthood. Furthermore, those who have reported being abused three or more times were twice as likely to abuse their own children (Johnson-Reid, Kohl, & Drake, 2012). This suggests that chronic maltreatment may result in an even worse outcome.

### Childhood Maltreatment

Childhood maltreatment has recently been studied extensively in order to better understand its long-term emotional, cognitive, functional, and social implications. Childhood maltreatment can be defined as any single or multiple set of acts that a caretaker inflicts on a child that result in harm, the potential for harm, or the threat of harm to that child (Leeb et al., 2008). Childhood maltreatment can be further divided into four major categories, including neglect, physical abuse, emotional abuse, and sexual abuse. Several studies have further defined these forms of maltreatment, and thorough reviews have been published (Hart & Rubia, 2012). The consequences of the various types of maltreatment include anxiety disorders, behavioral difficulties, academic decline, attention difficulties, substance use, dissociation, and affective instability, among other problems. Children who have experienced abuse are more likely to be diagnosed with another psychiatric disorder as a result of their behaviors and emotions, but if the cause is maltreatment, treatment for the identified diagnosis may not be as effective.

The World Health Organization estimates that 20% of women and 5% to 10% of men report being sexually abused as children. Furthermore, 25% to 50% of all children report being physically abused. Even more children are exposed to emotional abuse and neglect. An estimated 31,000 children under age 15 die annually by homicide, although this number is likely to be an underestimation of the role of neglect since it does not include deaths by fall, drowning, burning, or other tragic events that occur when children are left unsupervised. The high frequency of abuse increases the likelihood that a mental health practitioner will work with somebody during treatment who has a history of abuse or who is currently being maltreated.

## Genetics and Childhood Maltreatment

A mental health clinician likely does not immediately consider genetic implications associated with childhood maltreatment, since the child's genetics certainly do not cause the child to be abused. However, genetic variables certainly play a role in the clinical presentation of symptoms after a child experiences abuse. A functional polymorphism in the monoamine oxidase (MAOA) has been found to moderate the effects of antisocial problems that often occur in maltreated children (Caspi et al., 2002). The MAOA gene is located on the X-chromosome, and it metabolizes norepinephrine, serotonin, and dopamine. Higher levels of MAOA seem to better control some negative emotional responses to abuse. The results from this study led to the recognition that other genes also appear to play a role in responses to abuse during childhood, including the serotonin and tryptophan genes (Ciccheti, Rogosch, & Thibodeua, 2012). Identifying genetic risk factors that can increase the likelihood of specific symptoms might help guide treatment intervention strategies for children who have suffered maltreatment.

## Neurotransmitters and Childhood Maltreatment

Glucocorticoids (GCs) are adrenal steroids secreted during stress. It has long been established that overproduction of GCs can be extremely detrimental to the central nervous system (Krieger, 1982). It is clear that early stressors cause long-term changes in numerous areas of the brain and neural networks (Sanchez, 2001; Bremner, 2003). The HPA axis plays a critical role in stress response, and children's neurobiological response to GCs is not yet fully developed. Excessive GC release can lead to neuronal degeneration of the hippocampus. It has been hypothesized that early stress causes long-term increases of GCs disproportionate to those of other stressors, a likely contributor to further neuronal changes (Ladd, 1999). It has also been suggested that decreased neurogenesis occurs within the hippocampus as a result of overproduction of GCs.

Beyond GCs, other neurochemicals appear to be influenced by early stress. Research has indicated that serotonin, gamma-aminobutyric acid (GABA), and noradrenaline are dysregulated as a result of early exposure to stress. This disruption is thought to contribute to difficulties with social attachment, mood disorders, and cognitive problems (Caldji, 2000; Beers, 2002).

## Neuroimaging and Childhood Maltreatment

In general, research has indicated that the earlier the age of onset and the longer the duration of abuse, the greater the structural brain changes that occur (DeBellis et al., 1999). It has also been hypothesized that certain brain regions are more vulnerable to stress during their own unique sensitive periods to the effects of stress (Teicher, Tomoda, & Andersen, 2006). Studies have suggested that numerous areas of the brain are affected in children who experience maltreatment. One consistent finding is that the hippocampus, which is part of the limbic system and is important for memory and learning, is affected by abuse and trauma. Interestingly, research has found that the hippocampus is volumetrically normal at the time of the maltreatment but has later been seen to be smaller in those who have sustained maltreatment. As discussed above, this is likely associated with an increase in GCs.

Studies examining the prefrontal cortex, the last part of the brain developed and a region of the brain important to higher-order cognitive skills, have found it to be the part of the brain most susceptible to maltreatment during childhood. Multiple portions of the prefrontal cortex, including the orbitofrontal, dorsal lateral prefrontal cortex, and medial prefrontal cortex, have been seen with decreased volume even in children who have been physically abused but who do not suffer from posttraumatic stress disorder (PTSD; Hanson Chung et al., 2010). A smaller prefrontal cortex is likely associated with the emotional, behavioral, and cognitive problems that often occur in this population. The corpus callosum, the main set of white fiber tracts connecting the left and right hemispheres of the brain to facilitate hemispheric communication, has also been found to have decreased volume in structural imaging studies. Furthermore, DTI has indicated alterations in the neural pathways of white matter tracts in the corpus callosum in children who have been exposed to parental verbal abuse (Choi et al., 2009). The cerebellum has also been seen to be of consistently decreased size in children who have experienced maltreatment (DeBellis & Kuchibhatla, 2006; Carrion et al., 2009). This smaller size might account for memory difficulty as well as problems with fine motor behaviors.

There is uncertainty regarding other parts of the brain thought to be affected by childhood maltreatment. One part of the brain where changes might be anticipated is the amygdala, in light of its role in emotional processing and safety assessment. Conflicting evidence about the amygdala's size likely rises from methodological differences across studies. A meta-analysis found no differences in the amygdala between children who suffered maltreatment versus those who had not suffered maltreatment (Woon & Hedges, 2008). Overall, it remains unclear whether amygdala differences occur in this population. The anterior cingulate cortex (ACC), which is responsible for many skills, including the ability to shift cognitively, has not been studied in children, but it has been found to have less volume in adults who have suffered abuse as children. In light of the known role of the ACC in psychiatric disorders, this finding might help account for many other disorders. It has also been suggested that the parietal lobes are reduced, although there are fewer studies examining this brain region than might be desired.

Functional imaging studies have also demonstrated decreased metabolic activity in the prefrontal cortex, as well as in the hippocampus and temporal lobe during periods of rest (Chugani et al., 2001). Interestingly, the increased activation of the ACC has been documented in this population during performance of a measure of inhibitory control, suggesting increased cognitive activity on these tasks. Similarly, increased activation of the amygdala has been found in children when they are presented with angry faces (Maheu et al., 2010). Event-related potential studies (ERP) have shown that angry and fearful faces and voices elicit altered responses (Shackman, Shackman, & Pollack, 2007). These findings suggest abnormal ACC and amygdala involvement at a regional blood flow level and electrical level, rather than at a structural level, is occurring. Furthermore, sexually abused children have also been found to have increased activation in the cerebellum during resting state, which is hypothesized to be associated with the high density of GC receptors, as discussed previously (Anderson, Teicher, Polcari, & Renshaw, 2002). Altered hippocampus activation has also consistently been observed in functional imaging studies in adults who have experienced childhood maltreatment (Croy et al., 2010).

## Cognition and Childhood Maltreatment

Several cognitive studies over the past decade have been conducted on children who have experienced maltreatment. Studies have demonstrated impaired academic performance, lower IQ, and significant working memory problems

(De Bellis et al., 2009; Loman, Wiik, Frenn, Pollak, & Gunnar, 2009; Pollack et al., 2010). However, many of these studies failed to control for psychiatric comorbidities such as PTSD or depression. Executive functioning has been found to be impaired in children who have mixed maltreatment histories (DePrince, Weinzierl, & Combs, 2009). Specifically, problems with inhibition and working memory have been found. Maltreated children have also been shown to have limited ability to correctly identify and discriminate facial expressions (Pollack & Tolley-Schell, 2003; Wismer-Fries & Pollak, 2004). Yet these researchers have also documented that abused children respond more quickly to angry faces.

## Therapy for Childhood Maltreatment

In light of the clear neurobiological implications of childhood abuse, treatment must consider the complexities, comorbidities, and long-term consequences of maltreatment. Clearly, prevention of abuse is the most important step for this population. Helping parents who are at an elevated risk of abusing to develop appropriate parenting skills is important. Teaching children about child sexual abuse through schools, in an attempt to prevent it, is also suggested by the World Health Organization. Watchful and observant adults should help protect their own and other children. The mental health practitioner plays a role as a mandated reporter of child abuse or neglect and should assess for abuse in all children with whom he or she works.

After neglect or maltreatment occurs, treatment should be made available as soon thereafter as possible. Early intervention and developing coping skills are the most effective forms of behavior strategies with which to decrease the short-term and long-term negative consequences of neglect and maltreatment. Trauma-focused cognitive behavioral therapy (TF-CBT) has been developed to help a child cope with such trauma and improve his or her emotional and behavioral functioning. TF-CBT is discussed in more detail in Chapter 8, which deals with anxiety. Individual therapy, along with combined individual and group therapy, has been demonstrated to be an effective method of treating childhood sexual abuse when used for children between the ages of 6 and 17 years old (Nolan et al., 2002). It is likely that repetition during treatment will help overcome memory and learning difficulty. Furthermore, reprocessing of the event and guidance in effectively coping with and developing skills useful for managing the stress are likely to be effective. The group components of treatment may help a child realize that his or her experiences are shared by others.

There is no medication useful for treating children who have experienced maltreatment. Medication may assist in managing co-occurring symptoms of inattentiveness, anxiety, and depression, but the neurobiological processing of the trauma and subsequent behavioral and emotional sequelae, although likely to overlap with ADHD, anxiety disorders, and mood disorders, differs from them, and medications may be less effective. Psychopharmacological treatment can be beneficial to help a patient who has severe symptoms.

## CONCLUSIONS

Emotional, cognitive, and behavioral disorders can result from a variety of factors, and risk factors can be present starting at conception. A variety of genetic, environmental, and neurobiological factors can contribute to the onset of symptoms. What was once thought to be a nature-*versus*-nurture debate in behavioral disorders has been clearly demonstrated to be nature-*and*-nurture with the advent of advanced neuroscientific techniques. The mental health professional working with children plays an important role in ensuring that proper diagnosis and continued evaluation occur, because symptoms often cross over multiple diagnoses. Consideration for testing, and even genetic testing, can prove effective in determining the etiology of childhood difficulties.

## SUMMARY POINTS

- Normal brain development is a complex procedure that is strongly influenced by genetic and environmental factors.
- Children are being identified more frequently and at earlier ages for many disorders, including ASD, ADHD, and other developmental disorders.
- Reading disorder, the most common learning disorder, often called dyslexia, has been found to have strong neurobiological underpinnings.
- Childhood maltreatment has a high prevalence rate and increases the incidence of other diagnoses. Treatment is important, but genetic factors also potentially play a role in the outcome.

CHAPTER 6

# Schizophrenia

Schizophrenia is a debilitating disorder affecting approximately 2.4 million adults in the United States and 51 million people worldwide (Regier et al., 1993), which is approximately 1% of the world population. The frequency of schizophrenia is approximately the same in men and women, although the age of onset is younger in men (Angermeyer & Kuhn, 1988). Schizophrenia has been associated with higher occurrences of obesity, cardiovascular disease, minor physical problems, sociality, cigarette smoking, and violent behaviors (De Leon & Diaz, 2005; Newcomer & Hennekens, 2007; Palmer, Pankratz, & Bostwick, 2005; Swanson, Holzer, Ganju, & Jano, 1990; Weinberg, Jenkins, Marazita, & Maher, 2007). Understanding the clinical symptoms and neurobiological factors helps determine how to increase the likelihood of a positive outcome. At the very least, a clinician's neurobiological and cognitive understanding of schizophrenia will allow for a better prognosis and help guide treatment planning.

In light of the unique characteristics of the disorder, schizophrenia has long been studied. Historically, psychoses, including hallucinations and delusions, were taken to be demonic beings controlling a person's body. In 1896, Emil Kraeplin proposed that psychiatric disorders are natural disease units, resulting in the theory that psychiatric illnesses could be studied in the laboratory (Hoff, 1992). Thereafter, the first research institution was created for psychiatric illness. Since then, science has attempted to shed light on schizophrenia. Neurobiological factors have been clearly documented as playing a major role in the disorder. In fact, schizophrenia is perhaps the best-understood psychiatric disorder. Countless research articles and books have been dedicated to learning about it. As noted by Tandon, Keshavan, and Nasrallah (2008), "several hundred thousand publications pertaining to schizophrenia to date describe thousands of discrete findings." Consolidating all the known facts regarding schizophrenia is challenging, but understanding the interplay of complex neurophysiological components and symptoms can help the clinician serve the patient's emotional and functional needs. Despite all this research, much as with other psychiatric disorders, our understanding of the neurobiological underpinnings is incomplete. This chapter aims to help us understand the clinical characteristics of schizophrenia by providing the most consistent as well as the most interesting neurobiological findings concerning the disorder. Examination of genetic, imaging, neurochemical, and cognitive findings in this population will be presented. Interwoven within the chapter is an attempt to understand the differences found within schizophrenia. Finally, to help readers apply this knowledge to treating patients who have been diagnosed with schizophrenia, both medications and psychosocial perspectives are provided.

## CLINICAL SYMPTOMS OF SCHIZOPHRENIA

The clinical characteristics of schizophrenia are distinguished from those of most other disorders by psychotic symptoms. Hallucinations, paranoia, disorganized behavior, delusions, cognitive decline, and significant social dysfunction are all symptoms of the disorder. These clinical characteristics of schizophrenia have been categorized into positive symptoms, negative symptoms, and cognitive deficit. The functional deterioration within subsets of patients who have been diagnosed with schizophrenia is remarkable when compared to other psychiatric disorders. For example, negative and cognitive deficits are more difficult to control, and they often result in long-term functional difficulty, even when active psychosis is controlled (Andreason, Arndt, Alliger, Miller, & Flaum, 1995). Understanding the behaviors associated with negative symptoms is important to understanding how psychosocial impairment is likely to happen.

Positive symptoms of schizophrenia include the active psychotic presentation. Specifically, hallucinations, delusions, and paranoia are all positive symptoms and often get confused with one another by the public. Hallucinations are distorted or inaccurate sensory experiences that appear to be real perceptions. These sensory experiences can be of any modality but are most often auditory. In comparison, delusions are fixed false beliefs that are resistant to reason or actual fact. Delusions can include features of grandiosity, feelings of persecution, and erotomania, among others. Paranoia, another common feature, is the feeling that others are out to get oneself. Paranoia can result in bizarre behaviors arising from the desire to avoid being caught by another person. These positive symptoms of schizophrenia, although most popularly associated with the disorder in media, and despite being the symptoms most easily identified by others, can be more reasonably well controlled by medications than can negative symptoms and cognitive impairment (Tandon, Keshavan, & Nasrallah, 2008; Keefe et al., 2007).

Neuroimaging findings correlating brain structural volume and positive symptoms in schizophrenia are complex and inconclusive. Higher baseline positive symptoms have been associated with more hippocampal and striatal volume loss over time (Ebdrup et al., 2011). On the other hand, it has been suggested that reduction of positive symptoms in first-episode, previously drug-naive patients results in increased gray matter in the putamen after only 6 weeks of medication (Li et al., 2012). Future studies might continue to help us understand neuroanatomical changes associated with long-term medication management of the positive symptoms of schizophrenia. Ongoing research using new imaging methods should help determine whether the control of positive symptoms through medication management is linked with neurophysiological changes.

Negative symptoms of schizophrenia include anhedonia, flat affect, thought poverty, and social withdrawal. Anhedonia is the inability to experience pleasure from enjoyable activities, and flat affect refers to a person's face not moving or to a person speaking in a monotonous voice. Negative symptoms are often accompanied by social limitations, including failure to attend to personal hygiene, care for oneself, or maintain social relationships. A large multisite data analysis has indicated that improvement in negative symptoms has a distinctive and independent effect on functional outcome in comparison with other symptoms and in fact plays a role in initial baseline functioning (Rabinowitz et al., 2012). In fact, there is strong evidence to support the belief that negative symptoms are inversely correlated with functional outcomes (Herbener & Harrow, 2004; Rocca et al., 2009). Understanding the severity of negative symptoms and attempting to improve upon negative symptoms in schizophrenia can guide the clinician in treatment.

Interestingly, the neuroimaging findings associated with negative symptoms are different from positive symptoms. It has been found that ventricular increase over time results in reduced improvement of negative symptoms (Edrup et al., 2011). Increased severity of negative symptoms has been found to be correlated with ventricular size across the lifetime. Patients who have long-term negative symptoms have been found to have decreased brain volume in the dorsolateral prefrontal cortex and right temporal lobe, as well as enlarged right caudate volume (Buchanan et al., 1993; Galderisi & Volpe, 2008; Wible et al., 2001). Negative symptoms are also found to predict a worse outcome, so much so that the term *deficit schizophrenia* has been used to identify subgroups of patients who present with primary and enduring negative symptoms (Carpenter, Heinrichs, & Wagman, 1988). The neurobiological changes associated with negative symptoms can bring greater challenges for the clinician.

Another important aspect of schizophrenia to consider when developing treatment goals is the presence and severity of cognitive symptoms. Cognitive symptoms are often not as overt as positive and negative symptoms in schizophrenia. Objective cognitive tests consistently demonstrate cognitive decline in most areas. Patients who have been diagnosed with schizophrenia often demonstrate limited executive functioning, poor memory, and problems with attention. Further research examining cognitive functioning in schizophrenia will be presented later in this chapter, as there is evidence that various subtypes of schizophrenia demonstrate different cognitive strengths and weaknesses.

## ONSET AND COURSE OF SCHIZOPHRENIA

The onset of schizophrenia is insidious, and the disease often results in cognitive and social disruption as early as early childhood. Typically, full manifestation of the disorder occurs between late adolescence and early adulthood. Gender appears to play a role in schizophrenia, as the first episode involving psychotic symptoms is found to occur earlier in men, often between late teenage years and early twenties. Women are more likely to demonstrate symptoms in their twenties and early thirties. Men are also found to have a greater lifetime risk of developing schizophrenia (Aleman, Kahn, & Selten, 2003; McGrath et al., 2004), and they are found to be more likely to have a poor outcome (Green, 1996; Perkins, Gu, Boteva, & Lieberman, 2005).

It has been hypothesized that three critical periods across the life span play a role in the development of schizophrenia: conception, early development, and later development (Karlscot, Ellman, Sun, Mittal, & Cannon, 2012). At conception, genetic makeup is determined. As we will discuss, genetic predisposition for schizophrenia has been clearly demonstrated. The risk for developing the disorder from genetic predisposition is then determined by prenatal and perinatal brain development. Disturbances during these time frames, including but not limited to exposure to illicit substances, malnourishment, influenza, maternal smoking, and fetal hypoxia, can interfere with the growth of neural circuitry throughout life, resulting in the brain abnormalities found later in schizophrenia (Buka, Tsong, & Lipsit, 1993; Ellman, Huttunen, Lonnqvist, & Cannon, 2007; Ellman, Yolken, Buka, Torrey, & Cannon, 2009). Several researchers have concluded that neural changes associated with schizophrenia are seen as early as birth. This might help explain social difficulties and cognitive deficits in children who are later diagnosed with schizophrenia (Cannon et al., 2002; Rosso et al., 2000). During adolescence, the brain continues to demonstrate changes, including synaptic pruning. It has been found that progressive anatomical deterioration occurs during adolescence, prior to the onset of the symptoms (Gur et al., 1998; Thompson et al., 2001). These critical stages might also be influenced by socioeconomic status, as research has consistently demonstrated a higher prevalence of schizophrenia in lower economic status populations (Saha, Chant, Welham, & McGrath, 2005).

The course of the illness has been consistently characterized by experts as "variable." Despite the initial trend of significant change of behavior, cognition, and functioning in the early stages of the disorder, the course of the illness has been traditionally thought to plateau, and some people might even demonstrate improvement later in life. There is heterogeneity in outcome in the later life of the disorder, likely caused by a combination of innate neurobiological factors and controllable social factors, treatment, and support. Earlier studies suggested a heterogeneous long-term outcome ranging from fully recovered to severely incapacitated (Linn & Flak, 1988; Angst, 1988). Several hypotheses exist to explain this heterogeneity. One explanation is that broad differences might exist in the population before the onset of the illness. Another explanation points to the severity and presence of negative symptoms and cognitive deficits, as it is known that those characteristics can pose a significant risk factor for worse recovery (Davidson & McGlashan, 1997). Based on that knowledge, along with the understanding that different neuropathways and brain structures are primarily responsible for various symptoms, it can be concluded that recovery is influenced by neurobiological processes.

## GENETICS AND SCHIZOPHRENIA

As with most psychiatric disorders, genetic factors play a role in terms of increasing the risk for being diagnosed with the disorder. According to the National Institute of Mental Health (NIMH), individuals who have a family history of schizophrenia within first-degree relatives have a 10% lifetime risk of also manifesting symptoms of schizophrenia and are at increased risk if a second-degree relative is also diagnosed with the disorder. Studies note that the risk of being diagnosed with schizophrenia jumps up to 65% if an identical twin is diagnosed with the disorder (Cardno & Gottesman, 2000), with up to a 28% chance for a dizygotic twin (Kety, 1987). Indeed, studies consistently indicate that schizophrenia is highly heritable and that heredity accounts for approximately 80% of liability for the illness (Cardno & Gottesman, 2000; Crowe 2007; Lencz et al., 2007). Some imaging studies have suggested that structural brain abnormalities occur, even in the prefrontal cortex, at least to a milder degree, within unaffected family members, suggesting vulnerability markers (Boos, Aleman, Cahn, Pol, & Kahn, 2007, Keshaven et al., 2007). Other studies have found no differences occurring in striatal volume in monozygotic twins discordant for schizophrenia (Ettinger et al., 2012).

Despite the clear genetic link to the disorder, studies have demonstrated that no one specific chromosomal region is responsible for the disorder, and consistent identification of specific susceptibility genes is proving very difficult. Meta-analytic studies reveal susceptibility genes on at least 10 different chromosomes

(Badner & Gershon, 2002; Lewis et al., 2003). Rather, it is likely that markers at multiple chromosomal regions are responsible for increasing the risk for the disorder. Candidate genes, such as dysbindin, neuregulin-1, D-amino acid oxidase, and catechol-O-methyltransference, all have strong evidence supporting their role in schizophrenia (Funke et al., 2004; Greenwood, Light, Swerdlow, Radant, & Braff, 2012; Numakawa, 2004; Zhao, 2004). The roles of these genes are different, and each is very complex. However, many of these genes play a role in brain development, including in neuronal migration, neurotransmission, and cortical function (Greenwood et al., 2012; Tamminga, Shad, & Ghose, 2008). For example, a mutation on the catechol-O-methyltransference gene increases dopamine activity in the prefrontal cortex (see Figure 6.1). Alterations in these genes and dopamine dysregulation would disrupt normal brain development and functioning. These genetic findings are consistent with the imaging and postmortem findings indicating neuropathological alterations in this population.

## NEUROCHEMISTRY AND SCHIZOPHRENIA

In 1963, Carlsson and Lindquist observed that blocking dopamine receptors in the brain reduced psychotic symptoms in schizophrenia. Since that time, studies that have explored the dysregulation of dopamine in the brain have found increased density of dopamine receptors in the caudate and putamen (Abi-Dargham et al., 2000; Reynolds and Mason, 1995; Seeman, Guan, & Van Tol, 1993). Dopamine agonists increase schizophrenia symptoms, whereas dopamine antagonists decrease symptoms (Kapur et al., 2000; Lieberman, Kane, & Alvir, 1987). Simply put, the striatum, which is known to play a role in dopamine production, has been shown to have a high density of dopamine receptors in schizophrenia (Seeman et al., 1993). Overproduction of dopamine in the striatum has in turn been demonstrated to cause psychotic symptoms.

Dopamine is not the only neurotransmitter to be found dysregulated in schizophrenia. GABA, serotonin, and glutamate have been found to be abnormal in this population as well (Benes, McSparren, Bird, San Giovanni, & Vincent, 2007; Lewis, 2004; Tamminga, 1998; Van Praag, 1983). It has also been demonstrated that hypercortisolemia and dysregulation of the hypothalamic–pituitary–adrenal axis occur in schizophrenia (Phillips et al., 2006; Yuii, Suzuki, & Kurachi, 2007). Taken together, several neurotransmitters throughout various portions of the brain have been found to be dysfunctional in schizophrenia. This might account for the heterogeneity of symptoms and variability of successful pharmacotherapy treatment. It might also help explain the other structural and functional alterations within the brain.

## NEUROIMAGING AND SCHIZOPHRENIA

Studies have long demonstrated structural brain abnormalities in patients diagnosed with schizophrenia. CT studies as far back as 1976 have documented ventricular enlargement (Johnstone, Frith, Crow, Husband, & Kreel, 1976). As technology has improved, there has been a better understanding of the significant neurobiological changes seen in these patients. In 2001, a review of 193 magnetic resonance imaging (MRI) studies examining schizophrenia was conducted by Shenton, Dickey, Frumin, and McCarthy. They found that structural MRI studies not only confirmed the increased ventricular size, but have also subsequently

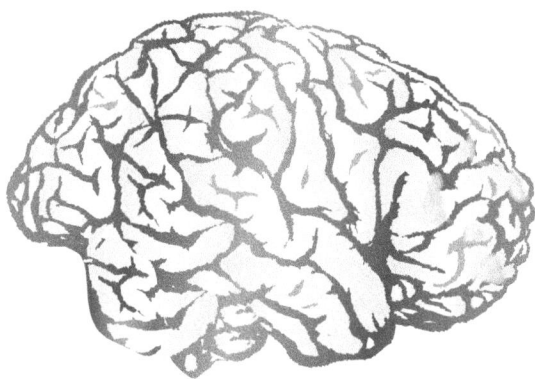

**FIGURE 6.1  Areas of decreased production of dopamine in schizophrenia associated with antipsychotic medication.**
Adapted from National Institute of Mental Health, National Institutes of Health, Department of Health and Human Services.

consistently documented increased third ventricle enlargement and decreased overall brain volume, with decreased volumes in the temporal-cortical structures of the hippocampus, amygdala, and other areas (Barta, Pearlson, Powers, Richards, & Tune, 1990; Lawrie et al., 2002). These gray matter deficits and the abnormal volume increase of the ventricles are present even in first-episode patients (Hulshoff-Pol et al., 2001; Steen, Mall, McClure Hammer, Lieberman, 2006). Specifically, gray matter reductions had often been found in the prefrontal and orbital frontal regions, and 60% of the studies indicated parietal lobe abnormalities.

Another consistent finding of abnormal imaging in this population is progressive striatal volume abnormalities. Volumetric deficits have been found in the caudate, putamen, and nucleus accumbens, with even greater reductions in the putamen and nucleus accumbens than seen in those of patients who have alcoholism (Deshmukh, Rosenbloom, DeRosa, Sullivan, & Pfefferbaum, 2005). Interestingly, postmortem studies had previously demonstrated a volume increase of bilateral nucleus accumbens (Lauer, Senitz, & Beckmann, 2001). Striatal dysfunction also leads to limited initiation and problems controlling motor behavior. Volumetric changes in the striatum are likely associated with not only the clinical symptoms of schizophrenia, but also the well-known dopamine dysregulation that occurs in schizophrenia. Striatal abnormalities are linked with overproduction of dopamine, leading to symptoms.

Temporal lobe studies examining schizophrenia have long demonstrated reduced cell size, dendritic density, and other pathological findings in postmortem studies (Arnold et al., 1995; Benes, McSparren, Bird, San Giovanni, & Vincent, 1991; Rosoklija et al, 2000). Structural imaging studies have supported temporal lobe abnormalities. Hippocampal volume reduction has been found, even in first-episode patients (Bogerts et al., 1990; Honea, Crow, Passingham, & Mackay, 2005; Wright et al., 2000). Reductions in the white matter found in the temporal lobe, including in the fornix, suggest white matter tract abnormalities that modulate functioning with the frontal lobes, specifically the prefrontal cortex (Fitzsimmons et al., 2009). This result might help us understand the memory impairments and executive function difficulties documented in these patients.

There is suggestion of cerebellar volumetric abnormalities as well. The cerebellum is important for gait, balance, coordination, fine motor control, and memory. This is consistent with a hypothesis that impaired coordination of senses and slower thought are associated with behavioral difficulties of disordered thought and language (Andreason et al., 1999). Andreason and colleagues' theory of the relationship between cerebellar impairment and schizophrenia is termed *cognitive dysmetria*. This theory hypothesizes dysfunction of the cerebellar–thalamic circuits. Newer MRI studies have typically confirmed the cerebellar-abnormalities, but these might be limited to certain patients (Bombin, Arango, & Buchanin, 2005) and to specific parts of the cerebellum (Bottmer et al., 2005). It is likely that these cerebellar abnormalities influence cognition by dysregulating sensory integration. Abnormal cerebellar volume could facilitate abnormal sensory integration, with subsequent emergence of abnormal cortically based cognitive processes found in hallucinations (Picard, Amado, Mouchat-Mages, Olié, & Krebs, 2008).

The anterior cingulate has been demonstrated to have lower volume as well (Wang et al., 2007). This brain region is thought to be important for self-monitoring, cognitive shifting, and verbal fluency. Clinically, patients who have been diagnosed with schizophrenia have a hard time caring for themselves and shifting thoughts, and they have diminished verbal output. Although volumetric differences in the anterior cingulate are not unique to schizophrenia, they do help explain the social and cognitive limitations that occur with the disorder.

Although there is limited evidence of white matter differences in first-episode patients (Sanfilipo et al., 2000; Hulshoff-Pol et al., 2004), diffuse tensor imaging has shown some matter differences to occur later in the disease.

This might provide evidence for neurodevelopmental changes throughout the course of the illness, evidence for heterogeneity in the patient population, or even evidence that various subtypes of the disorder might result in different neurobiological findings. White matter differences will continue to be examined as advanced imaging techniques are developed.

Though no brain region is entirely spared in schizophrenia, defining the core anatomical regions that are consistently irregular helps scientists better locate the areas of the brain that are important to examine from a functional and neurochemical regulation point of view. Although brain imaging studies are not yet used to differentiate or diagnose psychiatric conditions, using MRI in patients who complain of schizophrenia symptoms might be clinically helpful (see the Special Topics box). Structural abnormalities provide evidence that schizophrenia has preexisting neuroanatomical differences, which progress during the course of the illness in different ways, likely contingent upon predetermined genetic factors, symptom presentation, and treatment.

Numerous functional imaging studies have been done using positron emission tomography (PET) or functional magnetic resonance imaging (fMRI) in schizophrenia. Findings from functional imaging studies support the cortical dysfunction understanding of schizophrenia, with much evidence implicating the prefrontal cortex. Activation of various brain regions during executive functioning tasks is of interest in light of the role of executive functioning dysfunction in schizophrenia. Evidence from two decades of fMRI studies has found executive functioning impairment related to functional abnormalities of the prefrontal cortex, as well as within regions outside of the frontal cortex (for a review see Eisenberg & Berman, 2010). In fact, a meta-analysis found a slower frontal and limbic regional activation at rest, meaning that, even when the patient was not asked to complete a task while in an fMRI, differences were noticed (Hill et al., 2004; Taylor et al., 1999). Though studies have confirmed dysregulation of the frontal cortex, the alterations are not yet fully understood. Thus, the relationship of cognitive task and activation of brain regions in schizophrenia is somewhat complex. For example, though studies have consistently demonstrated working memory impairment within fMRI when examining schizophrenia (Barch et al., 2001; Driesen et al., 2008), the assumption of underactivation within the prefrontal cortex is inaccurate. Rather, it is likely that decreased working-memory–loaded tasks result in increased activation of the prefrontal cortex, but more complex working-memory tasks result in decreased activation. Callicot et al. (1999) proposed that when working-memory capacity is exceeded, activation decreases. It is thought that the function of dopamine response during various stages of working memory likely plays a role in the discrepant findings.

Abnormal fMRI studies have been documented in other brain regions as well. Both the anterior cingulate and hippocampus have been shown to be hypoactive, meaning that there is decreased activation in the regions. The neuronal connections between the anterior cingulate and prefrontal cortex further contribute to executive dysfunction by further disrupting neuronal transmission

> **SPECIAL TOPICS**
>
> **Clinical Imaging in Schizophrenia?**
>
> As demonstrated throughout this text, use of an MRI can help the clinician in a variety of different ways. Although it is not typically thought of as a primary recommendation for a mental health provider, there are a variety of reasons to consider encouraging an MRI in patients presenting with psychiatric symptoms, including depression, anxiety, somatic symptoms, and, particularly, psychosis. First and foremost, it can rule out an organic cause for the psychosis. For example, patients with certain forms of epilepsy, brain tumors, brain injuries, and other organic etiology can present with symptoms of psychosis. Typical treatment with an antipsychotic will not only be ineffective, but also might be harmful and could slow the timing of appropriate treatment.
>
> It has also been proposed that use of imaging in this population can predict the course of illness and its rate of progression. The extent of gray matter loss, as revealed by an MRI conducted at baseline and follow-up, has been associated with clinical outcomes (Cohen, Dembling, & Schorling, 2002). Also, determining which medication might be most effective based upon imaging findings has been proposed but has presented inconsistent results. For example, symptoms have been found to be more effectively decreased by use of Clozapine, rather than typical antipsychotics, when MRIs reveal sylvian fissure enlargement. At the very least, neuroimaging studies have helped decrease use of high-dose antipsychotic treatments. Specifically, dopamine receptor occupancy differs among patients, and exceeding a certain amount increases the incidence of side effects of tremors and other extraparamydial symptoms without increased clinical utility. Imaging study findings have helped provide support to decrease the dosage of antipsychotic medications due to the diminished therapeutic effects of high dosage. MRI has also been proposed to provide identification of risk factors for schizophrenia.
>
> Although only specialists who are trained in interpreting neuroimaging findings can provide insights into the findings, and only medical doctors can determine appropriate medications to prescribe, various levels of clinicians and other providers can be aware of these differences and can encourage the patient or suggest to the treatment team the clinical utility of these traditionally more neurological considerations.

(Winterer et al., 2004). Reduced hippocampal activity has been documented during a memory task (Heckers et al., 1998).

PET studies have also been conducted while studying schizophrenia. These studies have shown reduced metabolic activity in the limbic structures, thalamus, frontal cortex, and parietal lobe (Tammings et al., 1992). From these studies, the complexity of the illness has become increasingly clear. For example, it has been documented that some portions of brain activity, including in the prefrontal cortex, improve with symptomatic recovery. Other regions, such as the anterior cingulate and parietal regions, remain slowed when performing a complex motor task (Spence et al., 1997). Dopamine release has been measured by single-photon emission computed tomography (SPECT) and PET imaging as well. These studies have found that people who have been diagnosed with schizophrenia have an increased release of dopamine in the striatum during the acute phases of their illness in comparison with nonpatients (Abi-Darghan et al., 1998). Functional imaging has implied that excessive dopamine in the striatum is related to decreased glutamate regulation (Kegeles et al., 2000).

## COGNITION AND SCHIZOPHRENIA

No cognitive ability is completely spared in schizophrenia. In fact, neurocognitive deficits are a core feature of schizophrenia (Kenny & Friedman, 2002). Research has found that global cognitive changes occur and are affected at different times throughout the illness. Verbal and visual learning and reasoning have been demonstrated to be impaired from childhood through adulthood, whereas executive functioning limitations occur later in the course of the illness (Reichenberg et al., 2010). Deficits in various forms of memory and learning have been identified that subsequently negatively affect the person's livelihood. Working memory, which has been described by Baddely (1992) as concurrent storage and manipulation of information, has been found to be as impaired in first-episode psychotic patients as in patients with chronic schizophrenia (Zanello, Curtis, Badan, & Merlo, 2009). Interestingly, although working memory impairment has not been found to be influenced by length of illness, age of onset (Tuulio-Hendrickson, Partonen, T., Suvisaari, J., Haukka, J., & Lönnqvist, 2004), or duration of untreated illness (Rund et al., 2007), it has been associated with social functioning (Liddle, 2000) and quality of life (Fujii, Wylie, & Nathan, 2004). Semantic memory, information stored about a person's real-world experiences, has been found to be defective in this population as well (Titone, Holzman, & Levy, 2002). Thus, this may limit people from drawing upon their previous experiences when making future decisions. Long-term memory impairment for large amounts of verbal information has been found to occur in this population, which one might expect to negatively influence functional abilities (Gruzelier, Seymour, Wilson, Jolley, & Hirsch, 1988).

Studies have also shown attention and language to be negatively affected in patients with schizophrenia. Studies have shown verbal fluency to be more negatively affected than by other psychotic disorders, which in turn could affect an individual's ability to express herself or himself in a thorough and timely manner (Krabbendam, Arts, van Os, & Aleman, 2005). Limitations on sustained attention have also been found to limit an individual's ability to enter into and maintain relationships and exercise functional abilities. Widespread disturbance in attention and language processes in schizophrenia has been hypothesized to cause a disturbance in the internal representations of contextual information (Cohen & Serven-Schreiber, 1992). As a result, this disturbance contributes to functional limitations that the individual demonstrates.

As described in the imaging section, executive functioning has been widely studied in schizophrenia. Consistent with both structural and functional imaging findings that examine prefrontal functioning, executive functioning has been consistently found to be impaired in schizophrenia. Studies examining executive functioning in schizophrenia have demonstrated impairments in initiation, planning, problem solving, inhibition, concept formation, and cognitive flexibility (Chey, Lee, Kim, Kwon, & Shin, 2002; Velligan & Bow-Thomas, 1999). A meta-analysis comparing schizophrenia to affective disorders, including bipolar disorder and schizoaffective disorder, found that the schizophrenia group performed worse than the other groups in areas of executive functioning (Bora, Yucel, & Pantellis, 2009; Krabbendam, Arts, van Os, & Aleman, 2005). The widely known executive dysfunction found in schizophrenia is clearly linked to the functional impairments that the patients demonstrate. They account for difficulties in terms of caring for oneself, maintaining a job—even during periods of clinical stability—and creating and sustaining relationships. There is some evidence to support the notion that various subtypes of schizophrenia are associated with worse cognitive skills and subsequent functional variability. For example, people who have been diagnosed with paranoid schizophrenia have been found to have less executive impairment, as well as more functional strengths, than those diagnosed with the disorganized subtype of schizophrenia.

## TREATMENT FOR SCHIZOPHRENIA

Multiple factors play a role in the course of schizophrenia. Not surprisingly, prolonged periods of lack of treatment result in a worse outcome (Perkins, Gu, Boteva, & Lieberman, 2005). Family and patient psychoeducation (Lincoln Wilhelm, & Nestoriuc, 2007), social skills training (Pilling et al., 2002; Xia & Li, 2007), and early intervention in first-episode psychosis (Perkins et al., 2005) have all been found to improve recovery and increase the amount of time between a relapse. Also, there is evidence that early intervention in high-risk patients can be somewhat helpful in preventing the development of the disorder (Olsen & Rosenbaum, 2006; McGrorry, Killackey, & Yung, 2008). Identifying those who are at risk for the disorder might be beneficial in preventing the progression. The best predictor of those who are at highest risk for developing the disorder is family history.

### Medication

It has been found that antipsychotic medications are the only effective medications for schizophrenia, particularly for positive symptoms (Kapur, Mizrahi, & Li, 2005; Tandon et al., 2008). The antipsychotic medications are dopamine-2 antagonist medications—terms meaning that they decrease the production of dopamine (Figure 6.1). Yet the most potent dopamine blockers have not been found to be the most effective at treating positive symptoms (Kapur et al., 2000; Tandon et al., 2008). A common side effect of these classifications of medications is extrapyramidal symptoms. Common extrapyramidal symptoms include tardive dyskinesia, parkinsonism, akithesia, and dystonia. These symptoms are typically caused by the dopamine blockage or depletion in the basal ganglia. It has also been shown that extrapyramidal side effects of antipsychotic medications, including tremors, might increase cognitive and negative symptoms. In an attempt to decrease the severity of extrapyramidal symptoms commonly associated with typical antipsychotic medications that were first available in the 1950s, prescribers will often initially prescribe atypical antipsychotic medications that were developed more recently. (Table 6.1 provides typical versus atypical antipsychotic medications.) Yet, the outcome has not been as advantageous as hoped.

It is important to understand that the currently available antipsychotic medications are similar in terms of overall efficacy (Tandon et al., 2008). That said, negative and cognitive symptoms show less improvement than do positive symptoms. Traditional antipsychotic medications, such as Haldol, peraphenazine, and Thorazine

**TABLE 6.1 Common Antipsychotic Medication**

| a. Typical (First Generation) | | b. Atypical (Second Generation) | |
| --- | --- | --- | --- |
| Generic Name | Brand Name | Generic Name | Brand Name |
| Chloropromazine | Thorazine | Aripiprizole | Abilify |
| Halperidol | Haldol | Risperdone | Risperdal |
| Fluphenizine | Prolixin | Quetiapine | Seroquel |
| Pheraphenazine | Trilafon | Olanzapine | Zyprexa |
| Pimozide | Orap | Ziprasidone | Geodon |
| Loxapine | Loxitane | Clozapine | Clozaril |

## ETHICS

### Decision-Making Capacity in Schizophrenia

For a long time, decision-making capacity in patients diagnosed with schizophrenia has been an area of debate among professionals and the public. Indeed, this debate has been central to the ethical debate of research in schizophrenia (Annas, 1994; Levine et al., 1994). It has been argued that because most people with schizophrenia are not incompetent and are in fact able to handle their daily affairs, they should be able to self-determine whether they can consent to research. However, it has been pointed out that in research studies examining large groups of patients versus nonpatients, the patients who had been diagnosed with schizophrenia made poorer decisions regarding treatment than did other psychiatric patients (Grisso & Appelbaum, 1995). The initial decision-making capacity of patients who have been diagnosed with schizophrenia might be compromised by cognitive demands of informed consent (Carpenter et al., 2000). However, these researchers demonstrated that in many cases reduced capacity could be compensated for by education intervention during the informed consent process. A longitudinal study examining the consent abilities of participants in the CATIE study demonstrated that most participants had stable or improved consent-related abilities, but that almost one-fourth demonstrated a decline in abilities, with 4% no longer demonstrating capacity for enrollment (Stroup et al., 2011).

Attempts have been made to improve the understanding of the consent process in patients who have been diagnosed with schizophrenia. Although traditional methods of informed consent have included asking the participant what he or she can recall about the study, this approach might result in inherent limitations, considering the known poor performance on memory measures examining free recall (Goldberg et al., 1989). Providing cues, or, specifically, multiple choices, has been demonstrated to more effectively increase understanding of studies in this population than not doing so (Combs, Adams, Wood, Basso, & Gouvier, 2005). Applications of instruments for assessing the level of competency have been developed, such as the MacArthur Competence Assessment Tools, yet this instrument is not frequently implemented, as most researchers use their own criteria for determining competence. Another approach has been the use of multimedia tools in the consent process, rather than mere printed consent forms (Ryan, Prictor, McLaughlin, & Hill, 2008). Research has demonstrated that both DVD and web-aided multimedia consents have resulted in better comprehension of disclosed information than a routine consent form (Harmell, Palmer, & Jeste, 2012; Jeste et al., 2009). However, there are drawbacks to such approaches. Because the pace of DVDs is difficult to adjust, passive interaction between the researcher and the participant might result, eliminating opportunities to ask questions. Having the cognitive wherewithal to use web-based material might also be a limiting factor of this approach. Another possible way of ensuring appropriate enrollment in research studies is to use a proxy, another person who is the legally authorized representative of the study subject and who is able to make health care decisions for the study subject. This approach has limitations as well, though, as not only does it assume that the patient does not have the basic human right of autonomy, but it also assumes that the proxy has an understanding of the study and of what would accord with the patient's best interest and wishes. Research advanced directives that clearly identify the wishes of an individual before cognitive decline would be ideal, but these are barely used, not least because they are often impractical.

Obtaining informed consent in populations with impaired decision making is challenging. Though it is necessary that research continue in populations with clear

*(continued)*

> *(continued)*
>
> cognitive limitations in order to improve outcomes for these same or similar individuals, ensuring that the research is voluntary is absolutely essential. Research in this area needs to be carefully considered by investigators, treatment providers, and institutional review boards. Oversight of the determination of competency in this population is particularly important for preventing potentially harmful effects of investigational treatments.

are considered typical antipsychotic medications. In the early 1990s, newer atypical antipsychotic medications, such as Risperdal, quetiapine, and olanzapine were introduced with the intent of decreasing the extrapyramidal side effects and improving overall functioning. The results from the $42.6-million, multisite Clinical Antipsychotic Trials of Intervention Effectiveness (CATIE) study funded by the NIMH indicated that not only were the extrapyramidal side effects similar when appropriate dosages of perphenazine were prescribed, but also the newer atypical medications were linked to increased weight gain and potential diabetes, and they were approximately 10 times as expensive. Thus, the medications were similar in terms of discontinuation due to intolerable side effects. The most tolerated atypical medication was olanzapine, but it was linked with the greatest weight gain. Findings also indicated that patients who discontinued use of older antipsychotic medications were found to benefit more from quetiapine thereafter, although variability was seen. Taken as a whole, the NIMH concluded that the newer medications do not offer substantial advantages over the older medications, but also suggested that individual patient differences be taken into consideration. The NIMH website provides specific details regarding the studies that have produced these findings. Often, students question the capacity of patients who have been diagnosed with schizophrenia to participate in these types of studies (see the Ethics box). However, these studies are necessary to provide actual results, considering the cost of medications and decades of inaccurately perceived effectiveness. Without the thorough results from the CATIE study, providers would likely continue to prescribe the less cost-effective atypical antipsychotics, assuming that doing so achieved increased clinical effectiveness.

## Therapy

Traditionally, supportive approaches to assist with patients coping with psychotic symptoms and challenges facing them have been thought to be the most effective strategy when working with this population. Yet there has long been interest in psychological interventions that attempt to augment or improve upon cognitive processing (Flesher, 1990). Recently, cognitive remediation therapy (CRT) has been demonstrated to be useful in improving functioning in patients with schizophrenia. One goal of CRT is to help people to adjust their goals and expectations to match their own abilities and skills. Making such adjustments requires an objective understanding of one's own cognitive strengths and weaknesses. Another goal is to modify the goals of others, including relatives, to help provide support within the environment. CRT has been demonstrated to be beneficial to patients who have been diagnosed with schizophrenia by improving cognition and social functioning (McGurk, Twamley, Switzer, McHugo, Mueser, 2007). This meta-analysis indicated that the effect of improvement is increased with the use of adjunctive psychiatric rehabilitation as well.

Wykes et al. (1999) initially described CRT as a clinically based approach to teach cognitive skills of memory, cognitive flexibility, and planning through psychological principles of positive reinforcement and scaffolding. The goal is to teach self-monitoring techniques for developing adaptive thinking strategies. The results indicate improved memory performance, cognitive flexibility, and self-esteem. Wykes has demonstrated sustained improvement in cognitive functioning after 6 months (Wykes et al., 2003). Improvement in executive functioning has been shown to improve the quality of life in these patients through the use of CRT as well. Functional MRI studies suggest increased brain activation in these patients on measures examining working memory (Wykes et al., 2002). Challenges persist for CRT in schizophrenia, including generalizing processing strategies to new information in different environments, yet the treatment provides hope that psychotherapy can be effective beyond supportive techniques.

## CONCLUSIONS

It is helpful for the clinician to have an understanding of the neurobiological processes that underlie schizophrenia. Such an understanding should guide the clinician's determination of appropriate strategies for diagnosis and treatment. Initially, the clinician should be careful to conduct a sound clinical interview as a tool for understanding the hereditary factors that influence phenotype. Examination of family history is particularly important, given the clear genetic link, particularly in younger populations, of those who experience significant stress. During the initial clinical interview, positive and negative symptom evaluation is necessary for better understanding prognosis and potential barriers to more difficult treatment. Consideration and potential evaluation of cognitive factors can also assist in elucidating the clinician regarding treatment strategies and outcome possibilities. Specifically, determining the level of cognitive impairment in this population will be particularly important as it relates to identifying determinants that would prevent therapeutic progress. Being aware of the cognitive strengths and subtype of schizophrenia can lend to determining the goals of therapy, including the determination for specific medications and implementation of CRT. Working within a team to address the possibility of conducting MRI or other imaging studies of the brain in order to potentially rule out another organic reason for the presentation of symptoms could be beneficial. Similar to other diagnoses, the earlier the identification of accurate diagnosis, the more effective treatment is likely to be, leading to a better prognosis. The clinician should also remain up-to-date on the neurobiological studies that occur in this population so as to better understand treatment advances. For example, future studies, including DTI, are likely to consider even more modern strategies for understanding the deteriorating effects of the disorder.

## SUMMARY POINTS

- Schizophrenia is a debilitating psychiatric illness that typically has an onset prior to the mid-20s.
- Genetics, environment, and stress are important variables in the onset of schizophrenia.

- Schizophrenia's core psychotic features are the result of neurobiological dysfunction occurring at a neurochemical and neuroanatomical level.
- Dopamine, glutamate, and other neurochemical dysregulation are involved in the expression of psychosis.
- Treatment for schizophrenia needs to occur beyond medication management, and the mental health clinician needs to be actively involved in the care.

CHAPTER 7

# Mood Disorders

Fluctuations in mood states from happy to sad are normal within healthy individuals. It is human to have periods of time when one has increased energy, is more productive, and feels very happy about accomplishments, achievements, or even small events. At other times, a person might feel down, disappointed, or sad, often in response to stressful events or situations. The limbic system, our human emotion regulation system, is constantly integrating information from various resources, which result in different emotional states. Although it is not uncommon to have periods of happiness or sadness, it is a concern when these periods are prolonged and result in unhealthy functioning. In fact, when the fluctuations of these emotional states are excessive or are prolonged, the possibility of a mood disorder should be considered.

Mood disorders are collectively one of the most common psychiatric disorders. Of these, the most frequent disorders are depression and bipolar disorder. Mood disorders can vary in clinical presentations depending on age and gender, among other factors. Various subtypes of depression and bipolar disorder have been classified and can be diagnosed by a trained clinician. Despite the various diagnoses, the core characteristic associated with all mood disorders is an affective dysregulation resulting in atypical emotional states. The emotional dysregulation results in functional impairment in either occupational or social domains, including problems maintaining employment and relationships or attending school.

In general, mood disorders result in functional impairment, increased medical problems, and social impairment. The likelihood of having a mood disorder increases with a variety of medical and neurological problems, including traumatic brain injury, diabetes, stroke, heart attacks, sleep disorders, and various forms of dementia, such as Alzheimer's disease. Mood disorders are also associated with psychosocial stressors, including bereavement, loss of a significant other, major life stressors, and abuse. The link between mood disorders and various forms of medical, neurological, and psychosocial factors is associated with neurobiological etiology.

In light of the high prevalence and functional difficulty associated with mood disorders, as well as the strong correlation with medical and neurological disorders, having better understanding of the neurobiological factors that underlie these disorders is important for understanding which treatment strategies are most effective. The clinician will almost undoubtedly work during his or her career with people who experience some form of mood disorder. This chapter aims to help you better understand the two most common mood disorders—depression and bipolar disorder—by reviewing the genetic, neuroanatomical, and cognitive findings of the disorders. Though this chapter does not intend to exhaustively review the neurobiological literature on mood disorders—there are textbooks dedicated entirely to mood disorders—understanding core neurobiological features can inform psychotherapeutic intervention and pharmacological treatment, as well as how involvement of family and friends can be important in the disorder. Treatment of mood disorders has also expanded beyond traditional medication and therapy. Advancements in neuroscience have led to the development of technology, including deep brain stimulation and transcranial magnetic stimulation (TMS), as an effective form of treatment. This chapter aims to provide the clinician with a working knowledge of these forms of treatment as well as understanding of possible reasons for their effectiveness. In so doing, this chapter is intended to shed light on the neurobiology of mood disorder. Insight into the neurobiological aspects of mood disorders can help improve the outcome for these clients, which is every clinician's goal for treatment.

## DEPRESSION

Depression, also often referred to as unipolar depression, affects 121 million people worldwide and is responsible for 850,000 deaths annually (Centers for Disease Control and Prevention, 2011). The World Health Organization's World Mental Health Survey Initiative (Kessler et al., 2009) reported that 9% of the world's population experienced depression within the past year. The prevalence ratio between genders, female to male, is approximately 2:1. The average age of a first episode of depression is in the 30s, but depression can occur throughout the lifetime. Depression is common in older individuals—it has been estimated that depression affects approximately 40% of those living in nursing homes. However, depression is often difficult to diagnose in elderly populations in view of the concurrent increased risk for dementia. Indeed, depression can be difficult to identify at all times, owing to its wide range of possible symptoms. Understanding these symptoms is vital to the therapist being able to identify and treat depression effectively.

### Clinical Symptoms of Depression

The term *depression* is commonly used by lay people to encompass a variety of symptoms. Common clinical features of depression include feelings of helplessness, hopelessness, and worthlessness. An individual who has been diagnosed with depression feels sad and has decreased energy, amotivation, negative thoughts, little or no interest in daily activities, and poor self-esteem. Often, depression leads to sleep dysregulation, whether it is insomnia, which is disrupted or limited sleep, or hypersomnia, which is excessive amounts of sleep or sleepiness. Depression results in a decline in performance of daily activity, and it can negatively affect relationships and performance at work, among other things. Thoughts of self-harm and suicide can occur during periods of depression. Although depression can occur in anyone, the symptoms vary depending on gender, age, and socioeconomic status. In children, symptoms might include social isolation, anger, or poor performance in school. In elderly patients, changes in personality, verbal aggression, urinary incontinence, decreased weight, increased pain, reduced cognitive abilities, and an increase of care needs have been found to be symptoms of depression (Phillips, Rantz, & Petroski, 2011).

## Subtypes of Depression

Depression can be divided into various diagnostic categories, determined by the length of time a person has been depressed and the severity of the symptoms. For example, dysthymic depression is diagnosed when a person has symptoms of depression for a 2-year period, on more days than not, but never has enough symptoms to meet the criteria of a major depressive disorder. People might also experience recurrent brief depressive disorder, which includes symptoms of depression lasting from 2 days to 2 weeks; minor depressive disorder, which includes subclinical amounts of depressive symptoms for longer than 2 weeks; and premenstrual dysphoric disorder, which is associated with mood symptoms occurring during the last week of the menstrual cycle that interfere with activities of living. However, it is normal to experience bereavement—sadness after the loss of a loved one—as long as symptoms do not last longer than 6 months. After 6 months, a person might be experiencing complicated or prolonged grief. Individuals who experience prolonged bereavement are likely experiencing a different neurobiological response associated with the death of a family member or friend. When bereavement is prolonged, the person likely needs treatment. Postpartum depression occurs when a woman feels depressed for the first 4 weeks or so after giving birth. These and other various forms of depression that have been identified are clearly associated with a combination of chemical, hormonal, and/or other neurobiological factors contributing to their clinical presentations. Continued research is expected to shed light on the neurobiological differences among these forms of depression. Particularly, the most studied depressive disorder, and therefore the most understood in terms of its neurobiological factors, is major depressive disorder.

## MAJOR DEPRESSIVE DISORDER

Major depressive disorder (MDD) is often defined as having five or more symptoms of depression and at least one of those symptoms being either depressed mood or loss of interest or pleasure. For diagnosis, these symptoms must occur for no shorter than a consecutive 2-week period; must be a change from previous functioning; and must negatively affect social, occupational, or other areas of functioning. MDD has long been a public health concern, owing to its debilitating effects. It affects approximately 16% of people at some point in their lifetime, with high probability of recurrence (Judd, 1997; Kessler et al., 2003). There is as high as a 60% chance of having a second MDD episode after having a first. Approximately two-thirds of individuals who experience an MDD episode never recover fully. Thus, tremendous cost to society is associated not only with mental health fees, but also with medical bills, decreased work production, and functional difficulty within a person's immediate family.

## Genetics and MDD

Studies have clearly implicated the importance of genetic factors in depression. Studies examining genetics, including twin studies, have demonstrated heritability of depression to be between 33% and 50% (Kendler, Gatz, Gardner, & Pederson, 2006; Levinson, 2006). A review of genetic studies in depression found that early age of onset and recurrent episodic depression are strongly linked to familial risk (Levinson, 2006). Furthermore, first-degree relatives of patients with MDD have a risk of experiencing MDD that is almost threefold that of the general population's (Sullivan, Neal, & Kindler, 2000).

As mentioned in Chapter 2, there are various types of genetic studies, including linkage studies and candidate gene studies. Gene linkage studies begin by selecting a gene based upon its location within a linkage peak. The choice of a candidate gene typically "constitutes part of an investigator's best guess or hypothesis about etiologic mechanisms, and is informed by biological plausibility based on the limited knowledge we have gleaned from sources such as animal models, disease-correlated changes in clinical indices, or existing treatments for a disorder." Thus, hundreds of candidate genes and linkage genes have been studied. It is most likely that the interaction of genetic predisposition and environmental influences determines which genes are activated. For example, studies have shown that a low-efficiency–type serotonin transporter gene increases the risk of developing depression under environmental stress (Caspi et al., 2003; Lenze et al., 2005).

Genetic studies have identified regions on at least 10 different chromosomes that might play a role in the expression of depression. Lopez-Leon (2008) published a meta-analysis of nearly 200 studies showing that polymorphism in the apolipoprotein E (APOE) gene is the most associated with MDD, followed by SLC6A4, which is a serotonin transporter. Although numerous studies have examined genetic factors in MDD, unfortunately, replicating findings has been difficult, leaving investigators to continue attempting to understand the precise genetic attributes of MDD.

Imaging studies have also implicated hereditary factors in depression. Those with a family history of depression appear to have brains that are 28% thinner in the right cortex than those without a family history of depression (Peterson et al., 2009). In fact, familial risk for depression affects brain activity and functioning even before the onset of depression in girls who are at high risk. Specifically, teenage girls whose biological mothers experienced recurrent depressive episodes demonstrated reduced activation during reward processing and increased activation when processing negative images (Chen, Hamilton, & Gotleib, 2010).

## Neuroimaging and MDD

It is clear that underlying neurobiological processes are involved with the behavioral manifestation of MDD. Although the exact neurobiological mechanisms underlying depression might vary depending on the subtype of depression, symptom cluster, and age of onset, among other factors, studies have identified neurochemicals, neuropathways, and neuroanatomical components that seem to consistently play a role in the disorder. Clear brain pathology has been discovered within individuals who have been diagnosed with depression. Kempton et al. (2011) provide a database of the neuroimaging studies of MDD to allow researchers the opportunity to examine a meta-analysis of different brain structures that have been implicated in MDD. Studies included in that database have been relatively consistent in identifying the cortical and subcortical pathways thought to be involved in depression.

The prefrontal cortex and limbic system play a significant role in the clinical presentation of depression. The most common structural brain abnormalities associated with depression include decreased prefrontal cortex, hippocampus, amygdala, and basal ganglia volume (Sheline, Mokhtar, & Kraemer, 2003). There has also been some suggestion of decreased volume in the medial frontal cortex, which is thought to be important for self-monitoring. Negative cognitions associated with depression are caused by the imbalance of cortical control and subcortical functioning, including the dysregulated prefrontal modulation of limbic structures, among which are the amygdala and hippocampus (Beck, 2008; Disner, Beevers, Haigh, & Beck, 2011). Both structural and functional imaging studies have demonstrated consistent abnormalities in the dorsolateral, ventrolateral, and orbital frontal regions that have been linked to the symptoms of depression. These areas of the prefrontal cortex have been demonstrated to have decreased volume in structural studies and decreased blood flow during resting states (Goldapple et al., 2004; Videbach, 2000). In general, the prefrontal cortex is important for initiation and execution of movements and problem solving, among other abilities. The prefrontal cortex has been linked with irritability and agitation. The limbic system has been consistently implicated in emotion regulation. Functional magnetic resonance imaging (fMRI) studies have shown that decreased activity in the frontal lobes correlate well with overall depression score.

The amygdala, the most consistently studied brain structure in depression, has typically been found to have abnormalities in MDD. In general, decreased amygdala volume has been found in MDD within structural MRI studies. However, a recent meta-analysis indicated a lack of reliable difference in the amygdala volume in depressed versus never depressed people (Hamilton, Siemer, & Gotlib, 2008). The study further examined the effects of medication on MDD and found that with antidepressant medication no difference in the amygdala volume is apparent. It concluded that neurogenesis is promoted by the antidepressant medication, and that the medication might protect against glucocorticoid toxicity. In comparison, fMRI studies have demonstrated elevated baseline activity in the amygdala as well as an increase in amygdala response to affective stimuli (Hamilton et al., 2008). The dysregulation of the amygdala results in internal mood dysregulation as well as in difficulty interpreting social cues and environmental factors.

Another brain system known to be important in depression is the cingulate system. The cingulate gyrus, known itself to be important for cognitive flexibility, among other functions, has been found to be smaller in individuals who experience unipolar depression when compared with healthy volunteers (Hajek, Kozeny, Kopecek, Alda, & Höschl, 2008). This might suggest limitations on generating new ideas to resolve problems, perseverating on negative affect, and repeatedly trying the same solution to a problem despite a lack of positive outcome.

The primary neurochemical dysfunction in depression is of the monoamine neurotransmitter system, including serotonin, norepinephrine, and dopamine. A great deal of research has indicated that dysfunction of seroternergic neurotransmitters plays a particularly important role in the pathology of depression. Studies have demonstrated that lower levels of serotonin in cerebral spinal fluid, decreased serotonin receptor binding, and other serotonin abnormalities are present in patients who have been diagnosed with depression (Mann et al., 1996; Owens & Nemeroff, 1994). Furthermore, cerebral spinal fluid studies examining a metabolite of dopamine have been found to be decreased in depression (Reddy, Khanna, Subhash, Channabasavanna, & Sridhara Rama Rao, 1992). It has also been found that the HPA axis does not function in some patients with MDD. Research has suggested that alterations in cortisol regulation occur subsequent to early life trauma or abuse. Given the extent of literature on MDD, it is clear that a number of specific neurochemicals likely have a complex interaction with genes and brain structures, explaining the pathophysiology of MDD.

## Cognition and MDD

It is well known that individuals who have been diagnosed with depression have cognitive difficulties contributing to their depression. People with MDD complain frequently of memory and concentration difficulty. In fact, studies have demonstrated objective cognitive difficulty in areas of attention, processing speed, reaction time, memory, and executive functioning. However, the patient's complaints of memory problems are not always consistent with their actual performance on objective cognitive tests (Gualtieri, Johnson, & Benedict, 2006). In fact, performance on measures of attention frequently negatively affects performance on measures of memory. As a result, people with depression often feel that their memory is impaired when in fact poor attention is negatively influencing memory. This is one way to differentiate depression from dementia, as individuals in the early stages of dementia perform well on attention tasks but poorly on tasks measuring memory.

The performance by individuals who have been diagnosed with MDD on cognitive measures varies, but in general it has been found that the more severely depressed an individual is, the more likely he or she is to be cognitively impaired (Farrin, Hull, Unwin, Wykes, & David, 2003). That said, even mild depression results in decline of performance in selected areas of cognitive abilities (Porter et al., 2005). Generally, individuals with MDD have intact cognitive abilities in the areas of language skills, visuospatial and constructional skills, and intact orientation. The one exception with regard to possible language-based difficulties is verbal fluency, but that might be associated with executive impairment rather than with genuine language difficulties. Various areas of executive functioning, including problem-solving and cognitive flexibility, might be negatively affected by depression. Furthermore, individuals who have been diagnosed with depression will often have difficulty in terms of free recall of learned material, but when provided prompts or recognition conditions, the MDD individual will perform better. This suggests difficulty with free recall but less impairment in terms of encoding, different from what is found in dementia.

Studies have indicated that when MDD is effectively treated, performance on cognitive measures improves. Individuals who are clinically stable subsequent to administration of antidepressant medications have been shown to have improved cognitive flexibility, processing, vigilance, and complex attention (Gualtieri et al., 2006). Impairments in executive functioning, particularly in areas of initiation and perseveration, have also been linked with poorer outcome in depression (Fossati, Ergis, & Allilaire, 2002).

## Treatment for MDD

There are several possible effective treatments for depression. As with all psychiatric disorders, the skilled clinician must ensure that the patient has been recently evaluated by a medical doctor to rule out other possible organic etiology of the mood problems. For example, hypothyroidism, heart disease, and side effects from various medications have been linked to depression. It is the clinician's responsibility to inquire about and encourage an evaluation from a physician, most likely a primary care physician, to assist in the accurate diagnosis of MDD before implementing strategies for treatment.

The most common treatments, cognitive therapy and psychotropic medications, have been well validated and are highly efficacious. Indeed, these two types of treatments appear nearly equivalent in decreasing the acute symptoms for depression (De Rubeis, Gelfand, Tang, & Simons, 1999). However, several other forms of treatment can be used to ameliorate depression as well.

*Cognitive Behavioral Therapy*

Cognitive behavioral therapy (CBT) has proven to be effective for treating MDD to produce both short-term and long-term improvement. Several books have been written about how to use CBT effectively in depressed patients. The clinician is strongly encouraged to learn CBT and cognitive techniques if he or she plans to use therapeutic interventions. Briefly, CBT aims to alter faulty cognitions of individuals with depression and replace them with more adaptive and accurate cognitions. A person's mood arises in part from that person's thoughts. By using therapeutic strategies, people can alter their thoughts and behaviors to improve upon their emotional state. An aim of CBT is to cognitively restructure a person's thoughts and focus on specific problems in the here and now. It uses Socratic questioning, role-playing, imagery, and experiments to activate behavioral change. CBT has been found to be as effective as, and possibly more effective than, antidepressant medication. It encourages behavioral activation and provides an effective way to problem solve and decrease feelings of being overwhelmed. CBT has even been found to be effective for decreasing depression when used via the Internet, on a cell phone, or on a tablet or other mobile device (Kupfer et al., 2012; Watts et al., 2013). Accordingly, it might be possible to deliver treatment to people who have been diagnosed with depression but who have limited transportation or mobility or who are not living near a trained specialist.

*Psychotropic Medications*

Psychotropic medication became popular for depression in the 1980s with the popularization of Prozac, as described in Chapter 1 of this text. Since then, numerous selective serotonin reuptake inhibitors (SSRIs) have been introduced and used to treat depression. SSRIs have proven to be effective in up to 60% of depressed patients (Nelson, 1999). Table 7.1 provides the generic and formulary names of SSRIs. Another psychotropic medication class that has been effective in treating MDD is serotonin norepiniphrine reuptake inhibitors (SNRIs). Table 7.2 provides the generic and formulary names of SNRIs. SNRIs selectively inhibit norepiniphrine and serotonin reuptake. SSRIs and SNRIs are popular because they do not cause as many side effects as older medications. Another medication that does not fall in a category with

**TABLE 7.1 Common Selective Serotonin Reuptake Inhibitors (SSRIs)**

| Brand Name | Generic Name |
|---|---|
| Celexa | citalopram |
| Lexapro | escitalopram |
| Luvox | fluvoxamine |
| Paxil | paroxetine |
| Prozac | fluoxetine |
| Viibryd | vilazodone |
| Zoloft | sertraline |

**TABLE 7.2 Common Serotonin Norepinephrine Reuptake Inhibitors (SNRIs)**

| Brand Name | Generic Name |
|---|---|
| Cymbalta | duloxetine |
| Effexor | venlafaxine |
| Serzone | nefazodone |

any other antidepressant medication is Wellbutrin, which works on dopamine receptors to treat depression. If Wellbutrin, SSRIs, and SNRIs are not effective, other medications can be prescribed as well. For example, monoamine oxidase inhibitors (MAOIs), tricyclics, and tetracyclics might be more effective than the newer SSRIs and SNRIs in certain subgroups of depression. However, tricylcic medications can cause bladder problems, constipation, and dry mouth, often lowering compliance with medication. In comparison, MAOIs have a negative interaction with tyramine, which is found in many foods and medicines. Patients who are prescribed MAOIs must avoid over-the-counter cold medications, certain foods (such as cheese), and wines to take the medication safely. That said, the FDA has also warned that these medications can have unintentional side effects, especially in young people, including increased risk for suicide attempts. In 2007, the FDA proposed that the black-box warning label on all antidepressant medications be indicated for all people up to age 24, suggesting that those on the medication be carefully monitored for the side effects of suicidal thinking or behavior. Furthermore, the FDA warned against use of SSRIs or SNRIs with triptan, a medication commonly used to treat migraines, as this can result in serotonin syndrome.

### *Transcranial Magnetic Stimulation*

Transcranial magnetic stimulation (TMS) is a brain intervention thought to modulate activity in discrete cortical regions and associated neural circuits by noninvasively inducing intracerebral currents. When applied repetitively, it is referred to as repetitive transcranial magnetic stimulation (rTMS). rTMS has been used in a variety of capacities, and it has recently been found to have antidepressant properties when administered to the appropriate portion of the brain for a specified time. Specifically, daily left prefrontal rTMS has been found to have significant and clinically meaningful effects as compared to a placebo approach. Treatment is well tolerated, and adverse side effects are absent. Although the length of treatment is unclear, rTMS appears to be an upcoming form of treatment for MDD. See Figure 7.1 for an example.

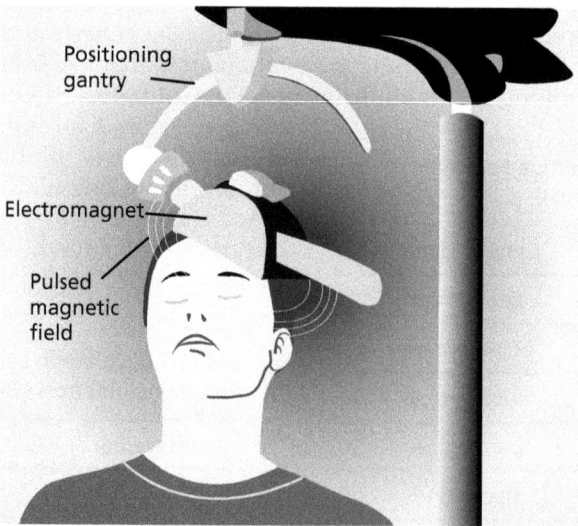

**FIGURE 7.1 A picture of transcranial magnetic stimulation (TMS).**
Adapted from the National Institute of Mental Health, National Institutes of Health, Department of Health and Human Services.

## Deep Brain Stimulation

Deep brain stimulation (DBS) is another new, experimental, more invasive treatment for refractory patients who have been diagnosed with MDD. DBS involves implanting electrodes in the ventral striatum, with four separately programmable contacts extending to a neurostimulator implanted in the clavicle area under the skin of the individual. Studies have shown that connection to white matter tracts in the cingulate cortex can also improve mood symptoms (Mayberg et al., 2005). The exact mechanism behind the effectiveness of the treatment is unknown. Although the FDA has not yet approved DBS for treatment of depression in the United States, it has been approved for obsessive compulsive disorder, and it might soon be approved for depression.

## Vagus Nerve Stimulation

Vagus nerve stimulation (VNS) also uses a device implanted in the body that sends electrical impulses to the vagus nerve. A small battery-operated device about the size of a half-dollar is implanted in the chest from which wires run to the vagus nerve of the neck, sending electricity to the nerve on a regular basis. The nerve then passes the signal to the brain. The effect of the stimulation can take weeks or months, but this treatment has been found to decrease symptoms of depression. VNS treatment has been approved in Canada, in Europe, and by the FDA in the United States.

## Electroconvulsive Therapy

Electroconvulsive therapy (ECT) is an older, typically effective treatment of MDD (see Figure 7.2). It is frequently administered to medication-resistant depressed patients and is performed more than 100,000 times annually in the United States. Despite its historically negative perspective, perpetuated by media accounts, it is an effective treatment in which patients undergo electric shock in order to produce a controlled seizure event in an attempt to improve mood. Although how it works is unclear, it has been found to be particularly effective in older, otherwise treatment-resistant patients. Despite initial concerns, numerous imaging research studies have demonstrated that ECT does not have an adverse effect on the integrity of

**FIGURE 7.2 Electroconvulsive therapy (ECT).**
Adapted from the National Institute of Mental Health, National Institutes of Health, Department of Health and Human Services.

brain structures (Coffey et al., 1991). However, ECT appears to result in an overall diminution of brain activity. Although the precise neurophysiological effects of ECT are unknown, evidence exists that metabolic reduction occurs in the prefrontal cortex following ECT (Nobler et al., 2001).

### Nonmedical Techniques

Practitioners often encounter questions from patients and their families about nonmedical approaches to treating depression. Various studies have examined the effectiveness of strategies such as exercise, yoga, and over-the-counter products to help improve upon mood difficulties associated with depression. A good clinician will familiarize himself or herself with the effectiveness of these strategies. In general, many of the strategies are behavioral in nature (yoga, running, journal writing) and can be attempted as supplements or as initial approaches to improve on mood, particularly in less severe states of depression. In fact, healthy lifestyle choices such as those described above should be a part of each individual's life, regardless of mood problems, as they might increase serotonin production, increase social interaction, and increase self-esteem. Aiming to increase physical activity, eat a healthy diet, and write a journal have been shown to be effective in improving mood in some, but is not likely to effectively treat long-term MDD symptoms.

Use of over-the-counter medications, vitamins, and supplements should be discussed with a physician. The clinician should strongly recommend that his or her patient discuss the possible implications of these approaches before using them, or as soon as the clinician discovers that they are being used. Although herbal medications have been used for centuries to treat depression, the effectiveness of these techniques seems limited at best. St. John's Wort is widely used in Europe as a form of treatment for depression. In a double-blind study of 340 people diagnosed with MDD, people who were prescribed St. John's Wort were found to improve at a rate similar to subjects taking a placebo sugar pill, as compared to a substantial increase of functioning in individuals prescribed an SSRI (National Institute of Mental Health, 2002). Different vitamins might improve mood and should be discussed with a physician.

Another strategy that can potentially improve mood is light therapy. Exposure to increased amounts of bright light can be particularly effective for those who have been diagnosed with seasonal affective disorders, a type of depression that occurs in fall or winter, when there is less daylight. Light therapy requires the client to use a device called a light therapy box, which shines a light that simulates outdoor light. It is thought that the bright light activates the suprachiasmatic nucleus (SCN), a circadian pacemaker of the brain, through specialized light-sensitive ganglian cells in order to release glutamate. Compared to a placebo, light therapy has been shown to improve mood just as well as an SSRI in seasonal affective disorder, including in elderly people (Lieverse et al., 2011). Although studies examining the effectiveness of light therapy on other people with MDD are unclear, there are no side effects, and the cost is inexpensive, which makes the treatment promising and certainly worth considering.

Various forms of treatment have been found to be effective for MDD. Though it is unclear which treatments are most effective for which patients, and why some forms of treatment are effective for some patients but not others, this is being researched. The Special Topics box examines this area thoroughly, as identification of these factors could substantially improve treatment efficacy and overall outcome. Neuroscience hopes to develop more consistent strategies for stabilizing mood through empirically sound techniques that are individually tailored in terms of psychotherapy, medication, behavioral, and other medical interventions, and by doing so to help decrease depression, thereby lowering health care costs and improving overall quality of life.

## SPECIAL TOPICS

### Neuroscience in Prediction of Success: MRI/Genetics/Cognitive

Imagine being able to scan a patient's brain, complete a genetic analysis, or conduct a thorough cognitive evaluation before implementing treatment strategies when working with patients who have been diagnosed with mood disorders in order to find out with near certainty that the treatment regimen will be effective. This knowledge would decrease patient frustration with failed techniques, substantially guide treatment providers wondering which medications or therapy protocols to use, and decrease medical costs. Patients, providers, and insurance companies alike would agree that being able to do so would be in everybody's best interest. Neuroscience is attempting to help achieve this goal.

Research determining predictors of success for various types of treatment is not new. Traditionally, predictors of success for therapy outcome have been limited to psychosocial, environmental, and other demographic factors. For example, studies have found that higher levels of family support, decreased amounts of environmental stressors, and increased access to treatment decrease the risk for future mood episodes. However, such factors can be limited in predictive value, and they are correlational in nature, not examining cause and effect. Identifying patient characteristics that highly correlate with successful treatment has been a recent subject of study in the neurosciences.

Studies have started to examine neurobiological factors that can play a role in determining effectiveness of treatment in an attempt to further tailor treatment goals for the individual patient. A study examining the predictive value of measuring activity at the subgenual anterior cingulate cortex (ACC) through fMRI to help identify patients who have been diagnosed with MDD is being conducted in order to determine who would benefit from cognitive therapy. It has been found that increased activation of the ACC was predictive of those who would demonstrate improved mood functioning with CBT.

A critical review conducted by Garriock and Moreno (2011) attempted to outline possibilities of using genetics to guide treatment in depression. Their results revealed that although progress has been made in predicting antidepressant response based upon pharmacokinetic genes, genetic data cannot yet provide a mathematical equation to guide treatment for depression or MDD management. Their review ultimately suggests that the best predictor of which medication will work for the patient is whether a medication worked for an immediate relative.

Some companies have started to offer neurotransmitter testing through urine analysis to evaluate for an imbalance of neurotransmitters. Although urinary monoamine analysis is not currently supported by modern scientific literature, attempts continue to use our neurobiological knowledge to clarify diagnosis with an eye toward helping with treatment planning in a consumer-friendly manner (Hinz, Stein, Trachte, & Uncini, 2010).

It can also be hypothesized that the effectiveness of CBT can be determined by cognitive-based testing. It is possible that individuals who have better executive functioning, memory, or attention might be better candidates for CBT. Use of neuropsychological testing to determine cognitive strengths and weakness has been a long-standing goal of evaluations. Doing so can help determine whether a person is nonverbally oriented

*(continued)*

> *(continued)*
>
> and whether treatment should focus on developing nonverbal strategies instead of on traditional CBT models. It is possible that individuals who have worse facial affect recognition might need increased focus on social interaction abilities if they are to improve functioning.
>
> Without doubt, neuroscience will continue to explore the best predictors of treatment success to guide future treatment of mood disorders in a bid to decrease cost while increasing treatment efficacy. Successful identification of neurobiological treatment would improve the quality of life for the patient and decrease the frustration of treatment providers.

## BIPOLAR DISORDER

Bipolar disorder (BD), formerly known as *manic depressive disorder*, is a mood disorder differentiated by periods of clinically elevated mood cycles. Bipolar disorder type I (BDI) is defined by the occurrence of mania on at least one occasion, even in the absence of depression. Mania symptoms include excessive energy, euphoria, irritability, grandiosity, lack of the need for sleep, and agitated behavior. During periods of mania, the individual will demonstrate cognitive limitations, including memory problems, poor attention, and ready distraction. According to the *DSM-5* (American Psychiatric Association, 2013), these symptoms must occur for at least 1 week and cause functional impairment to relationships, employment, or other such primary aspects of life. In comparison, bipolar disorder type II (BDII) need occur only for up to 4 days. BDII is differentiated from BDI by the presence of hypomania, rather than full-blown manic episodes. Hypomania is elevation of mood symptoms, but not to the same extent or severity of mania occuring in BDI. It is not uncommon for people to experience mixed episodes of BDI. In a mixed episode, the fluctuation between mania and depression occurs. Psychotic symptoms, including hallucinations, delusions, and thought disorders (symptoms discussed in the schizophrenia chapter of this book), often co-occur during the course of BD.

It is estimated that upward of 3% of the population experiences either BDI or BDII (Hirschfield et al., 2003; Judd & Akiskal, 2003). BD substantially increases the risk for suicide and is associated with shorter life span owing to a variety of factors, including increased difficulty of self-care and increased incidence of risky behavior (Goodwin & Jamison, 2007). For individuals diagnosed with BDI or BDII, there is an increase in severely strained relationships and failed occupational and educational pursuits, along with a variety of other social impairments. Yet during the manic phase individuals can often be creative and extremely productive. Many individuals prefer not to lose mania, which can make consistent treatment of the disorder challenging. Understanding the early symptoms is important to preventing continued decline of functioning.

### Genetics and BD

Similar to other psychiatric disorders, genetics plays a role in the onset of BD and the behavioral symptoms that occur. Family, twin, and adoption studies have indicated a heritability factor of between 63% and 79% for BD (Smoller & Finn,

2003). BDI patients who have a family history of mood disorder have earlier age of onset, often occurring before age 21, as well as more mood episodes and increased likelihood of hospitalization, than those without a family history (Mrad, Mechri, Rouissi, Khiari, & Gaha, 2007). Clearly, genetic factors play a role in the onset and expression of BD, although the exact mechanisms remain uncertain (Bienvenu, Davydow, & Kendler, 2011).

Numerous genes have been identified as playing a role in BD, and no single gene is completely responsible (Patel & DeUgiannidis, 2010). Patel and DeUgiannidis suggested that 56 genes likely contribute in combination to the expression of BD. Describing these genes in this context is beyond the scope of this book, and the interested reader is encouraged to review their article. Briefly, the genes thought to play a role involve numerous functions, including neurodevelopment, glutamate control, and regulation of calcium-channel modulation. Because such mutations, even when ever so slight, can increase the risk and development of BD, it is likely that coupling with stress or other environmental factors makes expression of BD likely.

## Neuroimaging and BD

Neuroimaging studies on BD have spanned nearly four decades. Early CT studies revealed an overall decreased volume, but identifying specific brain regions was difficult, owing to technological limitations (Adler & Cerullo, 2012; Pearlson & Robinson, 1981). As technology has progressed, the underlying neurophysiology of the disorder has expanded, with some consistent findings spanning that entire timeframe. Structural MRI studies confirmed a decrease in whole brain volume and atypically enlarged ventricle size (Nasrallah, Coffman, & Olson, 1989). However, it is clear that specific regions of interest have emerged in understanding the neurobiology of BD. Structural abnormalities have consistently been found in the prefrontal cortex as well as throughout the limbic system. Much as with depression, a website exists where MRI data is made publicly available for examination and further studies; it can be found at www.bipolardatabase.org. Several books, including a recent textbook edited by Stephen Strakowski (2012), go into great detail regarding the neurobiological findings of BD. Strakowski poses a working model of the underlying neurobiology of BD. For more detailed information, the reader is encouraged to review that or another bipolar-specific text. This section of this chapter is provided as a means of briefly summarizing the neurobiological findings of BD to conceptualize BD at a clinically useful level that will assist in understanding and applying the knowledge to benefit the patient and his or her family.

Not surprisingly, the amygdala, the emotion-control center of the brain, has consistently been found to be implicated in BD. Recent meta-analysis focusing on the limbic structures has found enlarged amygdala in adult bipolar patients (Hajek et al., 2009). The replication of this finding provides consistent evidence of the abnormal neuronal growth and lack of synaptic pruning, thereby likely decreasing the efficiency and effectiveness of this emotion-regulation area. Interestingly, the same analysis has shown that there are no differences in the amygdala volume in medicated and unmedicated BD patients. Although studies have been somewhat inconsistent, the most evidence of morphological differences in BD, other than an enlarged amygdala, include enlarged lateral ventricles as well as the smaller corpus callosum and anterior cingulate. From a behavioral and emotional perspective, this difference results in slower processing, limitations in terms of emotional control, poor flexibility, and dysregulation of emotional modulation.

Over the past decade a plethora of functional MRI studies have examined BD. These studies often focus on areas of the brain thought to be associated with the cognitive impairment found in BD. Studies have documented both the overactivation and the underactivation of various brain regions during specific tasks, which can change during the course of a single task. For example, problems in areas of attention have been found in BD. The cingulate system, left striatum, and left thalamus have shown decreased activation over time on sustained attention tasks, whereas the left amygdala has shown increased activation (Fleck et al., 2010) during periods of mania. The increased activation during simpler tasks and decreased activation with time or on various tasks suggests abnormal processing on various tasks. In comparison, hyperactivation of the amygdala appears to be consistent during manic states when the subject is asked to demonstrate emotion recognition. The amygdala shows no abnormalities on fMRI studies during periods of clinical stability. The orbitofrontal area has shown activation while observing emotions during all three mood states of BD. Otherwise, few studies have examined fMRI activation across various mood states in BD. Future longitudinal studies must better examine the unique activation pattern of BD across mood states and during clinical stability.

Another unique and clinically unclear finding in BD found on MRI, typically in the frontal–temporal regions, is the presence of white matter hyperintensities in the older population (Beyer & Krishnan, 2009). Although the behavioral implications of these findings remain a mystery, it has been hypothesized that they might increase late-onset mania, pediatric onset, and the quantity of episodes.

More recently, diffuse tensor imaging (DTI) has been used to examine white matter changes in patients who have been diagnosed with BD. Studies have found abnormal white matter tracts connecting the frontal cortex with the temporal and parietal cortex (Lin, Wing, Xie, Wu, & Lei, 2011). Furthermore, there appears to be a decrease of white matter connectivity in the subcortical circuits, including areas of the thalamus and cingulate gyrus. These results might account for alterations in mood or reflect increased vulnerability to the disorder. It has also been suggested that abnormalities in these pathways in DTI studies might be associated with earlier onset of the illness (Lu et al., 2012). Neurodiagnostic tools might be implemented in the future, not only to assist in diagnosing the disorder early in the disease, but also to examine whether the white matter tracts' integrity improves with various forms of early intervention and treatment for the disorder.

## Neurochemistry and BD

As we have discussed with other psychiatric disorders, including depression, neurochemical dysregulation occurs in BD. It is hypothesized that excessive dopamine release to the prefrontal cortex and striatal areas occurs during periods of mania. It is thought that this occurs as a result of slower activity of serotonin and norepiniphrine, which leads to a dysfunctional modulation of the human emotional system (Cousins, Butts, & Young, 2009). Magnetic resonance spectroscopy (MRS) has consistently shown dysregulation of concentrations of glutamate, choline, creatine, and gamma-aminobutyric acid (GABA) in those who have been diagnosed with BD compared to non-BD subjects (Strakowski et al., 2012). These findings make scientific sense. For example, we have learned that glutamate is the primary excitatory neurotransmitter and that it plays an important role in the prefrontal and striatal pathways. Excessive glutamate might lead to neuronal death and cell loss. Elevations of this neurochemical result in functional and structural changes to those pathways, subsequently resulting in behavioral and emotional dysregulation. Similarly, an increase of choline in the striatum might result in cognitive problems during manic states, particularly relating to emotional memory. In comparison, creatine concentrations have been found to be decreased in BD during affective episodes, which might account for dysregulation of frontal activity (Murashita, Kato, Shioiri, Inubushi, & Kato, 2000).

Neurochemical dysregulation in BD continues to be examined through MRS and other advanced imaging techniques. Understanding this dysregulation can assist in the development of further treatment strategies for improving symptoms and potentially preventing further decline.

## Cognition and BD

Neurocognitive studies have demonstrated that numerous areas of cognitive functioning are affected in BD. Although not all areas of cognition are negatively affected, such as overall IQ scores and most areas of language abilities, specific cognitive domains are affected. These include poor attention, limitations on working memory, impaired areas of executive functioning, and worsened facial affect labeling in spite of intact (nonemotive) facial recognition skills. Although it is likely that the cognitive profile is different during different clinical phases of the disorder, cognitive impairment in various capacities occurs during periods of mania, depression, and euthymia (Martinez-Aran et al., 2004).

Studies examining various aspects of executive functioning, including planning, organization, and inhibition, have demonstrated impairment during all three phases of the disorder, with typically worse functioning occurring during mania (Depp, Lebowitz, Patterson, Lacro, & Jeste, 2007; Martinez-Aran et al., 2004). In fact, a meta-analysis conducted by Bora, Yucel, and Pantelis in 2009 revealed that executive functioning deficits occur in clinically stable patients as well. Similarly, working memory, the ability to hold and recall information after presentation of other information and mental manipulation of information, has been found to be impaired during periods of mania and depression (Bearden, Hoffman, & Cannon, 2001). Along with working memory, fluency deficits have been most consistently found to be abnormal in BD across the various stages of the disease (Balanzá-Martinez et al., 2008).

Cognitive impairment is likely associated with a number of mood episodes. Although executive functioning decrements are likely to be present in patients who experience their first manic episode, there are greater deficits in individuals who experience a higher number of mood episodes (Nehra, Chakrabarti, Pradhan, & Khehra, 2006). It is also possible that the earlier onset of bipolar symptoms increases the risk for greater cognitive decline, particularly in areas of executive functioning.

Very limited information is available that examines the difference in cognitive functioning between BDI and BDII. Although it can be assumed that cognitive impairment is not as severe in BDII, this might not be the case. Although verbal fluency appears more impaired in BDI, it has been suggested that the increased depressive features in BDII might contribute to increased cognitive decline in that population as well.

The most unique aspect and the best-studied area of cognitive difficulty in BD is the inability to recognize emotional cues and facial affect. Dysregulation of the limbic system is thought not only to be responsible for the expression of the disorder, but to also play a role in terms of interpreting other people's affect. During periods of mania, bipolar patients have been demonstrated to have limitations in their ability to label another person's affect and discriminate among affect, and they are more likely to report higher levels of happiness and surprise, in contrast to the fear and anger seen in depression (Getz et al., 2003; Getz, Shear, & Strakowski, 2003). Facial affect recognition impairments likely play a role in functional impairment in social interactions, as the inability to correctly identify social cues contributes to poor relationships and difficulty interpreting meaning and modulation of one's own behavior based upon external feedback. Although facial affect recognition deficits might not be unique to BD, as it is also often seen in autism, schizophrenia, and unipolar depression, it is well documented that the impairments occur in this disorder, a fact of which the clinician should be aware as it relates to treatment and overall functional difficulty.

## Treatment for BD

The clinician's awareness of the complex genetic, biological, cognitive, and environmental influences in the expression of BD should lead to a better understanding of how to treat this disorder. It is not uncommon for individuals with mania to not want treatment, as the associated excessive energy and increased activity can create a perception of increased productivity. Although there might in fact be increased productivity, mania is also accompanied by poor decisions, increased risk-taking behavior, and relationship difficulty. Therefore, it's not uncommon for people to seek treatment for mania only with the assistance of family members who are concerned with the erratic behavior, or after psychotic behaviors accompany the disorder. Additionally, the patient might not seek treatment for himself or herself during the manic phase, but rather seek treatment only during the depressive phase of the disorder. Therefore, a thorough evaluation of each patient, including an evaluation of historical emotional and behavioral functioning, is important for appropriate diagnosis and for implementation of behavioral treatment. This is particularly important, as providers often have limited time to assess for all areas of functioning and could overlook the possibility of a previous, untreated manic episode that in some cases might have occured before age 20. Use of medication aimed at improving depression during the depressive state of BD might not be efficacious. Compared with medications aimed to stabilize mood, which are often prescribed for BD, an antidepressant is not likely to be effective (Sidor & McQueen, 2011). A thorough evaluation of history of mood symptoms is necessary to accurately determine the presence of BD.

### *Psychotropic Medications*

If BD is diagnosed, then a combination of medication and therapy is typically warranted. Medication aimed at stabilizing mood is often necessary. Table 7.3 lists common psychotropic medications that have been found to be effective for BD. Although these medications have varying degrees and types of side effects, they are often necessary for therapy to be effective.

The long-standing gold-standard treatment for BD is lithium. However, valproic acid (Depakote) is the most commonly used mood stabilizer in the United States, with lithium only the second most popular (Baldessareni, Henk, Sklar, Chang, & Leahy, 2008). Despite this trend, a large 12-year follow-up study concluded that lithium remains a superior treatment to valproate. The precise neurochemical alteration caused by lithium is complex and remains unclear, although it is known to be widely distributed throughout the central nervous system. It likely contributes to increased serotonin and decreased norepinephrine. It also regulates glutamate functioning. As a result, the person who has been diagnosed with BD will experience a stabilized mood. Lithium has been shown to increase overall gray matter volume in

TABLE 7.3 Common Mood Stabilizers

| Brand Name | Generic Name |
| --- | --- |
| Lithobid / Eskalith | lithium |
| Tegretol | carbamazepine |
| Depakote | divalproex |
| Lamictal | lamotrigine |
| Depakene syrup | valproate |

bipolar patients after only 4 weeks of administration, particularly in the left prefrontal cortex (Moore et al., 2009). Although widely known to be effective, lithium needs to be monitored closely with blood draws, as dysregulation of lithium can lead to lithium toxicity. Furthermore, the side effects of the medication are not tolerated by all people, so other alternatives are often initially prescribed.

Valproic acid (Depakote) enhances GABA by inhibiting an enzyme that breaks down GABA in the brain. Valproic acid is known to control seizure disorder, being an anticonvulsant. However, in the United States it is approved by the FDA for treatment of BD. Although less common than lithium, mismanagement of valproic acid could also lead to death, as it can result in liver or pancreatic dysfunction.

*Psychotherapy*

In light of the associated severe psychosocial stress, it is not surprising that CBT is often implemented for those who have been diagnosed with BD. Treatment aimed at developing strategies for controlling impulses and negative cognitions and maintaining healthy relationships is important for BD. In fact, BD patients have been found to be treated effectively with CBT as an adjunctive treatment to medication (Szentagotai & David, 2010). Although Szentagotai's meta-analysis indicated that CBT has been found to decrease symptoms, correlate with improved quality of life, and increase treatment adherence, CBT did not show a decrease in relapse or reoccurrence. Use of mood-monitoring checklists, thought records, activity logs, and behavior tracking can aid identification of early signs of a mood episode and thus assist attempts to decrease the severity of such an episode. Mood charting can help identify periods of time or triggers that can increase the likelihood for a change in mood in the patient. Use of CBT techniques should assist in psychoeducation and recognizing the importance of remaining on medication as well.

One potentially unique form of CBT that can be implemented with BD is family-focused treatment. The burden on families dealing with a loved one who is experiencing BD is tremendous. Therapeutic interventions aimed at decreasing family stress associated with BD have been developed in an attempt to decrease the overall burden and improve patient function. Caregivers who were provided CBT and psychoeducation experienced a significant improvement in their own mood in comparison with caregivers who were not provided treatment. Interestingly, patients' moods have been documented to improve with decreased depression in their caregivers, suggesting that caregivers who are burned out might have a harder time coping with BD, exacerbating mood problems in the patients themselves (Perlick et al., 2010).

## CONCLUSIONS

The neurobiological factors involved in mood disorders are very well studied. Although MDD and BPD have overlapping neurobiological components that are important for both disorders, including limbic system involvement, the exact underlying mechanisms involving both are complex and different. In fact, the

> # ETHICS
>
> ## Suicidal Behavior: When to Break Confidentiality
>
> The ability of clients to freely and openly reveal their thoughts is essential to successful treatment. Underlying all forms of talk therapy is the client's ability to be open and honest about his or her thoughts, feelings, and cognitions. Therefore, clinicians are bound to the sacred code of confidentiality, which embodies the inability to share any clinical information regarding the client under any circumstances. The Counseling Code of Ethics B.1.c. states: "Counselors do not share confidential information without client consent, or without sound legal or ethical justification." Similarly, Standard 4 of the American Psychological Association's Ethical Principles of Psychologists and Code of Conduct says that "psychologists take reasonable precautions to protect confidential information." It is clear that the ethical code of behavior and principle of confidentiality can be broken only under extreme circumstances. The most common instance in which confidentiality is broken is when the client presents an imminent threat of self-harm or appears to be at highly elevated risk for committing suicide. It is the clinician's duty to help the patient remain safe. No clinician desires to have a patient under his or her care commit suicide. Yet determining when to break that standard of confidentiality is often very difficult, even for the most skilled clinicians: doing so will result in ramifications to the therapeutic relationship throughout the course of treatment thereafter.
>
> It is often difficult to determine when to break confidentiality when working with patients who are depressed. Because depression is associated with sadness, low self-esteem, and feelings of despair and helplessness, and because in men it can be associated with aggressive behavior, a major risk factor for mortality in mood disorders is the increased risk for suicide. It has been estimated that individuals who have been diagnosed with mood disorders have a 17 times greater risk of suicide than the general population. Although there is increased risk, most people who have been diagnosed with mood disorders never attempt suicide. Attempted identification of increased risk factors for suicidal behavior should be of interest to any clinician in view of the code of ethics regarding confidentiality and the likelihood of a clinician's working with depressed individuals.
>
> Several contributing factors have been identified for increased suicidal behavior. Risk for both children and adults who have a first-degree relative who has attempted suicide is increased up to sixfold (Brent et al., 2002; Mann et al., 2005). Although this might suggest environmental factors, it also highlights the importance of genetic factors. Furthermore, neurochemical imbalance has been found to play a role. For example, serotonin dysregulation has been found in the cerebral spinal fluid and prefrontal cortex (Currior & Mann, 2008; Joiner et al., 2005). It has been suggested that low serotonin transmission might contribute a combination of aggressive and impulsive behaviors, thereby increasing risk for suicide (Mann, 2003). As mentioned previously, certain medications, or combinations of medications, can increase risk for suicidal behavior in children and young adults. Cognitive studies have demonstrated that suicidal behavior is linked with problem-solving limitations (Pollock & Williams, 1998) and poor decision making (Jollant et al., 2007). Imaging studies have found structural or volumetric abnormalities in specific regions of the brain. For example, reduced corpus callosum, the main subcortical white matter connector between the left and right hemisphere, has been found in people who demonstrate suicidal behavior (Cyprien et al., 2011). Furthermore, decreased orbital frontal volume in the presence of increased amygdala volume has been found in depressed individuals who exhibit suicidal behavior (Wagner et al., 2010).

*(continued)*

> *(continued)*
>
> Given our knowledge of high familial rates, serotonin dysregulation, limited executive functioning, reduced corpus callosum connectivity, small orbital frontal volume, and increased amygdala volume, it should be considered that neuroscience might be able to help the clinician decide when to break confidentiality to protect the patient from himself or herself. The goal of ensuring that the patient is safe is paramount for the clinician. Although it is impossible to predict suicide in any patient, knowing whether a patient meets the criteria for these factors can ultimately decrease the rate of successful suicide attempts in patients who have been diagnosed with mood disorders. Obtaining all this information might not be feasible for the clinician in our current health system, but it is possible that it will be more likely in the future. With increased resources, advances in genetic and brain-based technology, and more available and efficient cognitive testing, it might someday be possible to be aware of whether the depressed patient demonstrates these neurophysiological risk factors.

*DSM-5* has separated depressive disorders from BDs, which future clinicians will need to consider when evaluating, diagnosing, and treating these conditions. Neuroscience has helped the development of a variety of efficacious treatment methods now available for mood disorders.

## SUMMARY POINTS

- The neurobiology of mood disorders is complex, but they are among the better-understood psychiatric conditions.
- Though serotonin is a known neurochemical involved in mood dysregulation, other neurochemicals are involved in the expression of the symptoms.
- Limbic system involvement is central in the presentation of mood symptoms.
- The mental health clinician should be familiar with new neuroscientific techniques that have been developed to help treat mood disorders.

CHAPTER 8

# Anxiety Disorders

Anxiety is a basic human response to various experiences and stress. Anxiety itself is not a negative state of being. Some level of anxiety is essential to accomplishing difficult tasks. For example, most people reading this book have experienced anxiety in healthy amounts to help them reach this point in their education. At times, anxiety is necessary for safety. From an evolutionary perspective, appropriate anxiety response has aided survival. However, when anxiety is overwhelming, ongoing, ignored, or coped with ineffectively, it can lead to debilitating and harmful effects. Anxiety can result in cognitive, physiological, and emotional dysregulation, including symptoms of inattention, increased heart rate, and agitation.

Concisely defining anxiety is challenging—entire textbooks have been written on the topic. There are various forms and responses to anxiety. The goal of this chapter is to provide a basic overview of the neurophysiology underlying the innate human response of anxiety. Knowledge regarding the brain's response to normal levels of stress and how it responds in a healthy manner will help clarify the various ways the brain can misfire and produce debilitating symptoms and outcomes. Unlike mood disorders and psychotic disorders, there are a wide variety of different subtypes of anxiety disorders. Each subtype, although sharing in some overlap of neurobiological similarities, is also distinct in its neurobiological dysregulation. Thus it is necessary to understand the symptoms of the different anxiety disorders in order to decide which types of treatment will be most effective.

After providing the general neurobiological framework for how the body deals with stress, this chapter discusses the symptoms associated with the major anxiety disorders, as well as the best understood neurobiological aspects of the disorders. It also discusses implementing our understanding of the disorders in an attempt to provide the most scientifically grounded treatment strategies. The special section will discuss why people enjoy being scared at times, such as by watching certain movies or during Halloween. The Ethics box highlights the importance of understanding how anxiety and stress need to be considered in research and the clinical setting, particularly because

## ETHICS

### Stress in Clinical and Research Environments

It is well understood that stress activates the hypothalamic–pituitary–adrenal (HPA) axis, resulting in the production of glucocorticoids by the adrenals. Because that receptors for glucocorticoids are found throughout the brain, the glucocorticoid release can have long-standing effects throughout life. A review by Lupien, McEwen, Gunnar, and Heim in 2009 provides a thorough approach toward examining how glucocorticoids that occur as a result of stress affect people throughout one's life span. A mental health professional must consider this as it relates not only to his or her treatment for patients in terms of diagnosis and conceptualization, but also to his or her treatment of patients in general.

Stanley Milgram's series of obedience-to-authority studies in the 1960s are well known and discussed in introduction to psychology courses and in ethics courses alike. Milgram's studies used confederates who pretended that they were getting shocked by actual study participants, who were told by authority figures to continue to inflict the shock treatments on the confederates. This obedience behavior resulted in a high degree of stress for the participant. Although the study aimed to examine obedience, the study methods helped to lead to the formation of institutional review boards (IRBs). The ethics of invoking a high level of emotional stress for study participants were found to be problematic and unnecessary. The inducement of the stress in the patients is a large part of the reason why IRBs were used for a risk/reward analysis. Awareness that not all outcomes justify the means of obtaining the information is obvious in life-and-death situations, such as those studied in the Tuskegee syphilis study, but an appreciation of the effects that research studies have on emotional well-being is clearly indicated from the concern that was caused by Milgram's studies. The idea that researchers could continue to increase stress, even when participants no longer wanted to participate, has led to the requirement that any research participant be allowed to withdraw his or her informed consent to participate in a study, at any point, without repercussions. The right to withdraw from a study is as important as the willingness to consent to a study. This right to withdraw consent extends beyond the research world.

Although Milgram's studies were extreme, mental health practitioners often need to balance the challenge of invoking stress in their patients during the course of therapy and assessment of the benefits of the situation. For example, cognitive testing

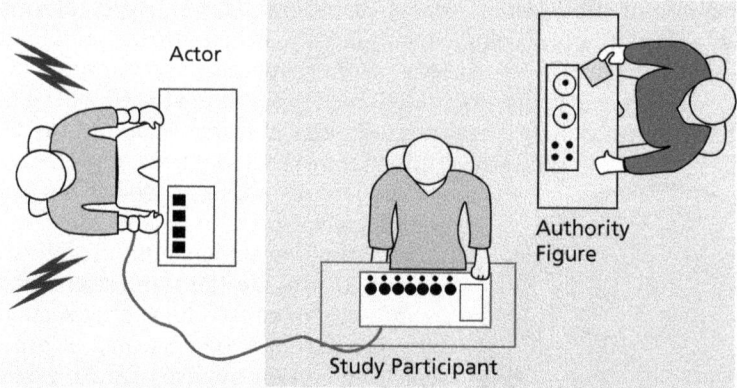

**FIGURE 8.1** Stanley Milgram's experimental design.

*(continued)*

> *(continued)*
>
> with people often leads to increased stress. Who wouldn't be stressed when asked to remember a list of words or to provide their general fund of knowledge? On the other hand, the stress is thought to be for the short term, and the neurochemicals released during that time are not likely to have long-term consequences. Yet if a person who is actively psychotic is completing a personality test and the test is producing increased agitation and anger, it is important to reassess the necessity of obtaining the information at that time in light of the influx of symptoms. Similarly, an increase in anxiety and the release of stress hormones will occur during the course of exposure and other types of treatment in an attempt to teach better coping strategies for the anxiety.
>
> Short-term periods of increased stress undoubtedly result in neurochemical releases that occur in other stressful situations; however, testing and therapy should be conducted in a controlled situation, in an attempt to improve emotional responses. When a mental health practitioner fails to use best practice strategies or implements a new or experimental technique to cope with stress, this can cause unanticipated reaction in the patient. Establishing that the patient has the right to withdraw from treatment, assuming that he or she is not at risk for harming himself, herself, or another person, is vital in providing ethically sound treatment.

anxiety produced in participants in certain psychological experiments has contributed to the creation of IRBs and helped us to develop standards for ethical treatment of participants.

## BASIC NEUROBIOLOGICAL RESPONSE TO STRESS: FIGHT OR FLIGHT?

Anxiety can occur in a variety of situations. People commonly experience anxiety when they feel threatened. As a result, the neurophysiological symptoms of anxiety can help people protect themselves from harmful situations. Anxiety can serve as an indicator of an unsafe situation. In these situations, people will respond with what is commonly referred to as a fight-or-flight response. The fight-or-flight response refers to when people experience an immediate increase of anxiety due to an unexpected environmental change that results in the brain's thought process of either warding off danger or fleeing from the situation. The autonomic nervous system becomes activated, and people respond without much cognitive control. For example, if a motor vehicle drove through the front window of your favorite coffee shop while you were sitting near the window, you would flee to safety without hesitation lest you be severely injured. Without conscious thought, your fight-or-flight response would be activated, and you would instantly protect yourself. This immediate fight-or-flight response is the result of multiple neurochemical and neurobiological systems combining to work effectively.

The autonomic nervous system (ANS) is activated when neurochemicals are transmitted from the brain. As a result, the sympathetic nervous system (SNS) and parasympathetic nervous system (PNS) are directly responsible for how we tolerate the environment. The SNS and PNS work in conjunction with each other in order to control the body's idle energy level, and they are responsible for fast reaction when necessary. When people encounter high-stress situations, the SNS is activated,

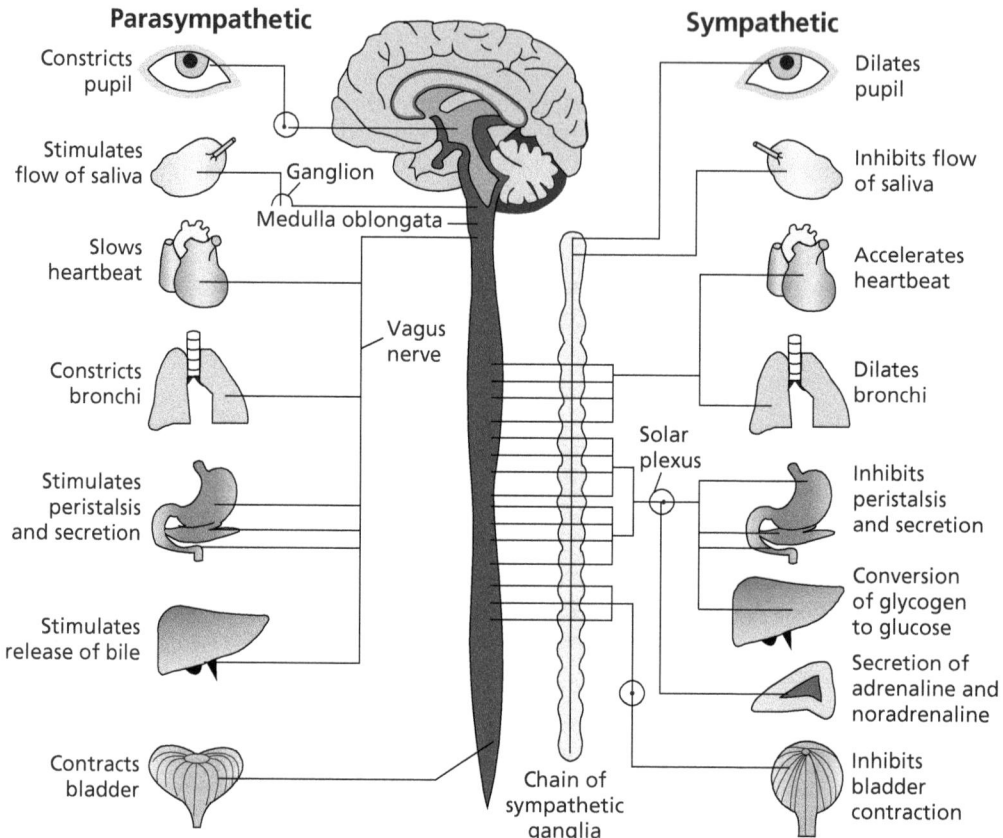

**FIGURE 8.2 Role of sympathetic and parasympathetic nervous systems on bodily functions.**

releasing adrenalin and noradrenalin from the adrenal glands on the kidney. As a result, heart rate increases, shortness of breath occurs, and an increase of bodily activation ensues. The SNS will be completely activated, rather than partially activated, preparing the person for action. The chemicals used in the SNS to maintain high levels of activity are eventually diminished in the central nervous system when the PNS becomes activated and restores the body to a relaxed state. When the person returns back to a normal state of safety, the PNS eventually levels the heart rate, restores breathing regulation, and resets core body levels. This restoration of idle body state is not immediate, as it takes some time to eliminate adrenaline and noradrenalin from the body. This delay explains the reason why people continue to feel keyed up or on edge for some time after a dangerous event has occurred. Furthermore, a dysregulation of this process can lead to continued high levels of anxiety and activation of adrenalin and noradrenalin well beyond a dangerous event.

## Physical Effects of SNS Activation

When the SNS is activated, a variety of physiological events occur to help maintain safety. From an evolutionary perspective, the activation of these different systems helps to keep the body safe and protects the body from perceived or real danger. For example, increased heart rate results in improved blood flow throughout the vascular system. The increased heart rate results in an increase in oxygen, and blood is directed to large muscles such as the calves, hamstrings, and biceps to

## SPECIAL TOPICS

### Why People Like to Be Frightened

Have you ever wondered why you or somebody you know enjoys watching a scary movie, riding a thrilling rollercoaster, being enthralled with a suspense-filled novel, or celebrating Halloween by attending a haunted house? Innately, it would appear that these situations should be avoided at all costs. There would appear to be no benefit to exposing ourselves to these or other perceived threatening situations. In fact, some people will avoid these and other scary situations at all costs. Yet, despite our innate desire to remain safe and avoid actual threatening situations, many humans also enjoy being scared in certain situations. As long as people know that they are in a safe environment, the increased arousal and short-term anxiety provoked in scary situations is tolerable, if not actually enjoyable, for many people.

When a person encounters a fearful situation, the release of cortisol and adrenaline activates the PNS. When that release takes place, people feel more alert, with a rush of stress hormones that increase excitement. This activation provides results in increased activation and arousal. It's a natural, stimulating activity that many try to reproduce through things such as medication and exercise. These activities allow the individual to explore highly anxiety-provoking situations secure in the knowledge that there is no actual threat to their well-being. People have the ability to experience both positive and negative emotions simultaneously. The ability to encounter fearful situations in a safe environment allows for a euphoric sensation at the conclusion of the situation.

After watching a scary movie, dismounting a roller coaster, reading a thrilling novel, or finishing a walk through a haunted house during Halloween, you might have experienced an immediate sense of relief or happiness. The neurochemical release that occurs in conjunction with the conclusion of these situations leaves many people wanting more. The natural high of endorphin release helps to explain why people are willing to momentarily frighten themselves. It is likely that genetic factors play a role in determining who enjoys these situations and how much. Specifically, given that females are consistently documented to have higher risk for anxiety disorders, it is possible that chromosomes influence the likelihood that males will enjoy scary situations more than females.

The next time you are presented with the opportunity to become frightened in a safe situation, reflect upon your physiological reaction and note the SNS activation with subsequent PNS activation as you continue your normal routine. This will help you understand the activating systems involved in anxiety.

prepare for action. Simultaneously, blood to the skin and extremities is diminished in order to decrease the likelihood that a wound to the body will result in excessive blood loss. This redirection of blood flow helps explain the cause of pale-colored skin during periods of anxiety. When this system is dysregulated, an individual might feel a significant increase in heart rate. For example, during a panic attack, a person might misjudge this natural increased heart rate as a heart attack.

Along with redirection of blood flow, breathing is altered with activation of the SNS. The SNS activates respiratory response, speeding and deepening breathing. This allows for more oxygen to flow to tissues. Once again, a dysregulation can lead to feelings of shortness of breath, dizziness, and feelings of hyperventilation when a person experiences increased levels of anxiety without the ability to exert the increased energy for productive purposes found in actual dangerous situations.

Other physical effects of SNS activation include decreased digestive activity, increased sweat, pupil dilation, and decreased salivation. These activities are all necessary, from an evolutionary perspective, to prepare the body for optimal activity during a threatening situation. This physiological reaction is invigorating at times, and it awakens the person to an extent in fearful situations, which explains our enjoyment of certain frightful situations (see the Special Topics box). For example, a person begins to sweat in order to make the skin more difficult to grasp when confronted with a stressful experience. There is also an increase in heart rate when one is scared, which increases adrenaline. The digestive system decreases activity in an attempt to prevent the organism from having to excrete unnecessary waste at an inopportune time. Pupils widen in order to allow for more light to assist in seeing more clearly in a dangerous situation. When these innate physical responses occur beyond the time of the stressful situation, anxiety symptoms of nausea, stomach pain, constipation, dry mouth, and blurred vision can occur. Similarly, headaches can occur as a result of the combination of multiple physical responses. Although the SNS allows for the person to be safe in specific situations, the continued activation during safe times, or overactivation during periods of only minor stressful events, or mentally stressful periods, contributes to the collection of bodily symptoms that occur in a variety of anxiety disorders.

## PANIC DISORDER

Panic disorder (PD) is the culmination of multiple unexpected panic attacks. These panic attacks are an intense, severe feeling of fear or extreme discomfort that occurs within a very short period of time ranging from seconds to minutes. During a panic attack, people experience a range of symptoms, including increased heart rate, shortness of breath, tremors or trembling, shaking, dizziness or lightheadedness, fear of losing control or dying, stomach pain, and numbness. The person will often experience sweating and might experience numbness in the limbs. These attacks can occur at any point, for no apparent reason, or when an environmental factor elicits the onset of anxiety. With PD, the symptoms will lead the individual to refrain from doing normal activities in an attempt to avoid situations, leading to negative alterations to functioning.

It has been estimated that the prevalence of panic attacks is approximately 2% within the United States, with an average age of first onset in the early 20s. It is highly unlikely to occur in people over age 45 or in children. The effects of a PD can severely reduce functional abilities, including maintaining employment, attending school, and participating in social activities—particularly novel activities.

### Genetics and PD

PD is considered the most heritable of the anxiety disorders (Martin, Ressler, Bender, & Nemeroff, 2010). In fact, studies have demonstrated that the risk of PD is increased sevenfold if a first-degree relative has PD (Smoller, Gardner-Schuster, & Covino, 2008). The exact area of genetic influence is complex, and numerous genetic sites have been discovered to play a role. The two most influential genetic sites appear to occur on 2q and 15q. Furthermore, polymorphisms in several genes have been implicated in PD. Polymorphisms have been identified in serotonin transporter genes and monoamine oxidase. These neurochemicals likely contribute to the behavioral manifestation. It has been suggested that multiple polymorphic presentations result in the increased likelihood of PD.

## Neurochemistry and PD

Serotonin dysregulation, along with amino acid neurotransmitters, including GABA and glutamate, are thought to play an important role in the presentation of PD. Serotonin projections throughout various areas of the limbic system and basal ganglia modulate neurocircuitry associated with PD. The combination of the dysregulation of the noradrenaline and serotonin at various sites leads to disruption of healthy neurocircuitry, resulting in the intense fear found in panic attacks.

## Neuroimaging and PD

Neuroimaging studies support the dysregulation of serotonin and noradrenaline in basal ganglia and limbic system circuitry, as well as in the cingulate system. A decrease in prefrontal cortex activity, with increased amygdala and cingulate gyrus activation, has been found in functional magnetic resonance imaging (fMRI) studies when compared to healthy controls. Panic attacks that have occurred during an fMRI have demonstrated increased right amygdala activation, consistent with the notion that elevated anxiety will increase emotional arousal for visual stimuli. Single-photon emission computed tomography (SPECT) studies have reported lower metabolism in the left parietal lobe and decreased blood flow throughout the brain in PD patients. Positron emission tomography (PET) studies have suggested elevated glucose activity in numerous areas of the brain in PD patients. Of note is the fact that completing imaging studies in this population can be challenging, as inducing a panic attack has ethical implications, and a medication-induced panic attack is likely to result in an altered neurobiological presentation. Structural imaging studies have demonstrated frontal, amygdala, and hippocampal abnormalities.

## Cognition and PD

Neurocognitive studies in PD have been complicated by the known cognitive deficits caused by benzodiazepines. Studies of PD typically occur after medication treatment is administered in an attempt to control the attack. Recent research, however, has attempted to separate the cognitive effect of PD from the cognitive dulling associated with benzodiazepines. It has been found that nonverbal short-term and episodic memory are impaired, and so are visuoconstructive abilities in PD in the absence of medication (Deckersbach, Moshier, Tuschen-Caffier, & Otto, 2011). Co-occurring PD in mood disorders has also been shown to decrease cognitive performance on objective measures of memory and executive functioning.

## Treatment for PD

### Therapy

Mental health professionals play an important role in assisting their patients in coping with PD. Cognitive behavioral therapy is the only psychotherapeutic treatment that has evidence supporting its effectiveness for PD (McHugh,

Smits, & Otto, 2009). In fact, CBT has proven effective at eliminating panic attacks in 70% to 90% of patients (Choy et al., 2007). CBT should help develop strategies for more effectively coping with situations that evoke feelings of panic. CBT strategy can include invoking panic attacks in session in an attempt to retrain the individual with effective coping strategies, including by implementing strategies while in session. In vivo exposure techniques can be implemented after the individual has practiced strategies for coping with feelings of panic. It is thought that the repeated exposures, along with learned strategies for altering cognitive processes, result in decreased amygdala and hippocampal activation.

## Medication

Psychopharmaceutical agents have also been demonstrated to be effective for treating PD. Specifically, selective serotonin reuptake inhibitors (SSRIs) are the first choice for PD. Furthermore, venlafaxine is the only serotonin norepinephrine reuptake inhibitor (SNRI) that has been proven to be effective for PD. Although monoamine oxidase inhibitors (MAOIs) have proven to be effective, dietary restrictions make them less desirable. Despite the known long-term effectiveness of psychotherapy and antidepressants, many patients prefer the immediate effectiveness of benzodiazepines. Although they have been proven clinically effective in decreasing the immediate symptoms, the long-term side effects of sleep disturbance, cognitive dullness, and memory decline, along with the high risk for tolerance, addiction, and abuse, make these agents less attractive to the treating medical provider. Informing the disgruntled patient about the long-term negative consequences might help the patient become invested in learning CBT.

# AGORAPHOBIA

Agoraphobia is the intense fear of something happening in various areas outside of the home, including on public transportation, in enclosed spaces, in open spaces, or in crowds. It includes a fear of being alone and of something terrible occurring. Individuals who have been diagnosed with agoraphobia may also believe that they are unable to escape a certain specific situation and so will avoid such situations, leading to functional difficulty.

It is estimated that approximately 2% of the population meet criteria for agoraphobia, although diagnosis can be difficult owing to avoidance of public situations. The condition is highly associated with panic attacks, with up to 50% of people experiencing a panic attack prior to meeting criteria for agoraphobia.

There is a strong genetic link to agoraphobia, with it being identified as having the strongest and most specific association with genetic factors in anxiety disorders. Otherwise, limited research exists examining agoraphobia in the absence of a PD. This paucity of research might be associated with the challenge of identifying, and the ability to get consent from, individuals who have inherent fears of social situations in which there is a perceived or real inability to escape (i.e., completing magnetic resonance imaging [MRI] studies in this population is difficult enough, but identifying agoraphobic individuals without panic attacks is very difficult). Although agoraphobia and panic attacks are related, they are distinct. Future studies elucidating the neurobiological distinction will continue to occur.

## SOCIAL ANXIETY

Social anxiety (SA) is an intense fear surrounding social situations in which the individual might experience, or perceive the experience of, judgment by others. Also termed as social phobia, SA results when a person is so fearful that other people will scrutinize his or her performance that the fear or behavior driven by it severely disrupts functional abilities. Specifically, it can result in few or no social relationships or difficulty obtaining intimate relationships, and it often results in feelings of humiliation, rejection, and extreme fear. Social situations are avoided at all costs, and when a social situation is unavoidable, the individual feels intense anxiety. The anxiety is beyond what should occur in the situation, and it is not just feelings of shyness or healthy performance anxiety. Individuals who have been diagnosed with SA often recognize that their anxiety is unreasonable but feel powerless to alter their behavior.

The average age of onset of SA is younger than most other anxiety disorders (with the exception of separation anxiety, which is not discussed in this chapter), typically occurring at age 13, with three-fourths of individuals experiencing SA between ages 8 and 15. Although childhood shyness increases the risk, it is more likely to occur after an actual humiliating experience. SA is found to be more common in females.

### Genetics and SA

Although there is behavioral evidence that SA occurs in families, there are limited genetic studies examining SA. Having a first-degree family member with SA increases the risk up to threefold in individuals. A norepinephrine transporter gene near chromosome 16 has been found to be implicated in SA; however, replicated results have not been published. Similar to other anxiety disorders, it has been suggested that multiple genes on several chromosomes are involved in the disorder. It has also been demonstrated that being raised in an overprotective family can be linked to SA.

### Neurochemistry and SA

Neuropeptides are thought to play an important role in SA. Specifically, decreased levels of oxytocin and vasopressin, which limit bonding and attachment, have demonstrated an effect on amygdala activity in SA. Serotonin and dopamine play a vital role in SA as well. Animal studies have demonstrated that mice who demonstrate timid characteristics have decreased dopamine in cerebral spinal fluid, although decreased dopamine in monkeys leads to lower observed social status. Furthermore, diminished dopamine found in individuals who have been diagnosed with Parkinson's disease has been linked to social phobia as well. The HPA axis is also thought to be dysregulated in social phobia. Cortisol has been found to be elevated in patients with SA despite normal adrenocorticotropic hormone (ACTH) levels. This dysregulation of cortisol suggests that social concern itself might excessively release this stress hormone, or that individuals with elevated cortisol are at an increased risk for SA.

### Neuroimaging and SA

Structural and functional imaging studies implicate the amygdala in SA. Elective, elevated activation of the amygdala occurs when individuals who have been diagnosed with SA are exposed to fear-provoking situations, as well as to negative faces, but not when they are exposed to neutral or positive stimuli (Labuschagne, Phan, Wood, et al., 2010). Left hippocampus and right amygdala activation have been observed when an individual with SA is provided imagery to control symptoms. Functional imaging studies have consistently documented decreased blood flow in the temporal, parietal, and occipital cortex after successful treatment for SA. This suggests an overactivation of multiple systems and pathways caused by social phobia, which can be altered with successful treatment. Imaging research in this population also demonstrates that hyperactivity in the frontal lobe and anterior cingulate cortex (ACC) might result in misinterpretation of social cues, thereby resulting in the experienced anxiety (Engel, Bandelow, Gruber, & Wedekind, 2008). Orbital frontal cortex volumes have been demonstrated to be smaller in SA (Atmaca, Yildirim, Gürkan, Gürok, & Akyol, 2012).

## Cognition and SA

Neurocognitive studies indicate that individuals who have been diagnosed with SA are more likely to be focused on negative or fearful stimuli, with decreased attention on positive or neutral stimuli. Therefore, emotional content is important for attentional processes in SA. Otherwise, studies have demonstrated a lack of cognitive impairment in individuals with SA (Sutterby & Bedwell, 2012).

## Treatment for SA

CBT is an effective treatment for SA. Use of exposure techniques, in which the individual is gradually presented with an object and learns to be less sensitive to it over time, has proven to be very successful. Acceptance and commitment therapy (ACT) can be effective in helping people to cope with unwanted thoughts, feelings, and beliefs by focusing on the here and now. There has been some evidence indicating that social skills training, in which an individual learns how to initiate conversation and engage with other people, can be beneficial as well. SSRIs are the first line of pharmacological treatment for SA. SNRIs, MAOI inhibitors, and benzodiazepines are also used to treat SA.

## SPECIFIC PHOBIAS

Specific phobias are intense, irrational fears and anxiety centering on a certain event, object, or situation. The occurrence or imagined occurrence of the identified object or situation results in immediate symptom presentation of high levels of anxiety. This leads to avoidance. Examples include enclosed spaces, flying, animals such as spiders or snakes, heights, medical procedures, and blood. There is recognition that this phobia is irrational, but the patient cannot control it without intervention or purposeful strategies. It is estimated that nearly 9% of people in the United States have a phobia, with females being twice as likely to experience one. That said, research suggests that certain phobias, such as exposure to medical procedures, are as high in males as they are in females.

## Genetics and Phobias

It has been suggested that genetic factors play a role in phobias (Kendler, Karkowksi & Prescott, 1999). A region on chromosome 13 has been found to be a site likely associated with specific phobias (Abby et al., 2012). It is possible that different phobias have various genetic alterations, and future studies will undoubtedly explore this possibility.

## Neuroimaging and Phobias

Similar to genetic findings, neuroimaging findings support the notion that various specific phobias demonstrate different neurobiological findings. In general, studies have shown a greater activation in the amygdala, ACC, insula, prefrontal cortex (PFC), and orbital frontal cortex. However, to date, most phobia studies examine individuals who have a phobia associated with animals. It has been reported that in individuals who have blood-injection-injury type of phobias, there is an increased activation of the thalamus, and occipital, parietal, and temporal cortex, unique to this type of phobia (Caseras et al., 2010). Regardless, consistent with other anxiety disorders, the ACC and amygdala play an important role in specific phobias. Other general findings from structural imaging have indicated an increased thickness of insular cortex and ACC.

## Treatment for Phobias

People who have specific phobias often catastrophize, expecting the absolute worst outcome in any situation or encounter with the identified phobia. The goal of cognitive behavioral therapy is to develop an understanding of the incorrect cognition associated with the identified object. Exposure techniques are the most effective form of treatment for phobias. Exposure, often in vivo, aims to develop compensatory strategies and realistic expectations when encountering the identified phobia. Systematic desensitization works toward the individual gradually being exposed to the feared situation, while monitoring and controlling the anxiety. Newer techniques have implemented virtual reality technology in the presentation of the feared stimulus. Regardless of the presentation of the feared stimulus, successful treatment typically involves decreasing anxiety while exposing the patient to the stimulus in a safe environment. Although antidepressants are often prescribed for specific phobias, there is little literature supporting their use for this purpose. Benzodiazipines can immediately decrease anxiety about certain situations, but they are not a long-term solution. Recent experimental studies have demonstrated short-term effectiveness of cognitive enhancers, such as glucocorticoids (Khalil, 2013).

# POSTTRAUMATIC STRESS DISORDER

Trauma is unfortunately one of the most common events that a person will experience in life. It can occur in a variety of contexts, including extreme personal events, such as physical or sexual abuse and tragic regional or global events such as the September 11 attacks or other mass murders in a public arena. Trauma can also be very personal, including loss of a close family member, a personal accident, or an unexpected outcome from a relationship. It can occur collectively, as with those traumas seen in military personnel engaged in wartime activity, and it can result in co-occurring emotional difficulty along with physical symptoms. Traumatic brain injury can result from a trauma, which then contributes to even more difficult and complex symptoms. Some individuals who experience trauma have subsequent anxiety symptoms, whereas others do not. Two people can be a part of the same traumatic experience and have radically different emotional and cognitive outcomes. Unlocking the mystery to responses to trauma is important to developing the best forms of treatment and identifying those who are at risk for suffering emotional difficulty subsequent to trauma in an attempt to implement early intervention, thus increasing the likelihood for better outcomes. Although the *DSM-IV-TR* conceptualized posttraumatic stress disorder (PTSD) as an anxiety disorder, the *DSM-5* has PTSD under the separate, newly developed heading of trauma- and stressor-related disorders. Regardless of this distinction, PTSD includes high levels of ongoing anxiety symptoms.

Although trauma occurs in everyone's life at some point, PTSD does not unfailingly follow trauma. Rather, PTSD is an unhealthy emotional reaction to the experienced trauma. Symptoms of PTSD persist after 1 month post-trauma, and they include behavioral symptoms of nightmares, flashbacks, and intrusive thoughts regarding the trauma. The lifetime risk of experiencing PTSD is approximately 8% in the United States. Females are more likely to experience PTSD, and they experience symptoms longer. A factor in the increased risk for PTSD in women is that they are more likely to experience abuse or violence. Functional impairment accompanies PTSD, and it has been linked to poor relationships, less education, and reduced occupational success.

## Genetics and PTSD

Genetic factors are thought to play a role in the predisposition to PTSD subsequent to trauma. In fact, it has been estimated that the heritability of PTSD subsequent to trauma could be as high as 40%. Serotonin transporter genes on chromosome 5 appear to be linked to the onset of PTSD. The interaction between environment and genetic influences likely contributes to the expression of the disorder. Polymorphic gene expression has been linked with the overactivation of the HPA axis that occurs in PTSD.

## Neurochemistry and PTSD

Neurochemical studies examining PTSD indicate hyperactivation of neuroadrenergic function and sensitization of dopaminergic functioning. Abnormal serotonin binding has also been documented as occurring during PTSD. The HPA axis has been found to have decreased cortisol and increased glucocorticoids. Cortisol is decreased in the plasma of saliva upon awakening. It is also known that inappropriate glutamate processing distorts processing in individuals with PTSD. The HPA axis dysfunction results in neuronal damage, particularly to the hippocampus region. Structural imaging studies have demonstrated a reduction up to 8% of volume in the hippocampus in individuals who have PTSD. Exposure to overproduction of glucocorticoids is known to affect the hippocampus in both human and nonhuman species. Structural imaging findings also indicate a volumetric decrease of hippocampal gray matter along with more severe trauma exposure, increased severity of symptoms, and increased cognitive impairment. It is likely that the dysregulation of the HPA axis is associated with severity of cognitive, emotional, and behavioral sequelae of PTSD.

## Neuroimaging and PTSD

Functional imaging studies support the subcortical dysregulation that occurs during PTSD. Decreased activation has been documented in Broca's area, with increased activation in the limbic system. This might account for difficulty verbally processing a recurrent intrusive thought or flashback that elicits an emotional response. PET studies have demonstrated decreased activation of the prefrontal and orbital frontal cortex along with the temporal areas. The ventrofrontal region of the brain is more activated in PTSD when the patient is observing fearful faces. Neuroimaging has also been used to predict success of treatment. For example, ACC volume has been demonstrated to predict success of treatment in PTSD (Bryant et al., 2008). The ACC is highly involved in cognitive flexibility, and this result suggests that the greater the volume of the ACC, the higher likelihood of treatment success.

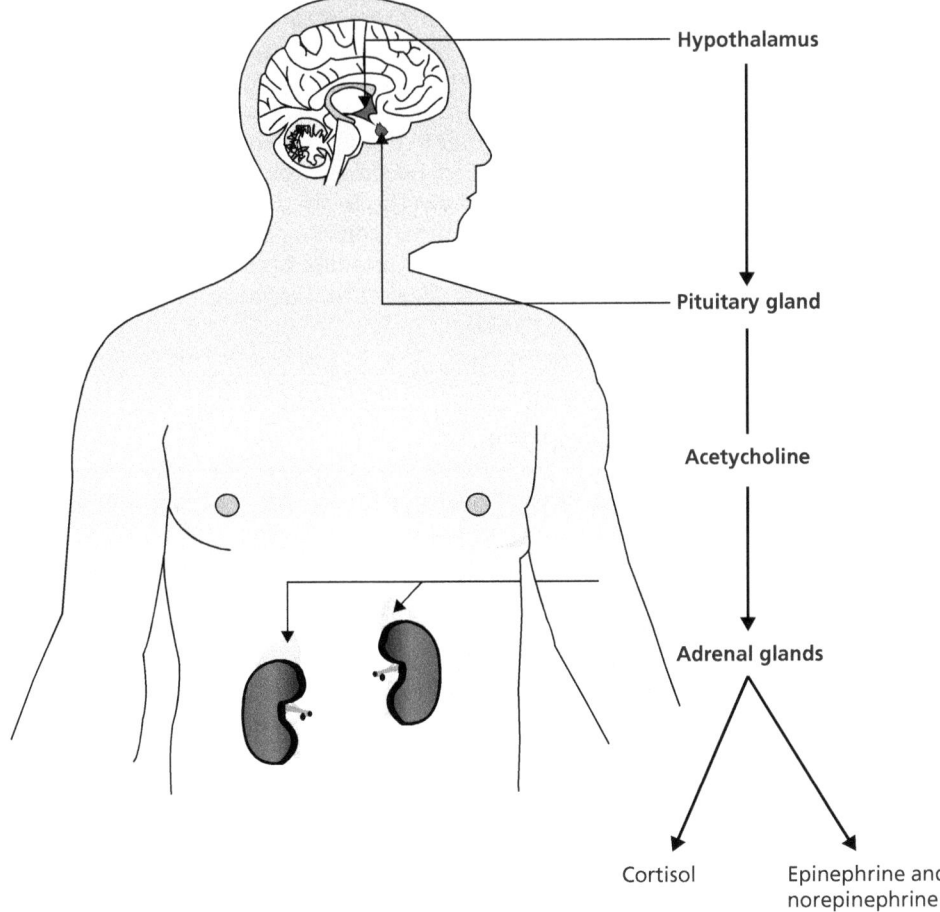

**FIGURE 8.3 The HPA axis.**

## Cognition and PTSD

Studying the cognitive effects of PTSD is more challenging than might be anticipated. Specifically, the cause, duration, and frequency of trauma can play a role in cognitive functioning of an individual. Furthermore, because many individuals who experience PTSD might also have experienced a blow to the head, traumatic brain injury might affect the results on cognitive tests. A recent review of the literature examining various reasons for and duration of PTSD indicates that attention, executive functioning, and memory are likely to be negatively effected by PTSD (Qureshi et al., 2011). Performance on measures examining cognitive flexibility has also been found to be impaired in PTSD. In comparison, visual spatial skills, language abilities, and constructional abilities remain intact in PTSD.

## Treatment for PTSD

There are several behavioral treatment strategies to cope with PTSD. Trauma-focused CBT (TF-CBT) has consistently been proven to be effective in treatment. TF-CBT aims to treat PTSD through psychoeducation, relaxation, affect expression and regulation skill development, cognitive coping and processing, as well as potential trauma narrative and in vivo exposure. Free online training is offered on TF-CBT at http://tfcbt.musc.edu, and mental health clinicians are likely to benefit from this training, which may help them understand treatment more thoroughly. Exposure therapy is another technique allowing the individual to deal with his or her fear in a safe way, and it uses cognitive strategies to help the person confront his or her memories surrounding the trauma. A form of exposure technique is eye movement desensitization and reprocessing, which has been reported to be effective in treatment for some people. Stress inoculation treatment can also be used to help the individual view his or her memories in a more effective capacity. Medication, including SSRIs, has proven to be of moderate effectiveness. Specifically, sertraline has been approved by the FDA to treat PTSD. However, studies have indicated that medication is not as effective as psychotherapy for treating PTSD.

## GENERALIZED ANXIETY DISORDER

Generalized anxiety disorder (GAD) is the most frequently diagnosed anxiety disorder in general clinics, but it is one of the least frequently diagnosed anxiety disorders in specialized anxiety clinics (Kessler, 2000). Much as with other anxiety disorders, GAD is used to describe intense feelings of nervousness occurring in broad situations. The main features of GAD are the cognitive symptoms associated with worrying. It is common for GAD to result in worry about the future, and this can subsequently lead to avoidance behaviors. Associated with this worry are feelings of being on edge or being keyed up. Furthermore, GAD results in restlessness, fatigue, agitation, and muscle soreness or tension. GAD results in functional impairment in terms of social or occupational problems. The age of onset slope of GAD is more similar to that of depression than it is to other anxiety disorders. Adolescence and early adulthood are the most common times for the onset of GAD, although the median age of onset is 30, much later than for other anxiety disorders. The lifetime risk of GAD is 9%, with females being twice as likely as males to experience this disorder.

## Genetics and GAD

The genetic contribution to GAD is thought to be less than that of other anxiety disorders (Martin et al., 2009), but it still accounts for one-third of the risk. There has been no replication of which genes or chromosomes are likely to play a role in GAD. Future studies will continue to examine and explore several possibilities.

## Neurochemistry and GAD

Despite this paucity of genetic studies, several neurochemical and neuroimaging studies have been conducted for examining GAD. Several studies have demonstrated that a dysregulation of GABA inhibitors occurs in GAD, and this dysregulation has been hypothesized to play a role in the etiology of the GAD (Nutt, 2001). GABA is the brain's most prominent inhibitory neurotransmitter, and it is widely dispersed throughout the brain, with a high concentration in the limbic system. This might explain the collection of symptoms that can occur in GAD. Serotonin involvement has also been demonstrated in several studies. It is not uncommon for depression and anxiety to co-occur. The overlap of similar neuropathways and neurochemicals helps explain this comorbidity.

## Neuroimaging and GAD

Neuroimaging studies reveal that a high ratio of gray matter to white matter exists in the temporal lobe, with increased amygdala volume in pediatric GAD. fMRI studies have demonstrated increased activation in the prefrontal cortex, even in resting states. Furthermore, there is increased activation of the amygdala in GAD when viewing negative facial expression and increased right amygdala activation when viewing angry faces. This increased activation is associated with acute awareness of social situations. The dysregulation of serotonin is known to play a role in GAD. Considering the overlap of symptoms with depression, this result should not be surprising. The ACC is also thought to play an important role in the expression of GAD. A recent diffuse tensor imaging study found abnormal white matter integrity in the areas of the ACC and amygdala (Shang et al., 2013) in first-episode GAD before treatment. Furthermore, the ventromedial prefrontal cortex has been found to play a role in GAD, as there is decreased activation in fMRI studies when the subject is exposed to fear stimuli (Greenberg et al., 2012). A combination of increased activation of the ACC and amygdala, coupled with decreased activation of the ventromedial prefrontal cortex, is likely to be a root neurobiological pathway in GAD.

## Cognition and GAD

Individuals who have been diagnosed with GAD have been found to have broad-based, but not global, cognitive decrements in areas of memory, executive functioning, and working memory, with increased difficulty in processing speed in the geriatric GAD population (Butters et al., 2011). Executive functioning difficulties occur in problem-solving, mental flexibility, and concept formation. These cognitive limitations might account for difficulty in terms of overcoming the cognitive components of GAD. Interestingly, cognitive improvement in these areas has been demonstrated with successful treatment in this population.

## Treatment for GAD

Treatment for GAD is similar to that of the other anxiety disorders discussed, as CBT is commonly effective in coping with the symptoms. Development and use of relaxation techniques, deep breathing, meditation, cognitive reframing, and exercise have also been found effective for GAD. Psychopharmacological treatment includes SSRIs, as all SSRIs have proven to be effective. Of note, however, is paroxetine, which has been the most frequently studied and has proven to decrease harm avoidance.

## CONCLUSIONS

Anxiety is a normal symptom that people experience in everyday functioning. The brain's response to certain stressful situations is adaptive and has been proven to aid many successful behaviors. Inability to control stress can lead to release of neurochemicals and alterations in the HPA axis, which can result in both CNS and PNS dysregulation. Long-term exposure to glucocorticoids can result in neuronal damage and brain volume reduction. The mental health provider should be familiar with various anxiety disorders, so to diagnose more accurately, because treatment differs depending upon the specific anxiety disorder diagnosis.

## SUMMARY POINTS

- Anxiety is the result of elevated levels of stress that induce activation of the HPA axis and the release of glucocorticoids, among other neurochemicals.
- Although there is overlap, genetic, neurobiological, and neurocognitive differences occur during the various anxiety disorders.
- Treatment often includes helping the individual confront his or her anxieties and learn new coping strategies.
- Although benzodiazepines are often preferred by the patient for their fast-acting response, SSRIs are the first line of pharmaceutical treatment for most anxiety disorders, with CBT found to be the most effective long-term treatment strategy.

CHAPTER 9

# Eating Disorders

Working with individuals who have been diagnosed with eating disorders can often be difficult, and it poses a unique challenge for the clinician. Individuals diagnosed with eating disorders have high mortality and comorbidity rates. A staggering 6% of individuals diagnosed with anorexia nervosa end up deceased from complications associated with the disease, including cardiac arrest, electrolyte imbalance, and suicide. Approximately 20% of people who have been diagnosed with the disorder remain chronically anorexic. Treatment is difficult, owing to the seemingly hard-wired behaviors of individuals with eating disorders. Not surprisingly, many eating disorders begin during adolescence, with adolescent girls being at high risk for these disorders. In fact, the prevalence of adolescent girls who demonstrate symptoms of an eating disorder has been reported to be approximately 44% (Ackard, Fulkerson, & Neumark-Sztainer, 2007).

Given the early neural connections, neurobiological underpinnings, societal pressures, and co-occurring emotional–behavioral characteristics, altering the behaviors of people who have been diagnosed with eating disorders can be extremely difficult and might take many years. It has been argued that understanding the neurobiology of eating disorders is an important step in developing appropriate pharmacotherapies (Avena & Bocarsly, 2012). Furthermore, a clear understanding of the neurobiological and cognitive components of eating disorders can help elucidate for the clinician the struggles of this patient population, more so than with most other disorders. Specifically, understanding the neurobiology behind these disorders can help the clinician develop the patience to accept slow change, tendencies, and desires for regression in behavior; it can also prepare the client for the challenges he or she will face and encourage families to aid treatment. Appreciating the biology behind eating-disordered behaviors can potentially decrease frustration and the likelihood that the disorder will be misconstrued as more easily controlled than it is. It also allows the clinician, client, and family to recognize that a variety of treatment techniques, including psychopharmacology, can be effective with this population.

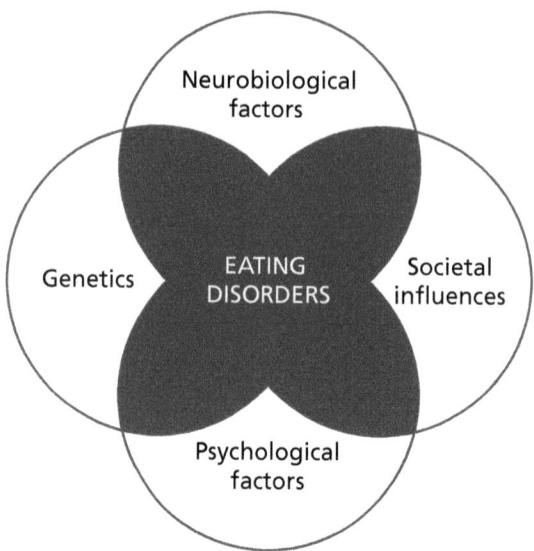

FIGURE 9.1 Contributions to eating disorders.

## ANOREXIA NERVOSA

The goals of this chapter are to understand the main characteristics of the three major types of eating disorders: anorexia nervosa, bulimia nervosa, and binge eating. After these characteristics are presented, each disorder is examined from a neurobiological perspective, including genetic factors when known, neuroimaging results, the understanding of neurotransmitter dysregulation, cognitive performance, and various types of treatment. Consideration of the unique challenges associated with comorbidity, societal pressure, and medical implications is presented. Finally, the effectiveness of specialized eating programs is presented.

AN is characterized by the refusal to maintain normal body weight, a pursuit of thinness, a fear of becoming fat or gaining weight, a distorted body image, and extremely disrupted eating behavior, with or without purging behavior. Individuals who have been diagnosed with AN will use compensatory strategies, including excessive exercise, ingesting laxatives and diuretics, and purging, to maintain a body weight attaining to less than 85% of the healthy body mass index (BMI). The individual who has been diagnosed with AN aims to maintain low body weight through seemingly any approach. This low BMI indicates that such an individual is underweight and medically unsafe. Denial of the seriousness of the low body weight is common, and in women, amenorrhea, the absence of a period during the menstrual cycle, often occurs. A recent review of the different classifications of eating disorders calls into question the specific criteria for the disorder and highlights the lack of consistent recognition of hyperactivity being a remarkable feature within the disorder (Uher & Rutter, 2012). These authors note that although hyperactivity is not present in every patient who has been diagnosed with AN, hyperactivity in AN is a differentiator from other causes of starvation. This characteristic is likely to be associated with the biological processes and is potentially relevant as it relates to treatment and misdiagnosis.

AN has existed for centuries and across cultures (Campbell, Mill, Uher, & Schmidt, 2011). Although societal factors play a role in the disorder, there are over 30 risk factors that have been identified for AN, including childhood eating and feeding patterns, overanxious parenting, perfectionism, and high levels of physical exercise (Jacobi, Abascal, & Taylor, 2004). People who have been diagnosed with eating disorders have increased anxiety disorders that predate the eating disorder (Raney, et al., 2008). Chronic stress has been implicated in the onset of AN. It is likely that maternal stress might affect the hypothalamic–pituitary–adrenal (HPA) axis in the offspring (Heim & Nemeroff, 2002). Identification of expectant mothers who are experiencing a high level of stress or anxiety is important in order to proactively decrease the likelihood of anorexia in offspring. Similarly, early life stress is likely to alter epigenetic factors that increase the risk for AN (Campbell, Mill, Uher, & Schmidt, 2011).

The average age of onset for AN is 19, with females being three times as likely to experience the disorder and an overall prevalence of approximately .6% in the United States (Hudson, Hirripi, Pope, & Kessler, 2007, from the National Institute of Mental Health website). The same researchers report that only slightly more than 33% of individuals who have AN receive treatment at some point over their lifetime. Other developed societies are similar, with the fear of gaining weight symptoms in AN occurring in a smaller proportion of non-Western cultures (Uher & Rutter, 2012). Only about 50% of individuals who have been diagnosed with AN will make a full recovery. Some people will demonstrate periods of recovery followed by relapse. AN has the highest death rate of any psychiatric disorder (Sullivan, 1995).

## Genetics and AN

For decades, genetic factors have been thought to have played a role in AN based on the results of aggregation studies and twin studies. Hereditability factors have been found to substantially increase the rate of eating disorders in individuals who have been diagnosed with AN. Twin studies have found a genetic contribution of 48% to 76% for AN (Striegel-Moore, & Bulik, 2007). First-degree relatives of individuals who have been diagnosed with AN have a risk of themselves having AN that is 10 times as great as for individuals who do not have such a first-degree relative (Lilenfeld et al., 1998).

Molecular genetic studies reveal that there are several regions of interest found on chromosomes 1, 2, and 13 (Pinhiero, Root, & Bulik, 2009). Although analyses of candidate genes have suggested several possibilities, the 5-HT2a gene has demonstrated consistent positive findings. This gene codes one of the receptors for serotonin and is thought to play a role in obsessive–compulsive behavior, perfectionism, and inflexibility (Kaye, 2007). Furthermore, people who have recovered from AN have persistent 5-HT dysfunction (Kaye, 2007). Though the genetic understanding of the disorder is in its infancy, evidence clearly exists for the importance of genetics in the disorder.

## Neuroimaging and AN

Neuroimaging techniques have been used to better understand the brains of individuals who have been diagnosed with AN over the course of time and treatment. A recent structural MRI study replicated consistent findings that people who have been diagnosed with AN have decreased gray matter in comparison with controls (Van den Eynde et al., 2011). However, whether changes occur within the brain before onset of the symptoms, and whether these reductions can be altered with treatment are not known, though signs are promising. Early studies revealed mixed results regarding structural differences in AN. Some studies indicated that a reduction in total gray and white matter volume in comparison to healthy control persists, even with increased weight (Artman, Grau, Adelmann & Schleiffer, 1985; Krieg, Pirke, Lauer, & Backmund, 1988). Other studies indicated that a decrease in cerebrospinal fluid, increase in brain matter, and decrease of ventricular enlargement occurred after the person diagnosed with AN restored his or her weight (Dolan, Mitchell, & Wakeling, 1988). Later longitudinal studies confirmed that treatment of symptoms resulted in a volumetric increase in AN. Structural imaging studies have since shown that the onset and timing of treatment of AN can make a difference in changes to various regions of the brain. In female adolescents ranging between ages 12 and 17, changes in the cortical midline ranging from the occipital to the frontal regions appear to have decreased gray matter (Mainze Shulte-Ruthe, Fink, Herpertz-Dahlmann, & Konrad, 2012). Weight gain in this study resulted in overall increased gray matter, with increased volume in the cerebellum, amygdala, hippocampus, and thalamus. Duration of illness likely plays an important role as well, as it has been found that significant reduction in total white matter volume and focal gray matter atrophy in the cerebellum occur with longer duration of symptoms (Boghi et al., 2011). Atrophy of the cerebellar gray matter might thus play a role in the chronic phase of the illness. Co-occurring mood difficulty, memory problems, and inattention during periods of AN are likely associated with these structural abnormalities; improved functional abilities occurring with increased nutrition and weight result in potentially increased brain volume. These brain-based changes support the importance of encouraging weight gain in this population in order to potentially increase and improve cognitive, emotional, and behavioral functioning.

Functional magnetic resonance imaging (fMRI) studies have also supported cortical changes associated with successful treatment. A positron emission tomography (PET) study of recovered AN women revealed that regional blood flow normalizes with long-term recovery (Frank et al., 2007). One overwhelmingly consistent finding in fMRI studies examining AN is that activation patterns in areas of the cingulate cortex are different from those in healthy controls (McCormick et al., 2008; Zastrow et al., 2009). The cingulate system has several functions but is largely responsible for cognitive and behavioral response shifting. The functional differences in this brain area help explain the desire of AN patients to be perfect, as well as their difficulty in altering unhealthy behaviors that are established. It remains unclear as to whether weight gain alters the functional aspects of the cingulate system. It appears that inhibitory control responses in recovered patients with AN are contingent upon the difficulty of the task, with less prefrontal cortical activation during more difficult tasks in recovered AN than in healthy controls (Orberndorfer, Kaye, Simmons, Stigo, & Matthews, 2011). Interestingly, even during resting states, the striatum, thalamus, and brainstem all demonstrate increased activation (even in individuals who are recovered from AN; Cowdrey, Filippini, Park, Smith, & McCabe, 2012; Uher et al., 2004). It is possible that altered dopamine signaling in the striatum may be altering the reward system, executive functioning, and behavioral routines in AN (Frank et al., 2005; Wagner et al., 2008). These findings help explain this population's difficulty in changing unhealthy habits and routines.

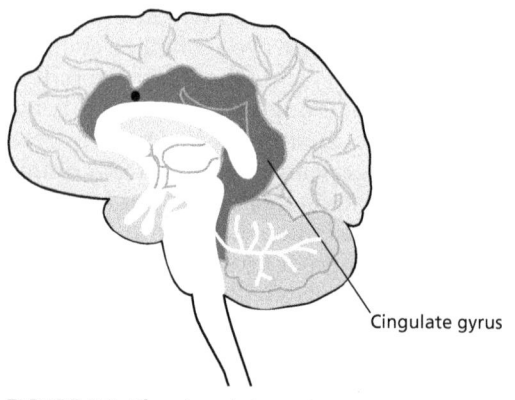

FIGURE 9.2 **The cingulate system.**

## Cognition and AN

Individuals who have been diagnosed with AN are often very high-functioning and intelligent, as well as mostly cognitively intact. Research has indicated specific areas of cognitive limitations on formal testing. The clinical characteristics of inflexibility of changing their unhealthy weight-related behavior has been conceptualized as limitations with cognitive and behavioral flexibility. In fact, a consistent neuropsychological finding in AN deficits is with set-shifting behavior. A meta-analysis found these deficits during periods of illness as well as during recovered states (Roberts et al., 2007). This cognitive deficit likely accounts for the behavioral tendency to get stuck or preoccupied with weight and high concerns over mistakes, as members of this population can't shift their mindset regarding lifestyle. These findings are not surprising in light of the alterations in the cingulated system. Another cognitive finding in AN is impaired global processing with inconsistent evidence of superior local information processing (Lopez, Tchanchuria, Stahl, & Treasure, 2008). This might account for limitations in seeing the big picture and instead focusing on minute, specific details. There is a suggestion of impaired social cognition, including processing of other people's thoughts and feelings, in individuals who have been diagnosed with AN (McAdams & Krawczyk, 2011). This limitation may also contribute to the behavioral manifestation of the disorder, particularly as it relates to distorted body image and weight.

## Treatment for AN

### Medication

There is no medication that has been created to improve the behavioral tendencies and cognitive misconceptions of individuals with AN. However, one recent study found that at least 50% of people diagnosed with AN were prescribed psychotropic medications (Fazeli et al., 2012). The two most common types of psychotropic medications prescribed are antidepressants and antipsychotics. This is not surprising, considering the likelihood of diagnosis of depression, likely caused by serotonin dysregulation. There is also overlap of symptoms between AN and depression. For example, distortions in body image are likely to be conceptualized as low self-esteem as a result of depression. Thus treatment aimed at improving self-esteem, decreasing anxiety, and stabilizing mood is often recommended.

Antidepressant medications, such as selective serotonin reputake inhibitors (SSRIs), specifically fluoxetine, also known as Prozac, have been used to treat AN. Some antidepressants have the side effect of weight gain as well. However, studies suggest that the effectiveness of such medication for AN appears limited both during intensive inpatient treatment and thereafter (Attia, Haiman, Walsh, & Flater, 1998; Walsh et al., 2006). Despite these findings, given the co-occurring depression and OCD, and lack of other options, SSRIs are still used to decrease psychiatric symptoms.

Antipsychotic medications have also been used to treat AN. The results, however, do not provide clear evidence of therapeutic success or failure (Court, Mulder, Hetrick, Purcell, & McGorry, 2008). A systematic review indicates that atypical antipsychotic medications, such as risperidone, olanzapine, and quetiapine, are effective for reducing depression, anxiety, and core eating-disordered psychopathology, but do not enhance weight gain (McKnight & Park, 2010).

Despite the high numbers of individuals diagnosed with AN who are being prescribed psychotropic medications, such medications' effectiveness appears limited. A review of pharmacotherapy literature for AN indicated that medication provides little benefit as it relates to weight gain for AN at the present time (Crow, Mitchell, Rorig, & Steffen, 2009). These medications likely improve mood, obsessive-compulsive disorder (OCD), and anxiety in this population, but they do not affect the core symptoms of the disorder. Therefore, nonpsychotropic medications and other interventions should be considered.

Nonpsychotropic medication can be prescribed if there are co-occurring medical complications in the patient. Estrogen has been found to decrease the likelihood for osteoporosis. It is possible that estrogen helps women who have been diagnosed with AN to remineralize their bones, thereby decreasing risk for fractures. Cardiac medications can be prescribed if arrhythmias or other heart-related difficulties are discovered. Medications used to improve other medical conditions occur, as determined by a medical provider, and they need to be carefully assessed and monitored in this population.

## Psychotherapy

Despite the challenges and complexities often associated with AN, a promising aspect of therapy is the documented changes in neurobiological processes, even occurring at the neurostructural level of the disorder. As stated, weight gain results in increased brain volume, changes in neurochemical activity, altered blood flow, and cognitive improvement. Unfortunately, the goal of the patient is often in direct contrast to the goal of therapy. The patient who has been diagnosed with AN often wants to lose weight. Altering the strong neural connections already created within individuals who have been diagnosed with AN, which have arisen by positive reinforcement from society, genetic predisposition, and brain-based changes, is challenging. Because of this, treatment needs to be consistent and multilayered.

Ideally, behavioral treatment begins with increasing weight. In extreme cases, the person with AN can be admitted to a medical hospital either for body weight issues or medical problems, or for both. In light of ethical guidelines, it is often difficult to determine when extreme measures are required, particularly if the patient is unwilling (see the Ethics box). When it becomes apparent that the patient needs direct medical care, the goal of medical units is often to immediately increase weight through diet or supplements. If the client refuses food intake or supplements, a nasogastric feeding tube (NG tube) can be used. An NG tube is a plastic tube that runs through the nose to the stomach via the esophagus. Though such feeding does not serve as a long-term solution, it assists in stabilizing severe cases.

There are several intensive residential treatment facilities throughout the United States and other countries that aim to increase weight in patients with AN. These treatment facilities often include a wide variety of professionals, including medical doctors, therapists, nutritionists, social workers, and other staff, who help patients make necessary diet and behavioral changes in a safe environment. However, these clinics are often expensive, and they might not be practical for a functional AN client. *A functional AN* is an individual who has a family and who is able to maintain employment and sustain functional capacities in numerous areas, but who still meets the criteria for AN. Rather, the functional AN individual, if treated, will likely be treated through an outpatient setting during the course of long-term therapy.

Another hypothesis regarding AN, which has implications for treatment, is that starvation causes mental problems in AN. This theory postulates that reduced food intake and enhanced physical activity are the two main risk factors for AN (Södersten, Nergårdh, Bergh, Zandian, & Scheurink, 2008). These experts suggest that AN patients need a high level of support. It is therefore important to reteach people with AN how to eat. Briefly, the model proposed by Soderston et al. suggests that learning normal eating patterns through behavioral strategies is important. For example, use of a scale that is connected to a computer can be used to monitor food intake by placing a plate of food on the scale. As the patient eats the food, the computer measures the eating curve at regular intervals, and the computer asks the patient to mark his or her level of fullness. The patient is then asked to modify his or her eating

# ETHICS

## Safety and Eating Disorders

Determining when a patient is at risk for self-harm can be difficult in any psychiatric population. Suicidal ideation and thoughts of death are not uncommon. Balancing the need for confidentiality and necessity to guarantee safety of the patient is without doubt a challenge for the clinician. It becomes even more difficult to determine when to break patient confidentiality to ensure safety in patients diagnosed with AN.

Evaluating patients' understanding of the potential self-harm they could inflict upon themselves with their behaviors is difficult. Continued behaviors that can increase risk for death do not necessarily create a need to break confidentiality and contact other professionals or family if the patient does not consent to these measures. Another challenge is to determine the patients' capacity to be aware of their behavior's effects on their health and well-being. Are patients intentionally trying to harm themselves or is the potential for harm a byproduct of the eating disorder? Although intentionality is a consideration in assessment, overall safety and well-being need to be addressed as well. A depressed AN patient might choose to slowly kill himself or herself through nutrition deprivation, rather than by overtly attempting suicide through violent means. It is often challenging for a therapist to make such a determination.

There are measures the therapist can implement to try to deal with these challenges. A therapist is encouraged to work closely with medical doctors in this population. Encouraging the patient to consent to open lines of communication with other providers can be helpful. Although this close collaboration might prevent the patient from initially establishing the trust required to share personal events and feelings, it can also serve as a safeguard for both the patient and the therapist. The patient should be encouraged to know that breaking confidentiality to discuss treatment with a physician will be limited to medical information, in order to maintain as much confidentiality as possible. In light of the known neurocognitive impairments with lack of nutrition and the neurobiological alterations, judgment is likely to be compromised in more severely undernourished individuals. This situation can increase the risk for suicidal action if depression is present.

Conceptualization of actual eating restrictions, including malnourishment or starvation, as being a means for suicide can be difficult to diagnose for outpatient therapists. If a patient is an imminent danger to himself or herself, then the therapist acts to take whatever means are necessary to prevent self-harm. For example, if a patient informs the therapist that she or he has purchased a gun and plans to shoot himself or herself, the therapist takes action to gain admittance for inpatient hospitalization. However, when the self-harm is medically intended and continues over the long term, it becomes more difficult to assess, even for therapists who have strong medical backgrounds. Lack of medical training, inablity to order and limited ability to understand lab results, and the lack of hospital privileges to admit an AN patient to a hospital can be significant barriers to the clinician. When assessing for imminent danger, the therapist should consider whether there are a decrease in weight, increased thoughts of death, changes in social support, decline in cognitive abilities, or emotional functioning and behavioral changes. If it is deemed that the patient is in fact in imminent danger, then the therapist is ethically and morally obligated to immediately ensure that patient's safety. The therapist is allowed to use whatever means necessary to ensure that the patient receives appropriate medical attention and/or increased psychiatric services. A challenge might arise, though, if the patient does not agree with the assessment, which is likely considering that patient's inaccurate perceptions. It is advised that the therapist carefully document his or her thought processes in these situations.

to appropriate levels generated by the computer during training schedules to adjust his or her level of fullness and eating rate. In doing so, patients learn to eat progressively more food in a shorter period of time (Soderston et al., 2008). This approach encourages healthier eating patterns to resolve other symptoms. Although such technology is not available to all clinicians, the general concept of altering understanding of fullness and eating rate can occur through behavioral interventions. Comprehensive treatment manuals for AN have been developed assuming the family's active involvement, including in implementing the strategies. The role of the environment is clearly a factor in treatment, as the brain is likely to perceive and react to familiar cues in a consistently unhealthy manner. Changing the environment is necessary to changing eating and other weight restriction behavior, and the family can assist in this goal.

Cognitive remediation therapy has also been suggested (Davies & Tchanturia, 2005; Tchanturia, Davies, & Campbell, 2007). This approach has demonstrated improved functioning on measures of cognitive shifting, a clear contributing factor to functional difficulty. Focusing on self-control, focusing on mood intolerance, and preventing avoidance of emotions is an approach of acceptance and commitment therapy (ACT), another therapeutic approach for AN (Fairburn, Cooper, & Shafran, 2003; Fairburn, Shafran, & Cooper, 1999; Schmidt & Treasure, 2006). The goal of this treatment is to help the client develop a different relationship with his or her cognitions and emotions by increasing cognitive flexibility. This treatment aims to help the patient reassess his or her values, as often the core values are replaced with a focus on eating-disordered behavior. It also focuses on mindfulness in order to stay in the present moment. Diffusion is also encouraged to help the patients recognize their thought patterns without restructuring the thought. Acceptance of oneself is a core tenet of ACT. The ultimate goal of ACT is to increase flexibility of one's thoughts as they relate to one's life and mentality, which is often a clear limitation in this population.

## BULIMIA NERVOSA

Bulimia nervosa (BN) has been described since 1979. It includes three main criteria: preoccupation with weight, recurrent binge eating, and recurrent compensatory behavior. Although these symptoms are clearly prevalent in BN, the exact amount of food eaten during a binge episode and the frequency of the behavior remains an area on which researchers do not agree (Le Grange, Crosby, Rathouz, & Leventhal, 2007; Schmidt, Morgan, & Yousaf, 2008; Uher & Rutter, 2012). For example, many people with BN report eating too much food, which might in fact appear to be objectively normal (Wolfe, Baker, Smith, & Kelly-Weeder, 2009). The subjective loss of control might be more relevant when defining the disorder, as people who experience this symptom have similar levels of use and effectiveness of treatment as those with objective overeating (Mond & Calogero, 2009). Subjective and objective binge-eating episodes have been shown to have similar levels of service use and response to treatment (Banasiak, Paxton, & Hay, 2005; Mond, Latner, Hay, Owen, & Rodgers, 2010; Niego, Pratt, & Agras, 1997). The frequency of recurrence for binge episodes is also unclear. Different diagnostic symptoms have various guidelines to diagnose BN, ranging from binge eating and compensatory behaviors in the ICD-10 to minimum frequency of two binge-eating or purging episodes per week for at least 3 months for the *DSM-IV-TR*. Subtyping of bulimia into purging and nonpurging has also been suggested in the *DSM-IV-TR*, but there was little evidence to support the validity or utility of such an approach (Van Hoeken, Veling, Sinke, Mitchell, & Hoek, 2009) and as such was not included in the *DSM-5*.

BN is a common disorder among women, with lifetime prevalence rates in women of 1.5% (Hudson, Hiripi, Pope, & Kessler, 2007). A review of 27 studies by Steinhausen and Weber (2007) indicates that nearly 45% of patients diagnosed with BN demonstrate full recovery, with another 27% on average improving considerably. However, nearly one-fourth of patients who have been diagnosed with BN have chronic protracted symptoms. BN is traditionally found to be far less fatal than AN, with a mortality rate of .32%, in comparison to the 5% mortality rate of AN (Steinhausen & Weber, 2009); however, one recent study indicated a mortality rate as high as 3.9%, with elevated suicide standardized mortality ratios (Crow, Mitchell, Roerig, & Steffen, 2009). Furthermore, there is also a high comorbidity of BN with other eating disorders and psychiatric disorders. Individuals who have been diagnosed with BN have increased risk of high impulsivity, sensation seeking, novelty-seeking, and traits associated with borderline personality disorder and substance abuse (Cassin, & von Rasson, 2005). This suggests that all eating disorders should be evaluated and treated cautiously.

## Neuroimaging and BN

Brain imaging studies in BN are less clear than in AN. Early structural brain imaging findings in BN conducted with CT reported findings less pronounced than but similar to those described for AN (Krieg, Lauer, & Pirke, 1987). Specifically, less cortical mass was found in early CT studies. More recent MRI imaging indicates that individuals who have been diagnosed with BN can have areas of the brain that have increased volume. It has been demonstrated that an increase in the medial frontal cortex, particularly the central striatal volume, occurs in BN (Schäfer, Vaitl, & Schienle, 2010). It has been hypothesized that limitations in synaptic pruning in this area are associated with dysfunction in food-reward processing and self-regulation.

fMRI studies have examined the prefrontal cortex in BN. BN has been found to be associated with hypofunctioning in the right insula in response to anticipated receipt of a chocolate milkshake, as well as activation in the left middle frontal gyrus and right insula after consumption of the milkshake (Bohon & Stice, 2011). This suggests decreased activation in the brain's reward system. The individual with BN might respond to this decreased activation by overcompensating with excessive eating. Increased caudate nucleus and dorsolateral prefrontal cortex have been found even in women who meet subthreshold criteria for BN (Celone, Thompson-Brenner, Ross, Pratt, & Stern, 2011), once again suggesting alterations in the brain's reward system.

## Cognition and BN

Although cognitive flexibility is clearly the main cognitive limitation in AN, the cognitive abilities associated with BN are not as well understood. A systematic review was conducted by Van den Eynde et al. in 2011. Their results indicated that sample sizes were small, diversity in methodology prevented meta-analysis, and neurocognition during the acute state was not yet well studied in this population. Their review suggested that BN does not have a clear cognitive profile. Inhibitory limitations, although not impairment, have often been found in BN (Kemps & Wilsdon, 2010; Rosval et al., 2006). This can be related to difficulty inhibiting the desire to eat. Interestingly, when modified Stroop tasks have been used in which words related to body weight and shape have been incorporated, a strong Stroop effect occurs.

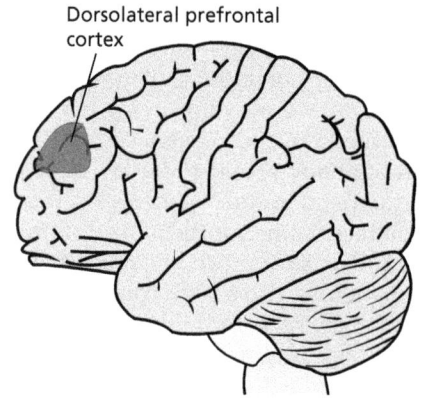

**FIGURE 9.3 Dorsolateral prefrontal cortex.**

This suggests less inhibitory control on measures intended to examine selective attention with body-related components. In comparison, it is less clear what effect food- and eating-related words have on inhibitory control. The systematic review also revealed that attention might be altered in BN, although the results are inconclusive. One study even found that performance on measures used to assess attention were even better in BN than in healthy controls (Ferraro, Wonderlich, & Jocic, 1997). Their review found that set-shifting results are also ambiguous. Similar to AN, detail-focused processing was found at the cost of global processing in one study (Lopez, Tchanturia, Stahl, & Treasure, 2008).

This phenomenon might be related to focus on specific eating- and weight-related issues, at the expense of understanding the larger contextual meaning. Performance on decision-making tasks indicates impaired functioning in several studies. Decision-making impairment might be associated with inhibitory control limitations and impulsivity (Waxman, 2009; Marsh et al., 2009 Liao et al., 2009). In comparison, there is limited evidence for poorer performance on measures examining memory, and results suggesting memory limitations can be associated with co-occurring depression, rather than resulting from BN. However, novel memory tests examining recall bias for disordered salient words have been found (Van den Eyende et al., 2011).

Taken together, there are few cognitive studies examining BN, but those that exist indicate potential limitations for inhibitory control, particularly as it relates to body weight and shape, global processing, impulsivity, and decision-making. Further studies examining this population might help the development of more specific treatment strategies, similar to those used to treat AN.

## Treatment for BN

### Medication

Much as with other eating disorders, there are no medications specifically designed to decrease the symptoms of BN. Rather, SSRIs have the most evidence for effectiveness and the fewest difficulties with adverse effects (Hall, Friedman, & Leach, 2008). Specifically, fluoxetine is the only medication currently approved by the U.S. Food and Drug Administration for treating BN. Interestingly, the dosages for effectiveness can be higher than those used for depression. In comparison, tricyclic antidepressants, monoamine oxidase inhibitors (MAOIs), and bupropion (Wellbutrin) should be avoided for their toxicity and lack of effectiveness (Hall et al., 2008). Much as with AN, SSRIs have the greatest effect on those individuals who also experience depression.

Determining which patients will likely respond to fluoxetine has been challenging. To date, there are no reliable relationships between patient characteristics and eventual response to medication. However, a study indicated that if patients had not responded to improved functioning of greater than 60% decrease binging or purging within 3 weeks of treatment, the functioning would not be likely to improve with the medication (Sysko, Hildebrandt, Wilson, Wilfley, & Agras, 2010).

*Repetitive Transcranial Magnetic Stimulation*

As discussed in a previous chapter, repetitive transcranial magnetic stimulation (rTMS) is a treatment option typically for refractory depression. The results for rTMS for BN have been inconsistent but promising. Target areas include the dorsolateral and dorsomedial prefrontal cortex (Downar, Sankar, Giacobbe Woodside, & Colton, 2012; Van den Eynde et al., 2010; Walpoth et al., 2008). Future studies will likely use techniques found to be effective in depression, considering the overlap of neurobiological findings.

*Psychotherapy*

Numerous studies have examined the effectiveness of various forms of therapy with BN. It has been found that interpersonal therapy and CBT are more effective than behavioral treatment without a cognitive component. Furthermore, multiple randomized controlled studies have shown that CBT is the most effective. Manual-based forms of CBT specific for bulimia have proven to be effective (Hay, Bacaltchuk, & Stefano, 2004). In fact, practice guidelines from the American Psychiatric Association indicate that CBT is effective as a short-term intervention when specifically directed at eating disorder symptoms and underlying maladaptive cognitions. CBT has been found to reduce binge eating and to also improve attitudes about shape, weight, and restrictive dieting. Intensive group therapy has also proven to be superior to medication (Mitchell et al., 1990). More recent studies have suggested that use of mobile phones and text messaging increases adherence to self-monitoring strategies (Shapiro et al., 2010). Ultimately, the combination of psychotherapy, particularly CBT, with use of increased self-monitoring strategies and antidepressants seems likely to be the most effective form of treatment for bulimia. Interestingly, similar to the response to medication, early response to CBT has been found to be a good predictor of long-term outcomes (Fairburn, Agras, Walsh, Wilson, & Stice, 2004).

For more severe BN patients, intensive treatment is recommended. A study comparing intensive outpatient day clinic treatment versus inpatient treatment for BN demonstrated comparable effectiveness immediately after treatment, as well at a 3-month follow-up, with both options proving to be effective (Zeeck et al., 2009). Consideration of the patient's safety and social support system is necessary in determining the most appropriate treatment in these situations.

## BINGE EATING DISORDER

Binge eating is characterized by rapid and excessive food consumption that is not due to metabolic need or hunger, occurring over discrete episodes that result in an uncomfortable full feeling (Brownly, Berkman, Sedway, Lohr, & Bulik, 2007). The *DSM-IV-TR* (APA, 2000) lists binge eating disorder (BED) as an area for further study, but the *DSM-5* includes it as an eating disorder and includes consumption of unusually large amounts of food, coupled with a sense of feeling out of control, occurring at least twice a week as its main symptom. Although there remains uncertainty as to the precise definition and criteria used for binge eating disorder and whether it is indeed an acceptable diagnosis, the psychological symptoms that accompany the behavior are numerous. Individuals who engage in these behaviors have feelings of distress and loss of control resulting from binge eating episodes. Binge eating is associated with depression, anxiety, and addictive behaviors.

# SPECIAL TOPICS

## Obesity and Therapist Role

It has been estimated that approximately 2% to 33% of U.S. children between ages 6 and 17 are obese or at risk of obesity. With the increasing rate of childhood obesity, it is expected that adult obesity rates will increase as well. These rising obesity rates are related to an increase in medical problems in both children and adults. The occurrence of type 2 diabetes, sleep apnea, asthma, orthopedic problems, and hypertension have been linked to obesity. It is also known that obesity is linked to psychological difficulties. For example, poor self-esteem, negative self-concept, depression, and withdrawal from peers are problems that have been associated with children who are overweight.

Research suggests that obesity can be linked to cognitive deficits. Studies examining adults have consistently demonstrated that depression, abnormal sleep patterns, and cardiac difficulties have also been associated with cognitive deficits. There has been much speculation that obesity might be linked to long-term cognitive deficits in adults, including by increasing the risk of dementia. Further evidence suggests that long-term obesity is linked to loss of brain tissue (Gustafson et al., 2004), providing a potential physiological correlate to the cognitive impairment. In fact, recent studies have shown that exercise is a protective factor against dementia in older populations.

Given the increased risk for mood problems and the potential for cognitive difficulties, it is important that the therapist consider a holistic approach when working with clients. Such an approach should examine dietary and exercise habits in all patients. Although a therapist might not be trained in these specific areas, behavioral strategies for improving behavior should be a strong consideration of the therapist. Treatment approaches used to develop behavioral modification plans to increase exercise and decrease sedentary behavior are strongly encouraged. Therapists might be reluctant to pursue such an approach with overweight or obese clients for several reasons, including lack of training, concerns regarding hurting rapport if weight is discussed as a treatment goal, and uncertainty regarding the therapist's role in developing eating plans and exercise routines, as specialists such as dietitians and physical trainers are explicitly trained in these areas. It is recommended that instead of refraining from intervening in this area, therapists should work with physicians, dieticians, and trainers to help improve their patients' functioning and quality of life.

Several steps should be implemented before developing a behavioral plan for controlling weight. Specifically, initially having the patient receive medical clearance from his or her primary care doctor or pediatrician is important for safety, as well as to rule out organic issues such as thyroid dysfunction that could contribute to weight gain. Furthermore, most medical doctors encourage exercise and healthy eating strategies, and reinforcement from the medical doctor will likely increase participation in the behavioral strategies. Consultation with a dietician to determine an appropriate diet and caloric intake goal, along with consideration of food allergy concerns, such as for gluten, might be beneficial as well. Having a dietician available for consultation should help the clinician better inform patients. Although referral to a physical trainer could be helpful, many clients might not have access to a physical trainer or an exercise facility because of restrictions on cost or location. However, online tools and smartphone applications allow for in-home, web-based training programs that can help users achieve exercise goals. Finally, ongoing emotional aspects and continued motivation and goal-monitoring will be important. Short-term, flexible, obtainable, and measurable goals

*(continued)*

> *(continued)*
>
> have been found to be an effective behavioral strategy to achieve success. From a behavioral perspective, less focus should be put on the actual weight of the client and more focus on implementing consistent, behavioral healthy lifestyle strategies that often result in weight loss. The neurochemicals released during physical exercise and the gains of healthy diet will be recognized throughout treatment from both a physical and emotional health perspective.

It is estimated that approximately 5% of the U.S. adult population has been afflicted by binge eating at some point (Hudson, Hiripi, Pope, & Kessler, 2007). Binge eating increases the likelihood for being overweight and for obesity. The physical, emotional, and cognitive effects of obesity can be overwhelming, even starting as young as childhood (see the Special Topics box). Therefore, understanding the behavioral manifestations along with the neurobiological explanation of the disorder can help the clinician identify it as a possible problem, and help him or her proactively treat people who are engaging in the behavior.

## Neurochemistry and BED

The neurobiological findings surrounding binge eating are consistent with those seen in individuals who have been diagnosed with substance abuse behaviors (Mathers, Sitch, Marsh, & Parry, 2011). Specifically, the neurochemical release of dopamine and endogenous opioids have been implicated in binge eating (Goodman, 2008). It has been found that eating desirable foods activates dopaminergic neurons within the nucleus accumbens (Kelly, Schiltz, & Landry, 2005). When rats are injected with high levels of sugar in a prolonged binge-like experiment, an increased dopaminergic receptor binding occurs in the striatum and nucleus accumbens (Avena, Rada, & Hoebel, 2008). These same researchers used a paradigm in which rats who ate sugar displayed decreased dopamine release within the nucleus accumbens after 36 hours of sugar deprivation and simultaneously increased food intake. When restrained from sugar and food, these mice demonstrated symptoms of anxiety. The striatum, which is thought to be important for food reward, satiety, and pleasure, was found to have altered dopamine levels (Michaelidis, Thanos, Volkow, & Wang, 2012).

## Neuroimaging and BED

In response to BED's recent recognition as a separate entity from BN, limited neuroimaging studies have been conducted. Often, these studies examine BED either in conjunction with BN or as a part of BN. Only recently have studies attempted to differentiate between the two eating disorders. A study conducted by Schäfer, Vaitl, and Schienle (2010) found that both BED and BN had greater cerebral volumes of the medial orbitofrontal cortex (OFC) compared to healthy controls. These structural abnormalities are likely associated with dysfunction in food reward processing as well as self-regulation.

Even fewer studies using fMRI have been conducted for BED. One study (Schienle, Schäfer, Hermann, & Vaitl, 2009) found that patients who had been diagnosed with BED demonstrate medial OFC activation while viewing images of food in comparison to overweight and healthy weight patients and patients with bulimia. In comparison, the patients with bulimia displayed more anterior cingulate activation and insula activation than the other groups when viewing images of food. There were no differences among the groups when examining fMRI responses to disgust-inducing pictures. This provides further neurobiological evidence for the uniqueness of BED as compared to BN. In fact, a more recent study completed by this research group indicated that evaluation of brain pattern activation can reliably contribute to differential diagnosis between BED and BN (Weygandt, Schafer, Schienle, & Haynes, 2012).

## Cognition and BED

There are very few studies examining cognitive functioning in binge eating. Typically, BED is examined with BN. As such, there are limited studies specifying cognitive functioning in BED. One study did not find inhibitory limitations such as those seen in BN, but did find difficulty in terms of set-shifting and problem-solving (Dueschne et al., 2010). Davis, Patte, Curtis, and Reid (2010) did find that BED patients were also more likely to perform worse on a decision-making task. Interestingly, in a study examining patients' self-reports, it was found that individuals who engaged in BED reported elevated difficulties in areas of executive functioning compared with those who do not have BED (Boeka & Lokken, 2011). Though understanding of BED is in its infancy, it is likely that studies will continue to identify cognitive differences with other eating disorders, shedding light on neurobiological underpinnings and thereby guiding treatment.

## Treatment for Bed

### Medication

Treatment for BED is challenging (Yager, 2008). SSRIs have been found to be effective for reducing binge eating (Brownley Berkman, Sedway, Lohr, & Bulik, 2007; Reas & Grilo, 2008). Given the weight gain in these patients, a goal of treatment is often weight loss. Unfortunately, SSRIs are not usually associated with clinically significant weight loss. Other medications, including sibutramine, topiramate, zonisamide, atomoxetine, and orlistat, are effective for decreasing both binge eating and body weight, but their administration often results in significant side effects and low adherence (McElroy et al., 2011). Challenges persist for this population, given the complexities associated with co-occurring symptoms and psychosocial problems.

### Psychotherapy

Cognitive behavioral strategies have been found to be effective at reducing binge eating episodes. Similarly, the role of therapist in individuals with obesity is expanding, whether it is due to binge eating or not (see the Special Topics box). Fairburn and colleagues have developed a fairly specific treatment protocol for binge eating disorder. Specifically, they have developed enhanced cognitive behavioral therapy (CBT-E). They provide details regarding three modules for treatment, including interpersonal psychotherapy (IPT) to improve upon interpersonal difficulties. Another module aims to improve upon coping strategies for negative affect, also referred to as "mood intolerance." There is also a module aimed at addressing body shape and weight. Treatment strategies should consider potential limitations with problem-solving and set-shifting, resulting in increased repetition in treatment as well as direct cognitive and behavioral strategies generated by the therapist. Insight-focused treatments are less likely to be effective in this population.

## CONCLUSIONS

Eating disorders are increasingly common, debilitating, and potentially life-threatening disorders that are clearly linked in their neurobiological basis. Mental health professionals should be aware of the signs and symptoms of eating disorders, as individuals might not disclose their eating habits as readily as their mood, anxiety level, or other symptoms. The co-occurrence of these symptoms along with eating disorders is likely linked to underlying biological mechanisms. Treatment is complex, as no medication has been shown to be consistently effective, and each eating disorder will bring with it specific goals. Furthermore, the medical complexities of these disorders require collaboration with physicians, nutritionists, and possibly other professionals.

## SUMMARY POINTS

- Eating disorders, which have long been thought to be mostly associated with societal pressures, have proven to be strongly linked to neurobiological processes.
- Fluoxetine is the only medication approved by the FDA for BN; there are no medications approved for other eating disorders.
- Successful treatment for AN has been linked to increased brain volume.
- Aiming to increase cognitive flexibility in patients with eating disorders will likely improve the chances of successful treatment.

CHAPTER 10

# Sleep Disorders

The potential for sleep disorders is often overlooked by mental health professionals as a root cause of problematic behavior and cognitive limitations. Sleep-related difficulties are a symptom of several psychiatric disorders, including depression, bipolar disorder, and post-traumatic stress disorder. Alterations in sleep are also often found in people diagnosed with dementia, attention deficit hyperactivity disorder (ADHD), generalized anxiety disorder, and traumatic brain injury. Sleep deprivation is also highly correlated with motor vehicle accidents and decreased productivity at work. Psychiatric classifications found within the *DSM-IV-TR* (American Psychiatric Association, 2000) and *DSM-5* (American Psychiatric Association, 2013) for sleep disorders include parasomnia, insomnia, breathing-related sleep disorders, and hypersomnia. From a developmental perspective, children are at risk for night terrors, enuresis, and encopresis, among other sleep-related problems. The prevalence of nocturnal seizures is high among children as well, including rolandic seizures and absence seizures. Because sleep abnormalities are often thought to be symptomatic of emotional or behavioral issues, they might not be thought of as a separate clinical issue. This chapter aims to help clinicians better understand normal sleep patterns and phenomena. Furthermore, it intends to help clinicians understand sleep in the context of emotional and behavioral disorders and to provide examples of common and interesting sleep disorders. Another goal of this chapter is to better relate sleep alterations to emotional and cognitive functioning, which can be mistaken for psychiatric diagnosis. Finally, the chapter applies this conceptualization toward clinical treatment from a behavioral perspective.

## PURPOSE OF SLEEP

According to the National Institute of Mental Health (NIMH), until the 1950s, sleep was viewed as a passive process in which our body lies dormant. During the past 60 years, researchers have identified that during sleep our brains are very active in terms of chemical release, electrical activity, and cognitive processing. Neurotransmitters, including serotonin, norepinephrine, and dopamine, are active at various stages of sleep. Different electrical activity is recorded on electroencephalographs (EEGs) and magnetoencephalographs (MEGs) throughout sleep stages as well. Similarly, cognitive activity is affected during sleep. Sleep has also been identified as being important for emotional functioning, bodily regulation, and behavior during periods of wakefulness. Limited sleep and sleep deprivation have been linked to inattention, irritability, moodiness, and physiological dysfunction (Durmer & Dinges, 2005).

These factors have led to a restorative theory of sleep, which postulates that for the nervous system to function effectively, sleep allows neurons that are active during the day to shut down and repair during the night. Other neurons that are not active during the day are activated during sleep, which increases neuronal connectivity. Without sleep, neurons become depleted of the necessary energy or polluted with by-products of normal activities, and this depletion results in malfunction. Another theory, evolutionary theory, notes that larger predators sleep more than smaller prey. For example, grizzly bears, which are relatively free from predators, are able to hibernate all winter and average 16 hours of sleep during other times of the year. In comparison, smaller animals, which are often prey, require far fewer hours of sleep. Evidently, although sleep might play a different role for different species, it is necessary for all species.

Theories ranging from restorative to evolutionary agree that sleep is necessary for survival. This is verified from animal studies. Specifically, rats who are deprived of the rapid eye movement (REM) sleep stage live for up to only 5 weeks, and those who are entirely sleep-deprived die approximately 3 weeks later. In comparison, the average life span of a well-rested rat is 2 to 3 years (National Institute of Neurological Disorders and Stroke, 2007). These sleep-deprived rats also develop abnormal body temperatures and lesions on their paws and tails, likely due to immune system dysfunction. In humans, prolonged sleep deprivation has been linked to psychosis. A person who sustained the longest period without sleep, nearly 19 days, experienced psychotic behaviors, including paranoia and hallucinations, as well as slurred speech, blurred vision, memory lapses, and concentration problems.

## NORMAL SLEEP

The average amount of sleep a person needs varies throughout the course of life. Humans sleep from approximately one-fourth to one-third of their lives. Understanding normal sleep processes is necessary to appreciate the effects of disrupted or limited sleep. Sleep is a general term used to describe a time period during which an individual is purposely not conscious. However, there are several different types of sleep, classified as sleep stages, defined by electrical activity within the brain. During these different stages, different neurophysiological activity, behaviors, and physiological changes occur. Sleep stages are traditionally differentiated based upon the neurophysiological recording of EEG characteristics. As described in Chapter 2, EEGs monitor the electrical activity of different brain regions during sleep studies (see Figure 10.1).

### Stage 0: Wakefulness

Stage 0, also known as wakefulness, is just that: when humans are alert, active, and functioning consciously. During Stage 0, EEGs record alpha activity, (medium-frequency brain waves, 8–13 cycles per second), interspersed with beta waves

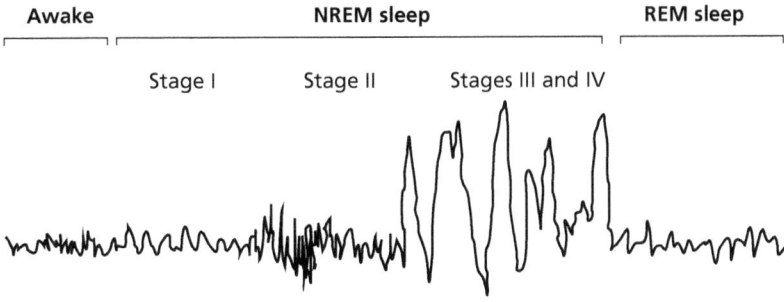

**FIGURE 10.1** EEG of sleep stages.

(irregular, low-amplitude waves, more than 13 cycles per second) throughout the brain. Alpha activity occurs less frequently in other stages of sleep, but it reappears in REM sleep, which will be discussed later. Beta waves are thought to be important for thinking, and they are present when an individual is attentive. During Stage 0, muscle activity is high in the body. Both rapid and slow eye movements occur throughout this stage. To enter into sleep, an individual must move from Stage 0 to Stage I. Falling asleep and losing consciousness gradually occur, and they are measured by changes in the brain waves. The stages of sleep are thereafter classified as either non-REM (NREM), which includes Stages I through IV, and REM sleep.

## Stage I Sleep

Stage I is an intermediate stage of sleep, occurring for 1 to 5 minutes per sleep cycle. Healthy sleep patterns exhibit approximately 2% to 5% of the night as being Stage I sleep. During this stage, the eyes begin to move more slowly, and the body can experience hypnic myoclonia, which are sudden muscle contractions. During Stage I, a person can be awakened more easily than from later stages of sleep. In fact, if awakened during this stage, individuals will often report not having slept at all. As sleep onset occurs, the person transitions to having fewer alpha waves. Specifically, alpha activity decreases to less than 50% as drowsiness ensues. Theta waves, on the other hand, are more prominent. Theta waves are longer brain waves (4–7 cycles per second), which also increase with calming activity, such as meditation and relaxation. Theta waves are slow-frequency waves, but with high amplitude. Delta waves (less than 4 cycles per second) also begin to occur in Stage I, but they increase dramatically throughout the various sleep stages, other than REM. Furthermore, during the onset of Stage I, serotonin and norepinephrine are secreted, and this secretion occurs throughout NREM.

## Stage II Sleep

Stage II sleep, also known as light sleep, comprises 50% of the sleep that occurs in healthy adults. During this stage, an individual's heart rate slows down, body temperature decreases, and eye movements completely stop. Delta activity still occurs, but less than 20% of the time. During this stage, theta waves continue to predominate, with two distinct EEG patterns. K-complexes and sleep spindles begin to occur on EEG recordings. A K-complex is a wave of delta activity on EEGs. A sleep spindle is a burst of wavelength that has a duration of one-half second. Though

TABLE 10.1 EEG Wavelength Description

| Wavelength | Most Active | Neurophysiological Activity |
|---|---|---|
| Beta | During daytime | Thinking, awake |
| Alpha | Awake | Relaxed, but awake |
| Theta | Near sleep and REM | Calm state |
| Delta | Asleep, NREM | Release growth hormone |

the function of sleep spindles and K-complexes remains uncertain, their presence indicates Stage II of sleep. As the hours of sleep increase throughout the night, an individual will cycle throughout the various stages of sleep and will often awaken for the start of the day from Stage II.

## Stage III Sleep

Stage III is the first stage of deep sleep. During this stage, breathing becomes slower, blood pressure begins to drop, and muscles become relaxed. It is characterized by 20% to 50% of delta waves, still interspersed by faster, smaller waves. These slow delta wave patterns predominate for the first third of a typical healthy night of sleep.

## Stage IV Sleep

Stage IV is the deepest and most restorative phase of sleep. Breathing continues to slow, muscles become more relaxed, and blood supply to muscles increases. During this stage, growth hormones are released, which is essential for muscle development. There is no eye movement or muscle activity during healthy Stage IV sleep. On an EEG, Stage IV is identified with more than 50% of delta waves. Mostly slow brain wave activity occurs during Stage IV.

## REM Sleep

Approximately one-fourth of healthy adult sleep is REM sleep. In comparison, approximately 50% of infant sleep occurs in REM. This difference is likely due to the 12 to 16 hours of sleep that infants are supposed to obtain. During this stage of sleep, there is no voluntary muscle movement, a state also known as muscle atonia, which results in no walking or talking. Breathing also increases during this stage. Indeed, the entire nervous system increases in activity, and the rapid eye movements that accompany Stage I of sleep predominate. An EEG is marked by sawtooth theta activity during REM sleep. During REM, serotonin and norepiniphrine are no longer secreted; rather, dopamine and acetylcholine are activated during REM. The activation of acetylcholine likely helps explain how a good night's sleep increases learning. Synchronization of oscillatory activity among different brain regions has been hypothesized as the mechanism by which memories and perceptions arise (Singer, 2001).

During REM, blood flow to the brain increases by approximately 50% to 200%, and sexual arousal increases. Eye movement increases as well and is similar to the pattern of functioning during wakefulness. For example, eyes move back and forth and up and down in the head, and they can appear to be focused on a moving object.

## Cycle of Stages

The stages of sleep continue to cycle throughout the night in healthy sleep. A person does not have one single block of a stage, but instead NREM and REM cycle throughout the night, with increasing amounts of REM sleep occurring as the person maintains sleep. These fluctuations occur approximately every 90 minutes, with 4 to 6 discrete periods of REM sleep throughout the night. As REM increases, deep sleep stages decrease. In a healthy amount of sleep, slow-wave sleep predominates in the first third of the night, whereas REM sleep predominates during the last third of the night (Yudofsky & Hales, 2008). Age is a factor in the various lengths of sleep and the type of sleep that occurs. Though the amount of REM sleep decreases from birth to adolescence before stabilizing, there is an additional decrease in the amount of REM after age 65. In comparison, the slow-wave sleep that occurs in Stages III and IV continuously declines after adolescence and can completely disappear in some elderly individuals.

In general, functional imaging studies have consistently found a decrease of thalamic and overall cerebral activation during NREM (Balkin et al., 2002; Kaufman, 2006; Kjaer et al., 2002). Furthermore, as NREM progresses, there appears to be a slower progression of brain activation, with light sleep being characterized by decreased activity in the frontal cortex, parietal cortex, and thalamus. In comparison, REM sleep results in increased activation of the limbic system and occipital cortex, with continued hypoactivity in the frontal regions of the brain.

## DREAMING

Attempts to understand dreams date far into history. Philosophers such as René Descartes and psychiatrists such as Sigmund Freud have attempted to understand and hypothesize about dreams. Although our understanding of the neural mechanisms associated with dreaming has advanced, an unknown quality of dreams remains a mystery to scientists. From an applied perspective, many clients will have vivid dreams and attempt to interpret them or have their therapist interpret them. Despite inherent concern regarding reliable recall of dreams, it is important to understand our current scientific understanding of dreaming.

Dreaming can occur throughout the various stages of sleep, but dreams are more likely to be recalled if the dreamer awakes during REM sleep. Though the exact neurocircuitry associated with dreaming remains unclear, a neurophysiologic association has been proposed at least since Hobson and colleagues did so in the 1970s. Hobson's theory, which has progressed from an activation–synthesis model to an activation, input, and modulation module, might continue to progress with advances in neuroimaging and other technology (Dawson and Conduit, 2011).

Positron emission tomography (PET) studies have shown that global brain metabolism is comparable during periods of wakefulness to REM sleep (Hobson, Pace-Schott, & Stickgold, 2000; Maquet et al., 2000). These studies have demonstrated occipital–temporal cortex activation during the periods of dreaming and wakefulness. This is not surprising: when people dream, they report seeing images and hearing sounds, including people talking—much like during periods of wakefulness. EEG studies have demonstrated brain patterns that are similar in areas of the brain during wakefulness and dreaming (Braun, 1998). Also, a large study conducted in 1997 indicated that dreaming does not occur in people who have lesions in or near the temporal–parietal system (Solms, 1997). A more recent study revealed total dream loss following posterior cerebral artery stroke as well (Bischoff & Basseti, 2004). REM sleep also increases limbic activity in functional MRI (fMRI) studies (Kaufman, 2006). These activation patterns likely account not only for the sensory experience of dreams, but also for the often intense emotional response to dreams.

## CIRCADIAN RHYTHMS

Circadian rhythms are defined as normal changes in mental and physical characteristics throughout the day. Circadian rhythm is controlled by the suprachiasmatic nucleus (SCN). The SCN is located in the hypothalamus above the place where the optic nerve crosses, and it is believed to contain approximately 20,000 neurons, which makes it approximately the size of a grain of rice. Light that reaches the photoreceptors in the retina generates neurosignals that travel along the optic nerve to the SCN, then subsequently to other brain regions, including the pineal gland. It is at this point that melatonin, a hormone responsible for drowsiness, is turned off. Secretion of melatonin peaks during nighttime and subsequently decreases during the day. As a result, the SCN plays a vital role in the sleep–wake cycle, including by facilitating alterations in blood pressure, hormone secretion, and body temperature.

The human body regulates itself to just over a 24-hour period, allowing human body temperature to be used as a common marker of circadian rhythm. As a result, when the body's temperature rises to its peak, humans are at their highest point of alertness, and as temperature falls, drowsiness occurs. As the body reaches is lowest temperature, drowsiness can reach the point of sleep and a significant decrease in alertness. As the body temperature increases once again, drowsiness subsides, and wakefulness occurs once again. The lowest body temperature typically occurs between 3:00 a.m. and 4:00 a.m., and once again between 3:00 p.m. and 4:00 p.m. Interestingly, if the body temperature is elevated due to an external mechanism such as a heat wave at night, restlessness often results. Similarly, the lack of synchronization between these circadian rhythms and consistent scheduled bedtime can cause less than optimal sleep, as well as decreased alertness during the day (Borberly & Acherman, 1992). The lack of parallel between circadian rhythm and work schedule for night-shift workers can contribute to long-term cognitive, emotional, and functional difficulty.

## LIFE SPAN AND SLEEP

The amount of sleep that is healthy throughout one's life alters. As previously mentioned, infants are expected to sleep 16 hours daily. Although this seems like a lot, it's not uncommon for infants to sleep 12 hours throughout the night and take two two-hour naps throughout the course of a day. As infants transition to toddlers and children, naps become unnecessary when enough sleep occurs throughout the night. In many industrial countries, including the United States, teenagers average around 7 hours of sleep due to school requirements, social activities, and extracurricular activities. Social media have also contributed to a decreased amount of sleep in teenagers. Regardless of these demands, research continues to document that teenagers require 9 hours of sleep nightly (Wolfson & Carskadon, 1998).

Similarly, adults have been shown to need 7 hours of sleep nightly, but they average approximately 6 hours nightly. In pregnant women, sleep increases by up to 2 to 3 hours per night, particularly in the first trimester of pregnancy. Studies have also shown that sleep deprivation in pregnant women increases the risk for miscarriage (Samaraweera & Abeysena, 2010). Often, after childbirth, women lose sleep. This deficient sleep can result in sleep debt. Sleep debt is the concept that the body will require extra sleep after too much sleep deprivation. This is commonly observed in teenagers who sleep for a prolonged amount of time during the weekends.

## SLEEP DEPRIVATION

Between 20% and 40% of Americans have sleep deprivation at any given time. Sleep deprivation occurs when an individual is unable to obtain sufficient sleep over an extended period of time. According to the American Academy of Sleep Medicine and Sleep Research, "sleep deprivation contributes to a number of molecular, immune, and neural changes that play a role in disease development, independent of primary sleep disorders." It has been documented that the alterations in biological processes as a result of chronic sleep deficiency can serve as causes for the development and/or increase of cardiovascular and metabolic diseases. Sleep deprivation has been documented to increase the risk for work-related injuries in various fields, including medical fields, and it increases motor vehicle and truck accidents, as well as the risk of other fatal accidents. It has been linked to aggressiveness, hostility, and irritability (Kampuis, Meerlo, Koolhaas, & Lancel, 2012). Long-term sleep deprivation has been documented to even decrease life span.

In general, the longer a person remains awake, the sleepier he or she becomes. The homeostatic regulation of sleep is similar to that of other motivated states in humans, including hunger, thirst, and sex. The hypothalamus is actively involved in all these drive states. As discussed in Chapter 3, the hypothalamus, which is about the size of a pearl and found deep in the brain, next to the thalamus and near the temporal lobes, has neural pathways running from the endocrine and nervous systems. During sleep deprivation, the hypothalamus releases less of the thyroxin and triodothyroine hormones (Everson & Nowak, 2001). This can contribute to emotional and cognitive decline in individuals who have sleep deprivation.

Identifying sleep deprivation in patients is therefore important if a therapist is to improve on emotional and cognitive functioning. Symptoms of sleep deprivation include falling asleep within 5 minutes. Although healthy sleep indicates that lying in bed for no more than 15 minutes is indicated, the sleep-deprived person is able to fall asleep too quickly. Also, it is not uncommon for people with sleep deprivation to demonstrate sleep debt, which is having periods of time in which they will sleep beyond a healthy amount of sleep during weekends or holidays. This is common in teenagers and other school-aged children on weekends. Evaluating for sleep deprivation and implementing corrective sleep strategies can improve on emotional, behavioral, and cognitive functioning.

### Cognition and Sleep Deprivation

Sleep deprivation also results in significant impairments in cognitive and motor performance. Studies have consistently indicated that sleep deprivation results in attention difficulties (Rocca et al., 2012). Even one night of total sleep deprivation results in an inability to learn new material (Yoo, Hu, Gujar, Jolesz, & Walker, 2007).

As mentioned earlier in this chapter, sleep spindles and k-complexes occur during Stage II sleep. It has been hypothesized that these occurrences are associated with cognition. Slow and fast spindle activity in Stage II was strongly correlated with performance on memory-related tasks (Schabus et al., 2006). Fogel, Nader, Cote, and Smith (2007) have correlated sleep spindles with scores on performance intelligence quotient tests. This study also found REM to be correlated with the verbal intelligence quotient index. Stage II sleep spindles have also been found to increase after learning new material before sleep (Schabus et al., 2008). Although sleep appears to affect intelligence through a complex interaction of physiology, individual traits, and mental states, it continues to be poorly understood (Kirov & Brand, 2012). It is important to evaluate for sleep patterns in one's patients as well as being cognizant of one's own sleep pattern as a mental health provider, which is further discussed in the Ethics box.

## Neuroimaging and Sleep Deprivation

Neuroimaging studies support the neurobiological effects of sleep deprivation. Structural MRI studies have demonstrated a variety of morphological changes in individuals with sleep-related disorders. There is hippocampus volume loss with chronic sleep insomnia (Riemann et al., 2007; Winkelman et al., 2012). Similarly, obstructive sleep apnea has been demonstrated to be associated with focal gray matter loss in the right temporal gyrus and the right cerebellum (Morrell et al., 2003). These authors conclude that this can be associated with motor dysfunction and motor deficit-associated motor vehicle accidents. A thorough review of the literature indicates that severe sleep disturbance in children results in neuronal cell death within a few weeks (Jan, Reiter, Wasdell, & Bax, 2009). Sleep duration during the weekdays has been found to be associated with bilateral hippocampus gray matter volume in healthy children as well (Taki et al., 2012). Their study examined 290 healthy children aged between 5 and 18. After controlling for age, sex, and overall brain volume, results indicated a positive correlation between the amount of sleep during weekdays and hippocampus volume in this age group. Functional MRI studies have also documented that neuropathophysiological changes are more consistently associated with sleep dysregulation than with structural brain abnormalities. These studies have demonstrated abnormal activity in the frontal lobes during executive functioning and attention tasks, suggesting that people who are sleep-deprived need to work harder to complete tasks than non–sleep-deprived people. Studies have also demonstrated decreased activation in temporal lobe areas during verbal learning tasks in sleep-deprived individuals than in non–sleep-deprived individuals.

## CLINICAL SLEEP DISORDERS

### Insomnia

Insomnia is the inability to maintain restorative sleep over a sustained period of time. It includes difficulty falling asleep or staying asleep. People who experience insomnia do not feel refreshed when they wake up from sleep. Insomnia is associated with a variety of medical and psychiatric disorders, and it increases mortality in adults (Chien et al., 2010; Hublin, Partinen, Koskenvuo, & Kaprio, 2011). Insomnia occurs in approximately one in four adults, and it can affect quality of life. Cognitive studies demonstrate mostly similar cognitive

> **ETHICS**
>
> **Are You Getting Good Sleep?**
>
> The first principle in the psychologist's code of ethics is that "psychologists strive to be aware of the possible effect of their own physical and mental health on their ability to help those with whom they work." Whether striving to be a psychologist or other mental health provider, the professional can often overlook his or her own well-being when taking care of others. This can happen subsequent to a variety of factors, including one's own personal life, stress associated with being a health care provider, and difficulty balancing work and personal demands. Disturbed sleep is a common reaction to periods of high stress. Decreased sleep, accompanied or not accompanied by increased sleepiness during the daytime, can contribute to significant difficulties in a profession that depends on the clinician's ability to concentrate on other people's problems. As mentioned previously, decreased sleep can contribute to attention problems, emotional dysregulation, and memory difficulty. Being able to concentrate and remember information from therapy sessions and to remain emotionally stable are necessary ingredients of successful therapy.
>
> The implementation of self-care strategies for the mental health provider is important in all areas, including exercise, social support, and diet. However, sleep can be difficult to assess. As discussed previously, the average amount of sleep ranges, and decreases, with age. Therefore, clinicians must carefully monitor changes in their own sleep patterns in order to determine whether there is cause for concern. Furthermore, determining the length of sleep-related alteration is also difficult. It is not uncommon for new parents to lose excessive amounts of sleep, particularly for the first year of a child's life. However, new parents are unable to predict when their child will be up throughout the night needing attention. It is unclear whether canceling appointments with patients who are in need of psychiatric care at the last minute or attending therapy while having less ability to concentrate throughout a therapy session is in the best interest of either the patient or the therapist. Clinicians need to evaluate themselves carefully in order to meet the goal of the first principle: taking care of themselves.

performance on tests (Basta, Chrousos, Vela-Bueno, & Vgontzas, 2007), although there is some suggestion of decreased reaction times and worsened sustained attention (Schneider, Fulda, & Schulz, 2004).

### *Neuroimaging and Insomnia*

There have been few neuroimaging studies examining insomnia. Individuals with insomnia have increased levels of a high frequency of EEG rhythm waves during sleep (Basta et al., 2007). Functional imaging studies have indicated decreased activity throughout the brain, including in the basal ganglia and the frontal, occipital, and parietal areas (Drummond, Brown, & Salamat, 2003; Smith et al., 2002). Furthermore, individuals who experience insomnia have higher cytokine levels during the day. Cytokines affect neuronal discharge in the hypothalamus, where melatonin is released.

### *Treatment for Insomnia*

Treatment for insomnia includes over-the-counter medications, prescribed medications, and behavioral strategies. Typically, individuals, particularly children, will be initially advised to use melatonin. Melatonin is a naturally occurring

TABLE 10.2 Commonly Prescribed Sleep Agents for Insomnia (with Generic Name)

| Benzodiazepines | Nonbenzodiazepines |
| --- | --- |
| Dalmane (flurazepam) | Ambien (zolpidem) |
| Halcion (triazolam) | Lunesta (eszopiclone) |
| Restoril (temazepam) | Desyrel (trazodone) |
| Valium (diazepam) | Sonata (zaleplon) |

hormone found in humans, plants, and animals that naturally helps regulate sleep and wake cycles in humans throughout the day. It is secreted in the pineal gland. When melatonin is ineffective, antihistamines, sedatives, and low-dose antidepressant medications are often used to treat insomnia. However, these can cause unpleasant side effects, including cognitive problems. Medications prescribed to help with sleep can be divided into benzodiazepines and nonbenzodiazepines (see Table 10.2). The benzodiazepines in Table 10.2 are anxiolytics. Typically the nonbenzodiazepines are not recommended for use beyond 3 weeks, with the exception of Lunesta, which can be effective for up to 6 months. In light of these limitations, cognitive behavioral therapy is also recommended for treatment to develop behavioral strategies for improving sleep hygiene (see Table 10.3). The National Sleep Foundation offers a variety of useful tools and information for clients and clinicians at its website.

## Hypersomnia

*Hypersomnia* is defined as excessive amounts of sleep. This can include extended nighttime sleep, unplanned daytime sleep, or inability to say awake during the day. As compared to insomnia, hypersomnia is far less frequent, but it is still fairly common. Less than 5% of the population complains of hypersomnia, and the symptoms typically occur between ages 15 and 30 (Sharon, 2006). Hypersomnia has been shown to result in cognitive impairment, selective attention, visual tracking and concentration, and worse performance on simulated driving, with improved functioning with successful treatment (Fulda & Schulz, 2001).

The neurobiology of hypersomnia continues to be explored and understood. It is likely different, contingent upon the etiology of the hypersomnia. Two more common causes of hypersomnia are sleep apnea and narcolepsy, discussed below. Otherwise, imaging studies in hypersomnia have not consistently demonstrated abnormalities. Cognitive measures do not demonstrate consistent impairment, but they do indicate poorer performances in attention and executive functioning, with intact performance on memory tasks (Bellebaum & Daum, 2010). There is evidence that a central histamine neurotransmitter is decreased with hypocretin deficiency (Kanbayashi et al., 2009). The hypothalamus has a large amount of histamine-releasing neurons that then project to the rest of the brain. This helps explain why histamine medications can improve sleep.

A common cause of hypersomnia is sleep apnea. Sleep apnea affects more than 12 million Americans and occurs when an individual repeatedly stops breathing throughout the course of the night. Risk factors of sleep apnea include being male, being overweight, and having low testosterone. There are two subtypes of sleep apnea: central sleep apnea and obstructive sleep apnea. Central sleep apnea occurs as a result of the brain's inability to consistently send proper signals to the muscles that control breathing. Obstructive sleep apnea is the result of the throat

TABLE 10.3 Behavioral Strategies for Improving Insomnia

- Set a schedule.
- Exercise.
  - Use activities such as yoga to help relax.
  - Try to exercise 20 to 30 minutes per day.
  - Try to get your exercise about 5 to 6 hours before going to bed.
- Avoid caffeine, nicotine, and alcohol.
  - Alcohol robs people of deep sleep and REM sleep, and it keeps them in the lighter stages of sleep.
- Practice deep breathing and relaxation before bed.
- Sleep until sunlight.
  - Sunlight helps the body's internal biological clock reset itself each day.
  - Recommend exposure to an hour of morning sunlight for people having problems falling asleep.
- Don't lie in bed awake for more than 20 minutes.
- Control your room temperature:
  - Extreme temperatures can disrupt sleep or prevent you from falling asleep.
- Allow no pets in the room.
- Have no television in the room.

muscles relaxing. Central sleep apnea and obstructive sleep apnea have different etiologies, but they result in the same increased risks for high blood pressure, weight gain, memory problems, headaches, and impotency. Untreated sleep apnea can increase the risks for motor vehicle accidents and job difficulty, and it can even result in death. Identifying sleep apnea is important, and helps increase the likelihood of improved functioning. Continuous positive airway pressure (CPAP) refers to a machine that treats obstructive sleep apnea. It increases air pressure in the throat so that the airway does not collapse during breathing. A CPAP machine uses a mask that covers the nose and mouth and prongs that fit inside the nose. If the patient can adjust to the invasiveness of the mask, sleep typically improves substantially, and the apnea is treated effectively. Other forms of airway pressure devices include the bilevel positive airway pressure (BPAP) and expiratory airway pressure (EPAP) devices. EPAPs have recently been FDA-approved, and they are promising, being less invasive devices that cover each nostril to help air move freely. More aggressive treatments include surgery of the jaw, tissues, nasal area, adenoids, tonsils, and even trachea, in more severe cases. Other forms of treatment for sleep apnea include hormone replacement therapy, exercise, mouth guards at nighttime, and psychotherapy, all of which can be beneficial for those who have sleep apnea, as well as other hypersomnias.

Narcolepsy, a well-known but rare hypersomnia, is characterized by frequent, brief, and uncontrollable sleep bouts. Narcolepsy with cataplexy (sudden loss of muscle tone) has been relatively well studied. Imaging studies indicate that narcolepsy is associated with reduced gray matter in the frontal lobes (Brenneis et al., 2005) and temporal lobe (Draganski et al., 2002; Kaufmann, Schuld, Pollmächer, & Auer, 2002), although there is dysregulation of neuronal activity and blood flow in the amygdala and hypothalamus during cataplexy when triggered by emotional response (Hong, Tae, & Joo, 2006; Schwartz et al., 2008).

Treatments of hypersomnia include neurochemical agents that are intended to stimulate brain functioning. Modafinil (Provigil), Ritalin, and other amphetamines have been prescribed to help increase wakefulness. Cognitive behavior strategies can also help maintain wakefulness.

## Parasomnia

Parasomnias are a collection of sleep disorders characterized by partial arousal. They are the result of the intrusion of sleep and wake states into each other. Parasomnia occurs when there are momentary or partial wakeful behaviors during non-REM sleep. Typically, parasomnias occur during the first third of the night, during slow-wave activity as registered on an EEG. Parasomnias should be diagnosed by a sleep specialist but can be identified by family members more effectively than by patients, who often have no recollection of the event. Parasomnias can result in dangerous activity and unhealthy sleep. Neuroimaging studies are difficult to obtain for parasomnias, considering the nature of the disorder and the infrequency of episodes. Sleep studies, however, help ensure that the unusual sleep behavior is not a result of a medical disorder, such as a seizure disorder.

Night terrors, also known as sleep terrors, are recurrent episodes of abrupt awakening throughout the first third of the night during non-REM episodes, in which the individual, often a child, demonstrates autonomic arousal, with eyes opening, screaming, and signs of extreme fear. The child is unresponsive to stimuli, will sit up in bed, and is amnesic to the event in the morning. The child is often unable to be awakened, and comforting the child is ineffective. Night terrors occur at least once in approximately 40% of children, with approximately 5% of children experiencing repeated episodes (DiMario, Francis, & Emory, 1987). There is a strong genetic link to night terrors. These are different from nightmares, which typically happen in the latter half of sleep, typically during REM sleep. The therapist is encouraged to remind the parents that although the child appears terrified, she or he will likely have no recollection of the episode, and the event is more stressful to the parent than to the child. Psychopathology is not associated with night terrors, but sleep deprivation can increase the risk.

Sleepwalking, also known as somnambulism, occurs when an individual gets out of bed and ambulates without awakening. It typically occurs during the end of the first or second slow-wave, deep sleep cycle. Sleepwalkers are often able to successfully navigate their environment, including by walking down stairs, not bumping into items or tripping on carpets, but they have also demonstrated increased risks, such as walking into a street.

Sleep bruxism, or grinding of teeth or jaw-clenching during sleep, is another form of parasomnia. It can be associated with movement disorders, substance abuse disorders, or developmental disorders—or it might have no clear medical cause. This involuntary and undesirable activity can result in abnormal tooth wear, create jaw pain, and cause disrupted sleep for a significant other. Neurobiological studies have demonstrated mild dysfunction of the basal ganglia, dopamine dysregulation, and high genetic influence (Alóe, 2009).

There is no evidence of high-level psycho-pharmaceutical treatment for parasomnias (Wilson et al., 2010). Although medications aimed at decreasing anxiety can be prescribed, there is no evidence of their efficacy. Rather, behavioral interventions are necessary. Specific strategies are contingent upon the disorder. For example, a mouth guard for sleep bruxism could help discourage the disorder,

## SPECIAL TOPICS

### Nocturnal Enuresis: A Behavioral Strategy

Primary nocturnal enuresis (PNE), bedwetting, is often a challenging and difficult parasomnia to treat. PNS is persistent bedwetting in the absence of any urologic, medical, or neurological anomaly in a child older than age 7. It is more common among males, and it also appears to have a genetic predisposition, as the incidence increases with the number of parents who also had PNE, as well as in monozygotic twins (D'Alessandro, Mason II, Pallone, Patano, & Marcus, 2005). It has been estimated that approximately 15% of nocturnal enuresis resolves spontaneously each year without treatment. The National Sleep Foundation indicates that nocturnal enuresis is developmentally normal before age 7. However, when these symptoms occur past that age, the symptoms can be frustrating for both the child and the parents. If a medical doctor has ruled out any medical causation for the enuresis, treatment is warranted. Learning effective strategies to help in this situation can be useful for the clinician to help his or her clients.

Pharmacological and medical interventions exist for this disorder, but they are not curative and can result in side effects. Research suggests that behavior interventions are more effective and have better efficacy than other forms of medical treatments. Initially limiting fluids and caffeine and maintaining a consistent sleep schedule are recommended. The most common treatment is the use of bell-alarm devices that indicate moisture level changes by setting off a bell that wakes the child to use the restroom. After time, the child learns to wake on his or her own to use the restroom before the alarm goes off. However, this strategy should be used as more of a behavioral last resort to avoid interrupting the sleep of others in the house.

A similar, less disruptive technique can be used by motivated parents: waking the child from sleep at a specific time, usually about 2 to 3 hours after bedtime. The child will typically become used to waking up and using the restroom on his or her own as needed. When the child demonstrates continence for approximately 2 weeks straight, he or she should be encouraged to drink more fluids before bed, in an attempt to increase bladder capacity. Eventually, the child learns to either wake during the night to use the restroom, or sleep throughout the night without wetting the bed. This strategy is more effective than an alarm, and it has lower relapse rates (Mellon et al. 2000).

The challenge involved in any behavioral treatments for nocturnal enuresis is that they require a high level of commitment from parents, motivation from the child, and a thorough understanding of the goals for treatment. Children who have been diagnosed with nocturnal enuresis may feel upset about their difficulty. Psychoeducation can help the child understand that he or she is naturally a heavier sleeper. These efforts should not be punitive in any way, and they should include positive reinforcement and consistency in both behavior and education. Motivational therapy can help eliminate guilt, provide support, and reassure the child and family that the issue will be resolved with consistent treatment (Cendron, 1999).

whereas locking windows, bolting doors, and sleeping on the ground floor might be helpful for individuals who sleepwalk. Avoiding frightening movies might help decrease nightmares. The Special Topics box discusses behavioral strategies for treating nocturnal enuresis. The clinician is encouraged to work with the patient to develop strategies for living a healthy lifestyle and effectively coping with stress, as many parasomnias are thought to be associated with anxiety and stress.

## CONCLUSIONS

Sleep is a vital human process. Sleep disorders are extremely common, and they can affect emotional, cognitive, and behavioral functioning. Many psychiatric disorders have sleep disturbances; likewise, sleep disturbances can trigger psychiatric difficulty. A mental health clinician should assess for quality and quantity of sleep. Referral to a sleep specialist might be more appropriate than treatment for psychiatric disorders in certain circumstances, in order to treat emotional and cognitive problems.

## SUMMARY POINTS

- Sleep serves many purposes, and the different stages of sleep are involved in different physiological processes.
- The purpose of dreaming is unknown, but scientists continue to examine the neurobiological processes that occur during dreaming.
- Sleep deprivation can result in significant consequences, including shortened life span and cognitive dysfunction.
- Clinical sleep disorders, including insomnia, hypersomnia, and parasomnia, have different presentations of symptoms, and various treatment strategies can be used to improve these symptoms.

CHAPTER 11

# Substance Disorders

## BACKGROUND

The effects of exposure to substances on brain development and functioning can be devastating throughout life. From conception on, exposure to illicit and controlled substances can substantially affect thoughts, behaviors, emotions, and social functioning. Though some substances might have only a short-term effect, other substances can have devastating long-term effects, even after exposure only once or on just a few occasions. Substance use can alter the central nervous system in a way that affects the remainder of the life span. From a social perspective, substance use can be harmful to relationships with spouses, children, employers, and peers. Of unemployed adults, it has been estimated that approximately 19% of them are illicit drug users (SAMSA, 2007). The rates of abuse tend to be higher among men than among women.

Despite substantial negative consequences associated with substance use, addiction continues to be very prevalent and difficult to treat. It has been estimated that 12.7 million people use illegal substances within any given month, for a total of up to 40 million people who use illegal substance within a year. Of those, it's estimated that nearly 3 million people would meet the criteria for being considered addicts; the rest would be considered casual drug users. According to the National Institute of Mental Health (NIMH), in 2012, approximately 8% of people older than age 12 had used some form of illegal drug. The NIMH has estimated that approximately 24% of all 18 to 20 year olds have used an illicit drug in any given month. Furthermore, legalized substances, such as alcohol and nicotine, have a substantially higher use rate and can also lead to abuse and dependence. For example, it has been reported that approximately 6.5% of Americans meet the criteria for either alcohol abuse or alcohol dependence. Another 25% of people between ages 12 and 20 have reported current alcohol use. There has also been a recent spike in the nonmedical use of painkillers, with most people who use them receiving them from a friend or relative. The most commonly abused illicit substances include marijuana, cocaine, methamphetamines, hallucinogens, and painkillers. Six percent of people abuse marijuana, although it has been legalized for medicinal purposes in some states, as discussed in the Special Topics box.

Addictions and substance abuse are commonly viewed as consequences associated with controllable behavior. This conceptualization can be empowering in terms of instilling confidence that strategies can be implemented to change both behavior and outcome. However, viewing substance use in this way also increases the possibility of overlooking the neurobiological role played in the addiction, as well as the relationship between the substance and the brain. For example, addiction is at least partially caused by biological changes that increase the brain's desire for the substance. Thus the central and peripheral nervous systems begin to experience changes that cause an ever-increasing need to ingest more of the substance. This process results in substance addiction.

Symptoms of addiction to substances include tolerance, sensitization, dependence, and withdrawal. Tolerance is progressive weakening of the effect of a substance after repeated use of that drug. For example, a person who drinks a case of beer every weekend for a decade will have a much higher alcohol tolerance than a person who drinks only once a year on his or her birthday. As a result of tolerance, a person must increase his or her consumption of the substance so that it will have the same effects on him or her as it did at the time of initial exposure.

## SPECIAL TOPICS

### Medical Marijuana—Helpful or Hurtful?

In March 2009, the U.S. federal government announced that it would not prosecute individuals who used or distributed marijuana for medicinal purposes if doing so was legal in their state. At the time of this publication, 18 states and the District of Columbia have legalized marijuana use for medicinal purposes. Sixteen of the 18 states require proof of residency in order to qualify as an eligible patient. Regardless of one's belief about the appropriateness of the treatment, the clinician should be aware of the implications involved in the use of marijuana for medicinal purposes in his or her clients.

Marijuana is thought to provide pain relief and to be effective for individuals who suffer from a variety of medical disorders. Prominent medical journals have published peer-reviewed studies supporting the use of marijuana for cancer, pain, glaucoma, and other medically disabling conditions. It has been argued that under the management and care of a physician, marijuana can be effective, safe, and controlled.

Others have argued, and studies have demonstrated, that marijuana may have several negative effects. Medical concerns can arise, given the known increase of lung problems, decreased immune system function, and increased infertility. There are also cognitive, emotional, and functional concerns. Regular use of marijuana increases the risk for depression, impairs cognitive ability to drive, slows processing speed, and is addictive.

Helping the client and/or his or her family to be aware of these possibilities and assessing for a decline in patients who are prescribed marijuana are likely to be a growing area of clinical utility for mental health providers. In light of the increased legalization of marijuana for medicinal purposes, it will be important for the clinician's understanding of the ongoing research in this area to be as comprehensive as his or her understanding of other medicines and treatment modalities.

*Sensitization* is the opposite effect, whereby repeated use of a drug results in an even stronger effect. *Dependence* results when continued exposure of the substance is necessary in order to avoid physical or behavioral disturbance when the drug is no longer used. *Withdrawal* is the term used to describe the behaviors and experiences of these symptoms. Though addictive disorders are historically thought to be a psychological dependence, alterations to neurobiology contribute to each of these processes of addiction. Similarly, there are neurobiological reasons why a person experiences withdrawal symptoms.

The goals of this chapter are to provide an understanding of the neurobiological processes involved in addiction to substances that are abused. The chapter provides a review of the general neurobiological response to exposure to substances. Although not intended to provide an exhaustive review of all possible addictive substances, the chapter does explore the commonality of the most prevalent substances that a general mental health clinician will encounter. If the clinician is interested in specializing in substance abuse, this chapter will provide a conceptual framework for considering the neurobiological factors playing a role in disease, rather than an exhaustive review of all possible substances. In fact, a general review allows the mental health clinician to better understand the effects of substances. An understanding of the neurobiological processes is important, considering the high comorbidity of substance use in other psychiatric disorders. The chapter also considers the ethical challenges involved in use of prescription medications to treat symptoms rather than disorders themselves, particularly amphetamines. As mentioned, the use of marijuana for medicinal purposes is examined in the Special Topics box.

## Behavioral and Neurobiological Response to Substance Exposure

The acute effects and chronic disruption from exposure to substances have significant effects on the brain at a cellular, chemical, and, eventually, structural level. The initial effect on the brain occurs at synaptic transmission. All drugs affect the brain by altering the amount of neurotransmitters present at a synapse, or by interacting with the neurotransmitters at the receptor site. In light of the high prevalence and comorbidity, a clinician should be able to recognize signs of substance use when treating patients, even if not specialized in substance disorders. Table 11.1 provides a list of various symptoms associated with substance abuse. Substances result in common behavioral effects, and they have positive reinforcing qualities that contribute to addiction. It is hypothesized that there are brain regions or systems that overlap with all substances to contribute to reinforcement.

**TABLE 11.1 Symptoms of Substance Abuse**

| Physical Symptoms | Behavioral Symptoms | Emotional Symptoms |
| --- | --- | --- |
| Tremors | Relationship changes | Argumentative/defensive |
| Bloodshot eyes or small pupils | Occupational difficulty | Easily irritated/agitated |
| New onset seizures | Stealing or borrowing money | Amotivation |
| Slurred speech | Suspiciousness/guardedness | Lethargy |
| Nose bleeds | Hyperactivity | Inattentiveness |
| Needle marks | Paranoia | Anxiousness |

It has been reported that individuals who become addicted to drugs have overwhelming and often uncontrollable cravings. These cravings are associated with alteration of synaptic transmission. Exposure to a substance influences the amount of neurotransmitter available at the synapse or binds with specific neurotransmitters to directly influence cognition and behaviors. Although the actions within the brain differ according to the substance used, they all have reinforcing characteristics that increase the risk for addiction. This reinforcing nature from the exposure of all substances suggests common underlying brain activity that plays a role in substance abuse.

One such neurotransmitter influenced by exposure to addictive substances is dopamine. Dopamine is important in reward-seeking behavior. When people expose themselves to the chemical properties of drugs along with the possible positive social interactions involved with substance use, dopamine is released in regions of the brain that assist in decision-making behaviors and mood. The prefrontal cortex and limbic system are sites of dopamine regulation that are altered when substances are introduced. Research has demonstrated that alcohol, methylphenidate, cocaine, opiates, marijuana, and nicotine influence the mesolimbic dopamine system (MDS), which includes the striatum. These drugs of abuse stimulate dopamine release, particularly in the nucleus accumbens. In light of this activation, it has been speculated that the nucleus accumbens is a primary site of the brain that is involved in substance abuse addiction (Di Chiara et al., 2004). Along with the release of dopamine, the MDS involves glutamate and GABA neurotransmission affecting brain activation. Simultaneously, the hippocampus is linked in the MDS, increasing short- and long-term memory of the exposure of the substance. This leads to drug-associated learning and memory, increasing desire for the substance (see Figure 11.1).

Imaging studies have demonstrated that the hippocampus is particularly sensitive to neurotoxicity. Both the right and left hippocampus volumes have been documented to be decreased in adolescents who have used alcohol versus those who have not (DeBellis et al., 2000). Recent diffuse tensor imaging (DTI) studies have demonstrated decreased myelination within the corpus callosum, suggesting that the integrity of white matter is being affected (Tapert et al., 2003). Bava and Tapert (2010) report that DTI has demonstrated frontal and parietal white matter circuits

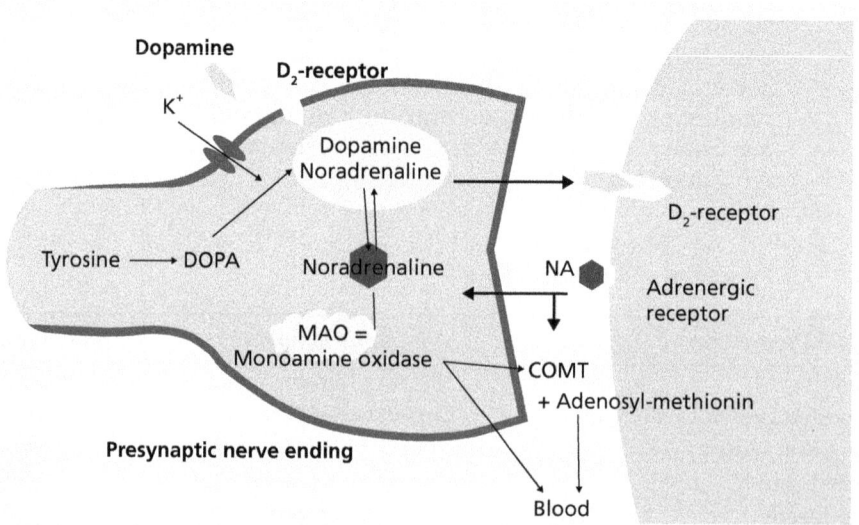

**FIGURE 11.1 Dopamine receptor.**

being negatively affected by substance use. Similarly, prefrontal volumes have been demonstrated to be reduced on structural imaging studies. Consistent with the imaging data, neuropsychological tests reveal worse performance on measures used to assess executive functioning in areas of risk-taking, problem-solving, and inhibition, as well as poorer performance on measures examining memory and processing speed, although there are individual differences that have a complex relationship with multiple factors (Bates, Labouvie, & Voelbel, 2002).

## Comorbidity of Substance Abuse and Psychiatric Disorders

Although mental health clinicians work with individuals who have co-occurring substance-related issues and emotional problems, it is important to recognize that one does not cause the other. Substance abuse issues occur either before onset of psychiatric issues or after onset of psychiatric issues. People who expose themselves to substances during adolescence appear to have a higher likelihood of later psychiatric problems. It has also been demonstrated that people who have psychiatric issues might self-medicate in an attempt to resolve emotional or cognitive difficulty. In fact, marijuana dependence has been noted to be most prevalent in individuals suffering from other psychiatric disorders (Gouzoulis-Mayfrank, 2008). It is also common for people who suffer from a substance abuse, or dependence on one substance, to have a co-occurring addiction to another substance.

This high comorbidity rate might in fact result from commonalities in neuropathways, neurochemical dysregulation, and cognitive impairment. As demonstrated in previous chapters, dopamine is important in a variety of disorders. The neuropathways involved in frontal activity, along with limbic system and striatal dysregulation, including the nucleus accumbens, have been found repeatedly to be associated with mood disorders, anxiety disorders, and psychotic disorders, among others. The overlap and earlier onset of symptoms are predicated on the neurobiology involved in the psychiatric disorders and dysregulation caused by exposure to substances.

Treatment is especially challenging when co-occurring substance and psychiatric disorders exist. Improved mental health is not likely to be achieved or sustained without elimination of the substance and control of the behaviors associated with the abuse. The amount and length of time that the substance has been used might have permanent effects on emotions, behaviors, and cognitions. The overlap of symptoms often requires specialized dual-diagnosis treatment facilities.

## GENETICS AND SUBSTANCE DISORDERS

It has been estimated that approximately 50% of an individual's likelihood of having a substance abuse problem is associated with genetic conditions. That said, a person does not inherit substance disorders, but rather a person might be genetically predisposed to abusing a substance. Whether a person suffers from abuse is contingent upon the central nervous system's exposure to the substance. Thus, a genetically predisposed individual might never demonstrate symptoms of abuse if she or he refrains from the substance throughout his or her lifetime.

Multiple genes, also known as *gene clusters*, interact with the exposure to the substance to help determine how strong a positive reinforcement the substance provides for an individual. This fact can account for why some people can have five alcoholic beverages on a holiday weekend and not feel the need to do so again, whereas other people have a craving the very next day to repeat the action. Various studies have examined possible links for genetic mutation that increase the risk for substance abuse. For example, a nicotine subunit receptor gene has been demonstrated to double the risk for nicotine addiction in smokers (Saccone et al., 2006). Genetic studies have also been used to predict which patients will benefit from treatment. Alcohol-dependent patients who were treated with naltrexone had a significantly lower rate of relapse if a specific variant of an opioid receptor was present (Ray & Hutchison, 2007).

## Prenatal Exposure to Substances

Prenatal exposure to various substances can have a variety of outcomes on the central nervous system. The effects of neurotoxins on the developing brain are likely associated with the type, frequency, timing, and amount of exposure. The threshold amount of substances that can be ingested during pregnancy before cognitive, emotional, and behavioral symptoms occur in children remains unclear. However, exposure to various substances is known to cause a variety of disorders in offspring, as well as to increase the risk of other disorders. For example, prenatal exposure to alcohol has been associated with impairments in executive functioning, learning, attention deficit hyperactivity disorder (ADHD), and oppositional tendencies, even in the absence of fetal alcohol syndrome. Prenatal exposure to cocaine has been linked to deficits in language development, self-regulatory difficulty, learning problems, and delays in executive functioning. Marijuana exposure in utero has been linked to worse performance on intelligence tests, ADHD symptoms, oppositional tendencies, and impulsivity. Even nicotine exposure has been linked to premature births, increased impulsivity, and working memory problems during childhood. There is even evidence to support that total brain volume and reduced head circumference are linked with in utero exposure to these substances (Rivkin et al., 2008).

## EXAMINATION OF COMMON SUBSTANCES OF ABUSE

Although there is overlap in the neurobiological processes associated with the various disorders of substances, a better understanding of the functional outcome and behavioral reactions to the substances appears warranted. Unlike previous chapters, this review will focus on the cognitive, emotional, and behavioral disorders that can occur as a result of exposure to substances. An understanding of the previous material describing the MDS and other neurobiological similarities in substance abuse is useful when conceptualizing the possible behavioral effects of exposure to these substances. Though treatment varies for the substances, the ultimate goal is abstinence. Cognitive training of frontal systems can be effective for abstinence from drugs of abuse (Garavan, Brennan, Hester, & Whelan, 2013).

## Alcohol Abuse

The most common form of substance abuse a mental health practitioner will encounter is alcohol abuse. Cognitive impairment, behavioral difficulty, and brain-based changes will occur long before liver damage, and often well before overt medical conditions associated with the alcohol use. Alcohol problems do not occur just in younger individuals; as the baby boomer generation has increased in age, alcohol-related problems have increased as well. However, identifying alcohol-related concerns in any population can be difficult, as individuals will underreport or not disclose the actual amount of alcohol consumed. Furthermore, language skills and social functioning can often be relatively well preserved. The mental health clinician should be familiar with common issues that occur in alcohol abuse.

Alcohol influences the central nervous system by acting as a depressant. It slows down the functioning of the central nervous system, including the brain. Although it initially improves mood or provides a euphoric sensation, it subsequently results in depressive symptoms. The initial euphoria leads to the reinforcing behavior of continued alcohol use, despite the times when people who use alcohol feel depressed, as people enjoy the initial improvement in mood. This paradox is a challenge for the mental health provider, as it reduces incentives for the patient to discontinue drinking alcohol. Alcohol, like every other neurotoxin, is poisonous. However, the metabolic activity inside the brain is different when using alcohol versus other addictive substances. It metabolizes at a consistent pace, regardless of how much of the substance is ingested. Though the metabolism differs according to weight, tolerance, race, gender, and height, it approximates one beverage per hour. For example, men metabolize alcohol more quickly than women do, and men therefore require more alcohol to become intoxicated. The amount of alcohol consumed by one person therefore results in different neurobehavioral and biological effects from those seen in another person. Such metabolizing at a consistent, steady pace regardless of the amount of alcohol is referred to as *zero-order kinetics*. Acute alcohol exposure leads to disinhibition, poor judgment, socially inappropriate behavior, poor attention, slower reaction time, and slurred speech. Aggression has often been linked to alcohol use, and alcohol use often increases incidence of violent behaviors (Leondard & Sinchak, 1993). This aggression is thought to be associated with the stimulating effect of alcohol. Aggression, along with the aforementioned cognitive difficulties, is modulated by frontal lobe activity. Alcohol intoxication can also result in lethargic behavior, sadness, anger outbursts, and other depressant symptoms—all behaviors indicative of limbic dysregulation. High amounts of repeated alcohol use increase the risk of depression and of cognitive impairment. With a decline in blood alcohol, reaction time, processing speed, and attention have been demonstrated to improve, even before complete metabolization of the substance (Schweizer & Vogel-Sprott, 2008). Schweizer and Vogel-Sprott also demonstrated that different cognitive functions are affected by alcohol when it is being ingested than when it is being metabolized, with verbal memory being impaired during the consumption of alcohol and visual processes being more impaired during metabolization.

Chronic alcohol use can result in severe cognitive impairment. Korsakoff syndrome can occur with repeated exposure to alcohol. It is characterized by impaired short-term and remote memory, apathy, passivity, confusion, flattened affect, and little insight, and the patient might demonstrate confabulation. *Confabulation* occurs when the individual makes up stories that can seem reasonable, but that upon further evaluation prove false, although the person is unaware that he or she is making up the story. Korsakoff syndrome leads to abnormal thought processes. Yet on cognitive tests, a person experiencing Korsakoff syndrome typically demonstrates intact intelligence and attention span. The difference between intelligence and memory functioning, though, is striking. The disruption in cognitive and functional abilities associated with Korsakoff syndrome comes from thiamine deficiency. Thiamine, also known as vitamin B1, is vital for metabolizing glucose from carbohydrates. It also assists in heart, muscle, and nerve functioning. With thiamine deficiency, a person experiences emotional, behavioral, and cognitive problems. Fortunately, if a person eliminates alcohol from his or her diet and supplements with vitamin B1, recovery from Korsakoff syndrome can occur.

Often accompanying Korsakoff syndrome is Wernicke's disease. Wernicke's disease is characterized by abnormal physical symptoms of gait irregularities and ocular symptoms, including unusual eye movement or gaze, ataxia, and confusion. The gait is marked by slow, uncertain movement and a wide stance. Ocular abnormalities include nystagmus or opthalmoplegias, weaknesses of gaze. Global confusion, with memory problems, persists. Alcohol dementia can occur, similar to other dementias described later in this text but which include progressive cognitive and behavioral decline. From a cognitive perspective, alcohol dementia typically includes the visual spatial and slower psychomotor speed also observed in chronic alcohol use even without dementia. However, memory impairment does not occur in all alcohol abusers.

Some literature suggests that small quantities of alcohol can have protective effects for the cardiovascular system. A systematic review indicated that limited alcohol use in earlier adult life can be a protective factor against dementia later in life (Peters, Peters, Warner, Beckett, & Bulpitt, 2008). In fact, low amounts of alcohol use have been associated with a 32% reduced risk of dementia. This is likely associated with decreased risk of vascular types of dementia and the vascular relationship found in Alzheimer's disease.

Alcohol use in adolescents is particularly problematic. It has been estimated that nearly half of all college students engage in binge drinking: the consumption of four or more successive drinks, for women, or of five or more successive drinks, for men, in a single sitting. Heavy consumption of alcohol has been linked to academic problems as well as to visuospatial limitations (Wood, Sher, & Bartholow, 2002). There appears to be a strong link to alcohol use and genetic predisposition. Also, long-term functional problems are more likely to occur in people who have used alcohol at an earlier age.

Treatment for any substance abuse must ultimately lead to discontinuation of the substance. Although the 12-step program is a well-known form of treatment for alcohol, it relies heavily on a spiritual component to which some people do not respond well. Cognitive behavioral strategies can be implemented to assist in discontinuing alcohol use. Developing coping strategies to deal with stressors as well as establishing behavioral goals to replace the alcohol consumption can be effective. Medication has also been used to decrease the use of alcohol. Antabuse is a medication prescribed for alcohol abusers, acting as a deterrent that makes the individual feel ill when ingesting alcohol. It works by blocking the breakdown of alcohol and thus causing vomiting, headaches, chest pain, stomach pain, and sweating. These occur even with minimal alcohol consumption. Consideration of environmental influences is necessary for the mental health provider as well. Elimination of reinforcers for the behavior can prevent the brain reward circuitry from releasing dopamine and by doing so activating the individual into resuming consumption. This leads to the notion that a person must stop ingesting alcohol completely in order to prevent another relapse from occurring. To increase the likelihood of a complete recovery, the person is encouraged to avoid situations, people, and events in which she or he might be encouraged or tempted to drink alcohol. Eliminating these high-risk activities decreases the likelihood of the behavior recurring. Although rearranging the social support system of a person who has alcohol issues sounds easy, the practice is difficult, as doing so means essentially asking the person to change his or her friends, routines, habits, and relationships. This also helps to explain the reason for the high relapse rates.

## Cannabis Abuse

Cannabis, also known as marijuana, is the biologically active compound tetrahydrocannabinol (THC) found in the plant cannabis sativa. This plant is native to Central Asia but can be grown easily in most environments. It has become the most common street drug in America, and it is often inaccurately thought to be a harmless drug from a cognitive perspective. It produces a change in mental state that is often described as a relaxed or calming effect. However, use can also lead to hallucinatory experiences, fluctuations in mood, and psychotic symptoms. Use of the drug has also been linked to the onset of schizophrenia, particularly in adolescents. A case study even linked it to a suicide attempt in a young female (Nussbaum, Thurstone, & Binswager, 2011).

Neuroimaging studies have demonstrated that the age at which an individual uses marijuana can significantly affect the biological process involved with the substance. It has been found that individuals younger than age 17 who have used marijuana have smaller whole brain volumes, less cortical gray matter, and a greater amount of cerebral blood flow than those who started using marijuana later (Wilson et al., 2000). This fact might account for the increased risk for schizophrenia and psychotic symptoms. Functional neuroimaging studies have also demonstrated an increase in the dorsal lateral and medial frontal regions, as well as in the occipital and parietal areas, during tasks requiring inhibitory control for adolescents who have a history of marijuana use (Tapert et al., 2007). This suggests that marijuana increases cognitive efforts for simple inhibitory tasks. From a cognitive perspective, a decrease in performance on measures during substance use and after sustained amounts of use has been documented. Slower processing speed, worse memory, and decreased attention have been documented. It is clear that even subtle brain abnormalities and cognitive problems from marijuana use in adolescents increases negative psychosocial consequences (Lisdahl, Gilbert, Wright, & Shollenbarger, 2013).

When withdrawing from cannabis dependence, individuals can experience cravings, anorexia, irritability, restlessness, and sleep disturbance. These symptoms begin to emerge 1 day after discontinuation, peak at about 1 week, and last for up to 3 weeks. It is important for the individual to refrain from marijuana during that time. The NIMH notes that the treatment for marijuana dependence is similar to that for other drugs of abuse, but that the long-term outcome might be less severe. Behavioral treatments have proven to be modestly effective. An abstinence of 1 year has proven to occur in only 10% to 30% of people who underwent treatment for marijuana abuse. In fact, only 50% remain abstinent beyond 2 weeks. Though cannabinoid agonist medication has proven to be a promising treatment in clinical trials, no medications have been FDA-approved for use for this population. Thus, treatment is often challenging and somewhat ineffective.

## Methamphetamine Abuse

Methamphetamine (meth) is an illegal psychostimulant that has an estimated 16 million users worldwide, a number thought to exceed the combined number of heroin and cocaine abusers, making it the second most widely abused drug after marijuana. Meth is relatively inexpensive to produce, has long-lasting effects, and produces a euphoric rush to the brain. Meth's chemical properties are similar to those of amphetamine, as well as to those of most other drugs, as it affects the dopamine receptors. In fact, meth blocks the reuptake of dopamine, leading to a rapid increase in motor activity, motivation, decreased anxiety, and hyperarousal. The effects can last for several hours, because the half-life of meth is between 10 and 12 hours. However, large doses of meth can also lead to acute toxicity, neurological damage, hyperthermia, renal failure, cardiac failure, cerebrovascular hemorrhages, seizures, renal failure, and stroke. Less severe consequences include functional disability to care for children and maintain employment, and such doses also cause relationship difficulties, among other significant behavioral difficulties. Meth abuse increases the risk for contracting HIV and hepatitis B and C, and it also increases the risk for long-term psychosis.

Neuroimaging in chronic meth users has demonstrated significant structural alterations and functional problems in areas of the limbic system and prefrontal cortex. Positron emissions tomography (PET) studies have demonstrated reduced striatal binding of dopamine, which has persistent effects. In fact, there is evidence that people who abuse meth have an increased risk for developing Parkinson's disease, owing to the striatal damage of dopamine receptors caused by meth (Callaghan, Cunningham, Sykes, & Kish, 2012). A thorough review of studies conducted by Krasnova and Cadet in 2009 indicates the damage of neuronal tissue in studies examining monkeys on meth. Large portions of the studies indicate striatal damage along with select other damage found within the brain. Cognitive functioning has been demonstrated to be impaired in memory, attention, selective attention, inhibition, and decision-making abilities. Furthermore, fine motor skills, gross motor skills, and processing speed have been documented to be impaired in chronic meth users (Simon et al., 2000; Woods et al., 2005). Even when abstinent, chronic meth users continue to demonstrate cognitive impairment.

Treatment for meth is challenging. A cognitive behavioral therapy (CBT)-based approach using family resources, individual counseling, group counseling, and drug testing has met with some success in this population. Though no medications are approved for meth treatment, recovery from meth has been observed in humans with increasing abstinence. PET studies have demonstrated a nearly 20% increase of dopamine-binding agents, increased thalamic glucose, and a decreased level of microglial activity (Wang et al., 2004). Although the mental health clinician might not understand the clinical implications of these activities, it could be anticipated that neurobiological recovery occurs spontaneously over time with abstinence, suggesting improved cognition and functional abilities. This finding will likely lead to the development of medication to enhance these neurobiological processes, thus enhancing the observed functional improvement.

## Cocaine Abuse

Cocaine is a stimulant drug made from leaves of the coca plant native to South America. Similar to meth, it provides a rush when inhaled. Taking this drug results in increased alertness, arousal, and confidence. This high will last 15 to 30 minutes when the drug is snorted and approximately 5 minutes when it is inhaled. This leads to increased amounts of time using the drug, thereby increasing tolerance. Much like other drugs, it results in an increase of dopamine on the reward circuits, which in turn results in immediately improved feelings followed by long-term effects of decreased libido, possible delusions, seizures, and other negative effects. Tolerance leads to an inability to experience pleasure. Neuroimaging studies have consistently demonstrated dysregulation of the frontal lobes. Structural imaging studies have demonstrated decreased gray matter in the prefrontal cortex. Functional magnetic resonance imaging (fMRI) studies have demonstrated that patients who are cocaine-dependent have demonstrated increased activation of the prefrontal cortex and limbic system when presented with cocaine-related videos depicting drug administration. Furthermore, PET studies have demonstrated decreased metabolism of the prefrontal cortex (Kim et al., 2005).

Consistent with prefrontal dysregulation, executive functioning in terms of poor decision-making, increased risk-taking behavior, and disinhibition have been documented to occur in this population.

Treatment is difficult. There is no pharmacological treatment for cocaine dependence. According to the Substance Abuse and Mental Health Service Administration (SAMSA), several studies exploring drugs that are able to reverse cocaine-induced neuronal changes have shown promise. The majority of cocaine abusers who seek treatment are those who smoke crack, which is a form of cocaine that has been processed in a certain way so that it can be smoked. Though treatment is difficult, motivational incentives have proven helpful for achieving initial abstinence from cocaine. This includes a prize-based system encouraging better decisions to avoid cocaine use. From an applied biological perspective, this result suggests that the dysregulation of the frontal cortex results in the individual needing guidance and motives to make better decisions. Behavioral strategies can help guide the individual in these areas. With time, it is hoped, the brain redevelops neurocircuitry to consistently make better decisions avoiding substances. Residential programs, community-based programs, and other support systems should be provided to help the individual consistently avoid cocaine use.

## Hallucinogen Abuse

Pure ecstasy, lysergic acid diethylamide (LSD), mushrooms, ketamine, and phencyclidine fall under the broad term *hallucinogens*, owing to their effect on hearing, smell, touch, and vision. These drugs are referred to as dissociative analgesics because of the pain-killing effects they have on the user, as well as the chemical property of the substance. Many hallucinogens have chemical structures similar to those of the natural neurochemicals serotonin, acetylcholine, and catecholamine. Dopamine is overreleased in the midbrain and prefrontal cortex, contributing to the effects caused by these drugs. It is likely that these drugs bind with other neurochemicals as well, contributing to the collection of symptoms. MDMA, also known as ecstasy or molly, has the chemical properties of both hallucinogens and amphetamines, with effects lasting up to 6 hours. Treatment for hallucinogen use includes use of a stimulus-free room devoid of distractions and noise. There is little published data on treatment, but supportive and CBT techniques are likely to be effective in long-term treatment, as well as coping with the short-term symptoms. Furthermore, benzodiazepines have been used to decrease the agitation that occurs during an acute state.

## Opiate Abuse

Narcotics, including heroin, codeine, and morphine, are synthetic forms of opiates that have strong addictive characteristics due to the decreased pain that an individual might feel along with increased euphoria. Opiates such as heroin selectively stimulate receptors, which generate increased dopaminergic response that typically provides inhibitory regulation. Opiates directly affect several brain regions, including the thalamus, amygdala, and hippocampus, releasing endorphins that underlie the positive feelings that occur. Chronic exposure to opiates leads to gray matter deficits in the frontal and temporal lobes, as well as to decreased activation in the cingulate cortex, basal ganglia, and frontal lobes. Recent DTI studies indicate that demylination occurs in the frontal and temporal lobes (Bora et al., 2012; Liu et al., 2008). Although the exact location of the neuronal alterations can be localized, based upon the type of opiate, it is clear that opiate abuse leads to structural and functional alterations. Abstinence from opiates has been shown to increase activation of the HPA axis, increasing corticotrophin-releasing hormones. Neuropsychological tests also indicate that marked cognitive impairment occurs in this population, including the impairment of memory, executive functioning, processing speed, and efficiency.

Treatment for opiate abuse is contingent upon the type of opiate being abused. For example, heroin-dependent patients often benefit from methadone maintenance to provide a long-lasting, oral opioid agonist in lieu of the heroin. Naltrexone has also proven to be effective for treating heroin dependence. Other longer-lasting oral opioid agonists can be administered as well. This medication blocks the euphoria that comes when heroin is administered, thereby decreasing the reinforcing emotional response. With abstinence, cognitive improvement occurs in this population.

## Caffeine and Nicotine Abuse

The mental health clinician should be aware that he or she, too, might be using a drug of abuse, such as nicotine and/or caffeine. More people use these substances than use any other addictive drug. Caffeine is a stimulant that is absorbed rapidly, with peak effects within 30 minutes of digestion. The effects last for approximately 2 hours, and loss of tolerance can last up to 2 weeks. Dependence can develop with high amounts of caffeine intake. Caffeine affects the neurotransmitter adenosine. Adenosine is important for sedation, and as a result a stimulating effect occurs with caffeine. This leads to increased cortical tone, with improved alertness, arousal, and energy.

Nicotine is the most addictive psychoactive substance, as it is also a central nervous system stimulant. Approximately 90% of nicotine is absorbed, making it one of the most toxic drugs. Nicotine acts much like acetylcholine by affecting cholinergic sites. This action reportedly has positive effects on memory and learning. When the nicotine is stopped abruptly, people who smoke report decreased memory functioning.

Although caffeine and nicotine don't typically have the same negative short-term effects on health and behavior, the long-term effects have been linked to physical health problems. Treatment for caffeine might be required to improve sleep and reduce anxiety and irritability, whereas treatment for nicotine might aim at improving breathing-related disorders and other lung diseases. Eliminating caffeine and nicotine from one's behavioral patterns is challenging. Both substances are widely available, are legal, and have social connections. People will take smoke breaks or coffee breaks at work or at school, developing peer relationships through the activity. The activation of dopamine receptors, as well as serotonin receptors, occurs with even the social cue prompting ingestion of the chemical. Though both substances have addictive characteristics and can affect health in a variety of other contexts, otherwise cognitively intact individuals have a hard time discontinuing use, owing to the substances' addictive qualities.

## CONCLUSIONS

Substance disorders are highly prevalent, and they frequently co-occur with other disorders. It is likely that almost everybody has been exposed to a substance of some kind that has resulted in cognitive, emotional, and behavioral alterations.

# ETHICS

## Use of Amphetamines as Performance Enhancers

Several studies and reports detail recent increases in the use of performance-enhancing medications in a variety of contexts, including steroids and human growth hormones (HGH). Professional athletes are frequently suspended, punished, and vilified for their use of drugs, to the point that athletes who are suspected of using performance enhancers are not being honored for their accomplishments. Although eliminating the use of anabolic steroids and other muscle-building substances is challenging thanks to imperfect drug testing, the use of cognitive-enhancing medications is even more challenging to limit and control. The use of cognitive-enhancing drugs is also likely to be more widespread and not limited to competitive athletics, but rather found in all aspects of life. In fact, federal funds are being poured into developing medications that reduce the decline of cognitive aging, particularly as it relates to dementia. Medications to alter cholinergic and glutamate functioning are ultimately intended to preserve or potentially enhance cognitive abilities, but this is being done for either clinically at-risk or clinical populations. An ethical debate arises when examining whether use of cognitive enhancers in nonclinical populations is appropriate. Although nonamphetamines, such as anticholinergic inhibitors, are currently being prescribed for the older population, the use of amphetamines for both prescribed and nonprescribed instances is becoming more common in many populations.

Individuals who are in the military, pilots, athletes, truck drivers, and academicians, among others, are thought to be using amphetamines to improve concentration and enhance performance. Perhaps the most popular capacity in which this issue is observed affects you, the reader—amphetamine use commonly occurs on college campuses in order to improve academic functioning. Although the rates of use in this capacity have ranged considerably, it has been indicated that up to 55% of fraternity students have used cognitive-enhancing drugs at least once (DeSantis, Noar, & Web, 2009).

There has been much debate regarding the use of cognitive enhancers for nonclinical purposes (Bostrom & Sandberg, 2009; Faraah et al., 2004). In fact, various forms of cognitive enhancers, such as transcranial magnetic stimulation, brain training games, certain diets, and even the use of caffeine and nicotine, have proven to be effective in improving cognition. However, the use of amphetamines has gathered increased interest and concern. In fact, the term *cosmetic neurology* has been used by Dr. Anjan Chatterjee, a neurologist, to describe the use of such drugs. Although there are clear medical concerns regarding off-label use, including increased risk for cardiac concerns, anxiety, and even addiction, from an ethical perspective the mental health practitioner who endorses this activity might not only be encouraging illegal activity, but also be promoting unfair discrimination. Socioeconomic status can dictate who has amphetamines available to them, and they might be competing against people who cannot afford such medications. Also, psychiatrists, nurse practitioners, and prescribing psychologists in a few states might be faced with the dilemma of determining whether an individual does in fact meet criteria for ADHD, indicating appropriate use, or whether attention problems exist at subclinical levels, in which case the medication might be of help but not be necessary. However, in this scenario, it appears that only those who can afford mental health benefits have this opportunity to further their cognitive functioning.

Another concern is competitive balance. Use of performance-enhancing medications by some individuals, but unavailability of these medications for others, seems unfair. It eliminates the notion that hard work pays off, and for some it appears to be a way

*(continued)*

> *(continued)*
>
> of cheating. On the other hand, use of the medication in itself is not going to lead to learning, but rather can only aid concentration in the service of learning. Although the distinction might seem a narrow one, it is an important distinction that continues to produce debate.
>
> Regardless of one's perspective, the use of prescribed medications for nonclinical populations occurs with high frequency. The clinician should be aware of such occurrence, evaluate for its possibility, and be familiar with the effects that it can have from a cognitive and physical perspective.

Although a single exposure to certain substances can lead to extremely harmful long-term effects, repeated exposures to high quantities of substances are found to have the strongest effects on the central nervous system. Treatment providers should evaluate for and be knowledgeable about additional resources, including but not limited to 12-step programs, that can assist their clients, considering the clear neurobiological effects that substances have on the individual.

## SUMMARY POINTS

- Dopamine is the main, although not the only, neurotransmitter involved in substance addictions.
- The mesolimbic system is highly involved in substance disorders.
- Genetics play an important role in substance disorders, particularly in adolescents.
- Treatment for substance disorders often combines behavioral strategies and medications; ultimately, abstinence results in improved cognitive functioning, as well as increased brain volume.

CHAPTER 12

# Medical Disorders

Medical conditions are often linked with emotional, behavioral, and cognitive problems. In fact, clear organic alterations of brain changes are classified within diagnostic systems as a separate entity. The disorders of dementia, such as Alzheimer's disease, are identified in the *DSM-5* because of the clear behavioral and cognitive manifestation of symptoms, but these differ from many other psychiatric disorders as they are often as likely to be diagnosed and treated by non–mental health providers, including neurologists and primary care physicians. Yet the mental health provider's role in working with individuals who have been diagnosed with these disorders is vital and often necessary for safety and a relatively positive outcome. There are other medical disorders that result from or co-occur with behavioral, emotional, and cognitive alterations, but they are not as well classified within the *DSM-5*. These include seizure disorders, strokes and vascular events, and tumors. Even relatively common medical procedures, including hip and knee surgeries, can result in psychiatric-like symptoms in older populations. With an ever-increasing life span and the influx of a geriatric population coinciding with the aging of the baby boomer population, the mental health provider should be aware of these disorders, as well as of how people might benefit from psychological therapeutic strategies for improving functioning.

The goals of the current chapter are to examine the more prevalent disorders associated with medical conditions. Specifically, several dementias and movement disorders often resulting in dementia will be examined, along with the effects of tumors, strokes, and seizures. The Special Topics box will examine how providers can assess for and assist in areas of determining capacity, particularly for independent living and driving. The Ethics box discusses the complexities of determining whether patients should use genetic and neurobiological knowledge to become aware of their own risk for inherited diseases, such as Huntington's disease, and the effects of these decisions.

> **SPECIAL TOPICS**
>
> ### Driving: When Is It No Longer Safe?
>
> Determining the level of safety for patients in a variety of contexts is an important responsibility for those working with older individuals, as well as for those experiencing medical events, such as strokes, tumors, and seizures. One of the most difficult decisions made by family, health care workers, and physicians alike is determining the most appropriate time to remove a driver's license from an individual. Calling the state and reporting that a driver is no longer capable of driving safely is an important responsibility, and for some professionals it is required by state law. Psychologists are ethically bound not to break confidentiality unless an individual is at risk for harming himself or herself, or others. Although driving in itself does not suggest that a cognitively impaired patient is going to bring harm to himself of herself, or to others, there is a substantially increased risk of such a possibility. In some instances for some states, the decision is simple. For example, Pennsylvania requires that if an individual sustains a seizure, his or her driving privileges are revoked for a 6-month period, during which the individual must remain seizure-free. Although this requirement can result in inconvenience and frustration, with adequate treatment and symptom-free functioning, the person will be able to resume driving in the future.
>
> Deciding to revoke a driver's license is more difficult when the decision is more likely to be absolute, particularly when it occurs in progressive dementias. The physician must make a judgment call and balance safety issues with the emotional well-being and independence of the patient. The challenge is that there are many bad drivers who are not demented who retain their licenses, and the diagnosis of dementia does not automatically preclude an individual from driving. Formal neuropsychology evaluation and information obtained from the patient's family can help the physician make the appropriate decision regarding driving. Though no one test is predictive of driving capabilities, tests that can aid this determination include the Trail Making Test, sustained visual attention tasks, and maze tests. Slower processing speed, poor planning, and limited attention can increase driving difficulty. Furthermore, virtual reality testing exists to help determine whether an individual is capable of driving, although there is concern regarding the novelty of such tasks for older individuals.
>
> A mental health clinician can assist in this important decision by not only providing formal assessment of cognitive abilities, but also working to help the patient deal with the emotional impact of such a change. The clinician is encouraged to address practical concerns with the patient while also helping the individual overcome a potential loss of independence. The emotional toll of losing one's license is often a burden on the patient and his or her family. Providing external resources that can help meet the needs of an individual who cannot drive can help the family deal with the increased burden. The clinician can also empathetically listen to the patient's struggle and help him or her recognize that it is not a loss of identity, but rather simply an inability to safely react to obstacles quickly enough when driving.

## DELIRIUM

*Delirium* is a sudden decline in alertness, attention, and cognition. It is often misdiagnosed, even in a hospital setting—an error that can have a substantial negative impact on short-term recovery and long-term functioning. The risk for delirium increases in older populations, has been shown to have hospital costs of $8 billion annually, and has been found to increase mortality up to 33% (Inouye, 2006). Delirium has been associated with decreased independence, increased nursing home placement, and expensive post-hospital costs. Yet, many mental health providers are not familiar with the symptoms of, causes of, or treatment for delirium. Indeed, health providers in general have a tendency to overlook delirium as a cause for acute behavioral and cognitive changes, often assuming progressive dementia, depression, or another reason for neurobehavioral changes (Voyer et al., 2010). In fact, it has been suggested that delirium be placed on equal footing with other medical disorders, considering its frequency, attendant distress to families, and ultimate outcomes (Teodorczuk, Reynish, & Milisen, 2010). It is important to be familiar with the symptoms of a disorder, and an awareness of the neurophysiological and cognitive findings associated with delirium can help the clinician determine treatment and capacity within this population.

There are several symptoms of delirium, but it is generally agreed that there are two principal factors of delirium: cognitive and behavioral (Jain, Chakrabarti, & Kulhara, 2011). From a cognitive perspective, delirium can include a reduced awareness of one's surroundings, resulting in perseveration of conversation, wandering attention, visual spatial difficulty, memory problems, and failure to respond to external stimulus. There is often a lack of response or an incoherent response to the environment. A hallmark cognitive feature includes disorientation, which helps differentiate delirium from the early stages of dementia and depression. Furthermore, speech-related difficulty is often seen, in which the individual will demonstrate random, nonsensical, or incoherent speech patterns. It often appears that the individual has comprehension difficulty and difficulty following directions. Another hallmark feature of delirium is the waxing and waning of cognitive functioning. For example, depending on when the person is evaluated, he or she might demonstrate significant differences in terms of clinical presentation, even by the hour. This clinical variability is also atypical in other cognitive disorders, which typically demonstrate either consistent cognitive difficulty or progressive decline. From a behavioral perspective, delirium can result in hallucinations, delusions, irritability, and agitation. Sleep is often disturbed, with significant sleep–wake cycle disruption.

Several possible causes for delirium exist, including substance intoxication, alcohol withdrawal, infection—particularly urinary tract infections, side effects of medication, critical illness, malnourishment, dehydration, and immediate effects of closed head injury, among other medical reasons. In children, delirium can occur when coming out of anesthesia, after severe burns, or during the course of high fevers (Saxena & Lawly, 2009). It is not uncommon for delirium to co-occur during the course of dementia. There are usually multiple factors accounting for the presence of delirium. In fact, it has been estimated that 90% of people with delirium have three or four identifiable factors, 27% have 2 factors, and only 16% have only 1 identifiable factor causing the delirium (Camus, Gonthier, Dubos, Schwed, & Simeone, 2000). Thus, the neurobiology of delirium is complex, and it is often associated with the etiologies behind the delirium. For example, old age in itself is a risk factor for delirium. It is thought that age-related loss of cholinergic reserve occurs (MacLullich, Ferguson, Miller, de Rooij, & Cunningham, 2008). As discussed previously, acetylcholine plays an important role in attention, memory, learning, and wakefulness, and it plays a role in dementia. As we will discuss below, acetylcholine dysfunction is also found in dementia.

The neurobiology of delirium remains poorly understood. It is clear that the imbalance of disruption of neurotransmitters leads to the experience of delirium. It has been consistently demonstrated that acetylcholine is diminished during periods of delirium. In fact, anticholinergic drug use has been shown to induce delirium. Taken together, reduction of acetylcholine often occurs during delirium. Dopamine, norepinephrine, and serotonin have also been demonstrated to alter alertness and affect the sleep–wake cycle. Unlike the deficiency of acetylcholine, it has been suggested that there is an increase of dopamine (Moyer, 2011). Not only have dopamine-blocking medications proven to be effective to treat delirium, but stress increases dopamine production, increasing the risk for delirium in medically unstable patients. This excessive dopamine might account for the psychotic symptoms often seen during the course of delirium. The dysregulation of serotonin, in which there is either too much or too little reuptake, can also increase the risk for delirium (Hshieh, Fong, Marcantonio, & Inouye, 2008). Electroencephalograph (EEG) studies have demonstrated a link between delirium and slower activity throughout the brain, but particularly in the occipital lobe. From a functional perspective, this is likely linked to general slowness in cognition, sleep disturbance, and perceptual disturbance. Furthermore, MRI studies have demonstrated thalamic frontal and limbic system pathway involvement. This is associated with the behavioral manifestation of the disorder, including irritability, agitation, and lethargy. Individuals who have experienced basal ganglia lesions have increased risk for delirium. Bilateral prefrontal cortex involvement has been shown—which links with executive dysfunction—including limited problem-solving skills during delirious states. The parietal lobe has also been implicated at times during delirious state.

Delirium needs immediate medical attention and can be resolved when identified. It needs to be differentiated from other medical concerns, such as depression, drug use, and dementia. Treatment often includes medication, such as neuroleptic agents including antipsychotic medication. Of note, if treatment with a benzodiazepine or a stimulant occurs, it will result in increased confusion and disorientation. Correct diagnosis is necessary for resolution.

## THE DEMENTIAS

Dementia is defined as cognitive impairment in areas of memory and at least one other area of cognitive ability, including language, visuospatial, and executive functioning. Individuals who suffer from dementia have apraxia, aphasia, ataxia, executive dysfunction, or a combination of those symptoms, as well as memory impairment noted as a decline in functioning from previous levels of abilities that negatively affects their daily life. There are several types of dementia, some of which are better understood than others. It is not uncommon for an individual to have more than one type of dementing process as well. For example, the postmortem neuropathology of Alzheimer's disease can find occlusions consistent with a vascular dementia. Understanding each of the biological mechanisms, cognitive capabilities, and course of illness can assist the mental health clinician with behavioral treatment planning, recommendations for other providers, and working with families.

### Alzheimer's Disease

Alzheimer's disease (AD) is the most common form of dementia and is estimated to have directly affected approximately 5.2 million Americans in 2013, with an estimated 5 million of those people older than 65. Across the world, it is estimated that more than 26 million people have AD. Within the next decade, it is expected that over 7 million people will be afflicted with AD. It has been recently estimated that by 2050, nearly 14 million people will have AD, if a cure is not found. Age is the greatest risk factor for the disease, and with medical breakthroughs extending people's lives, this rise in the numbers of those with this diagnosis is not surprising. Women are more likely to have the diagnosis than men, largely because their life expectancy is longer. The destruction of the brain from AD ultimately leads to death, and AD is the sixth leading cause of death in the older population (those older than 65). It is estimated that $203 billion dollars will be spent caring for those who have been diagnosed with AD during 2013. The cost of caring for these individuals extends beyond the financial, as caregivers and families often need to deal with the emotional toll of experiencing a decline and the ultimate death of loved ones. This experience, in turn, decreases the physical and mental health of caregivers.

There are many lifestyle risk factors associated with AD beyond age and gender. For example, high-fat diets and sedentary lifestyle can increase the risk of AD. It has been found that high cholesterol, high blood pressure, and a low amount of education can also increase the risk for AD. In comparison, physically and cognitively stimulating activities such as exercising, reading, and dancing are all protective factors from the cognitive decline associated with AD.

As with all dementias, the hallmark feature of AD is cognitive decline, specifically in areas of memory. Individuals who have been diagnosed with AD will often forget people's names, conversations, and events. They might not remember what they ate for an earlier meal or even whether they ate an earlier meal. There is often language difficulty, including word-finding difficulty and anomia. The later stages of AD are often marked by severe naming difficulties. Individuals who have been diagnosed with AD typically have intact attention skills until later in the disease, which often contributes to the family members' inability to identify the onset of the disease. They are able to maintain a conversation in the moment, but they cannot remember any portion of the conversation at a later time. There are visual, spatial, and constructional limitations that often result in functional impairment, including difficulty driving, balancing a checkbook, writing out bills, and, potentially, recognizing the need to care for themselves. Furthermore, executive functioning abilities such as problem-solving, planning, and cognitive flexibility can become impaired. Individuals with AD will often perseverate in conversation, tasks, and other behaviors.

It is also common for AD to be associated with mood-related difficulty. It has been estimated that approximately 50% of individuals with AD have co-occurring depression. Mood problems typically occur during the early stages of the disorder, as the neurophysiologic changes are occurring, and the individual might have insight into his or her increased functional difficulty. With time, insight into one's own problems decrease, and depression is likely to co-occur, although family members can also have an increased risk for depression.

Individuals who have been diagnosed with AD will have other behavioral disturbances and functional problems. For example, sleep is often disrupted in AD. The sleep–wake cycle can be disturbed to the point of sundowning, or waking in the middle of the night with confusion. People often have difficulty with their sense of smell as well. This can contribute to their forgetfulness to perform tasks, such as turning off the stove or oven, and it might explain dietary changes.

A common concern among children of people who have been diagnosed with AD, particularly as they age, is whether they will have the same fate as their parents. As is the case in almost all disorders, a genetic predisposition to the disease does not guarantee expression of the disease. The genetic contribution of AD is complex. Certain genetic mutations are rare, but some can lead to early onset. For example, mutations in genes that are known to play a role in encoding amyloid precursor proteins are rare, but they lead to early onset of the disease. The amyloid precursor gene on chromosome 21 is known to play an important role in the early expression of the disorder. Genetic mutation on the apolipoprotein-E (APOE ) 4 allele on chromosome 19 and the APOE 2 allele also increase the risk of the disease. Studies have suggested that different genetic variants are present in individuals who have late-onset AD as well. The NIMH Alzheimer's Disease Genetics consortium has allowed for thorough evaluation of other genetic factors that increase the risk for the disease. Continued research will help further elucidate the genetic factors contributing to environmental influences as well as to the timing of the onset.

AD is a neurodegenerative disorder that occurs as a result of an accumulation of amyloid-related peptides and tau proteins. As a result, neurotic plaques and intracellular neurofibrillary tangles are the hallmark features found within the brains of people with AD. Neurotic plaques are multicellular lesions containing amyloid peptide deposits. Neurotic plaques are a cluster of dead and dying nerve cells. Neurofibrillary tangles are twisted fragments of proteins that combine with neurotic plaques to clog the nerve cells. As a result, there is a loss of neurons that occurs throughout the brain. This destruction of nerve cells decreases synaptic communication and neurotransmission, ultimately leading to disconnection of neurons and cell death.

The plaques and tangles that occur in AD will often first appear in the temporal–occipital regions (Duyckaerts & Dickson, 2011). There is evidence that the hippocampus, amygdala, neocortex, and basal forebrain are severely affected. The neurofibrillary tangles have a hierarchical pattern of accumulation, initially affecting large projection neurons important for memory-related neuronal systems. Although amyloid deposition occurs at a constant and slow rate, clinical symptoms of AD are associated with neurodegeneration rather than with this amyloid deposition (Jack et al., 2009). Through animal studies, it has been recently demonstrated that tau protein is transferred downstream to neighboring cells, as well as to synaptically connected neurons (Calignon et al., 2012). Once tau protein has accumulated, synaptic destruction occurs in the brain, resulting in loss of cholinergic neurons. The breakdown of acetylcholine within the brain subsequently leads to progressive and severe memory loss. With time comes continued dysregulation of other neurochemicals, including glutamate, serotonin, and dopamine. This is not unexpected, given the cell destruction throughout the brain.

It is recommended by the American Academy of Neurology that a structural imaging scan, ideally a magnetic resonance imaging (MRI) scan, occur as part of a work-up for AD. Furthermore, the *DSM-5* (American Psychiatric Association, 2013) recommends that imaging techniques be used to aid diagnosis of the disease. Structural imaging findings might find atrophy in the temporal and parietal regions. Mesial temporal degeneration, hippocampal atrophy, and posterior cingulate, as well as the parietal regions, are clearly affected early in the disease (Choo et al., 2010; Pengas, Hodges, Watson, & Nestor, 2010). Gray matter atrophy can be observed with advanced imaging techniques as well. Recent functional imaging studies have tried to better link the cognitive and behavioral impairment observed in AD with neuropathology. These regions have been found to play a role in the topographical memory loss in AD as well (Pengas et al., 2012). Positron emissions tomography (PET) studies have demonstrated hypometabolism throughout the brain, with the exception in the basal ganglia and primary sensorimotor cortex (Silverman, 2004). In fact, PET studies demonstrate slower metabolism in high-risk individuals, even before the onset of AD for those who do subsequently meet the criteria (Kadir et al., 2012). A detailed review by Braskie, Toga, and Thompson (2012) indicates that a major advance within AD research has occurred recently with the ability to assess levels of amyloid plaques and tau neurofibrillary tangles in the living human brain through PET studies. Cognitive performance has been correlated with plaque and tangle findings in these PET studies. Greater ventricular volume and decreased hippocampal volume have been demonstrated to successfully predict cognitive decline (Fleischer et al., 2008). New automated imaging analysis is assisting in the efficiency and effectiveness of using scans to help diagnose this disorder.

The term *dementia* indicates that a person has problems in areas of memory as well as in one other area of functioning, including an aphasia, apraxia, agnosia, or executive impairment. People who have been diagnosed with AD will demonstrate decline from previous functioning. The cognitive impairment and decline from previous abilities result in functional impairments.

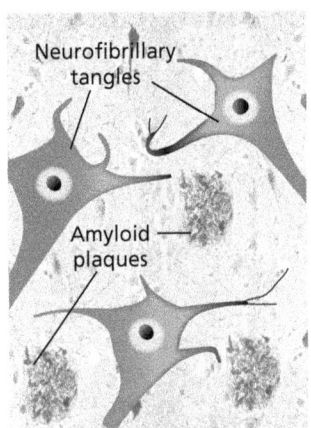

FIGURE 12.1 Neurofibrillary tangle and amyloid plaques.

Topographical memory impairment has been found in AD, and evidence suggests that the treatment for AD has been limited in both effectiveness and outcome. Medications that are currently prescribed have produced limited clinical benefit, and they do not treat the underlying cause of the disease. Cholinesterase inhibitors (ChEIs), donepezil, galantamine, and rivastigmine, are often administered to delay the breakdown of acetylcholine released into synaptic clefts and so enhance cholinergic neurotransmission. Memantine is approved for treating moderate-to-severe forms of AD. It aims to block the debilitating effects of low levels of glutamate that occur with the progression of the disease. Though treatment can assist in slowing the progressive cognitive decline, it does not reverse the process.

Behavioral strategies are often implemented as well for AD. Consideration of behavioral strategies to ensure safety in the home environment, including bedroom placement on the first floor, locked doors, GPS tracking devices, removal of driving privileges, proper nutrition, eating meals, and general health issues, need to be addressed. Other issues associated with end-of-life decisions are important to address, including health care proxy and power of attorney determination. Every effort should be made to make these decisions before the disease progresses to the point at which the person who has been diagnosed does not have the capacity to make informed decisions. Consideration of treatment for depression and anxiety needs to occur early during the disease, considering the high comorbidity. Furthermore, care and treatment for caregivers is necessary, because family members often feel anxiety and depression with the decline of functioning in their loved one, but also guilt for being concerned about their own future functioning. AD treatment is multifaceted, and it needs to be carefully addressed on each level.

Although desired, there has been no support for the use of cognitive or physical activity to stop the progression of the disease. Research has demonstrated that consistent physical activity, including dance, walking, and jogging, can decrease the likelihood of or delay the onset of AD. Similarly, cognitive activities such as reading; working crossword puzzles; and playing active computer games, card games, and sudoku can also slow cognitive decline in the elderly. Eating a healthy, protein-laden, low-fat diet has been suggested as well, to maintain cognitive health. Literature suggests that individuals with higher levels of education might have a lower risk of being diagnosed with AD, suggesting that the theory of cognitive reserve might help prevent AD. Cognitive reserve is the theory that the more cognitive abilities and learned material one obtains, the less affected that person is when neurobiological processes occur that are associated with decline. It is certain that maintaining general health, including heart health, decreases the risk of AD. The most effective treatment for AD comprises behavioral strategies aimed at prevention and healthy lifestyle choices.

## Vascular Dementia

Vascular dementia (VD), formerly known as multi-infarct dementia, comprises the long-term, progressive cognitive and behavioral effects following cerebral vascular and cardiovascular events–related occurrences, such as stroke, arteriosclerotic disease, cardiac embolism, vasospasm, and hypoxic–ischemic encephalopathy. Otherwise, seemingly healthy older individuals who undergo knee surgery or hip replacement surgery have an increased risk for VD. Similarly, hypertension, hypercholesterolemia, diabetes, and smoking increase the risk for VD. Though AD is more likely to occur in females because of their longer life expectancy, VD is more likely to occur in men, likely owing to their diet and generally lessened concern for their own general health care. Onset of VD is typically between ages 60 and 75.

VD is characterized by memory loss as well as one other area of low cognitive functioning. The neurocognitive functioning of VD varies according to the location of the lesion, but it is typically associated with retrieval deficits rather than with the encoding deficits found in AD. Furthermore, there is increased risk for psychomotor slowing and executive dysfunction. VD will often lead to language deficits that include oversimplified vocabulary and shorter sentences or slurred speech. Visual planning deficits and visual spatial impairments often occur early in the disease as well. Physical symptoms include upper body weakness—typically focal and opposite the side of the vascular event, rigidity bradykinesia, and loss of bladder control. VD is often associated with walking difficulties, including rapid or shuffling steps or gait abnormalities. Functional difficulties occur more abruptly, rather than the slow insidious onset of AD.

The evaluation of the role that genetics has on the onset of VD is starting to gain attention. Schrijvers et al. (2012) conducted a genome-wide study of VD. Their study indicated a novel genetic locus of VD on the X-chromosome. This might also account for the increased risk of VD in males, considering the presence of only one X-chromosome. The APOE gene has also been found to play a role in the presence of VD (Schmidt, Freudenberger, Seiler, & Schmidt, 2012).

Focal lesions that result in VD can often be found by using imaging techniques. Though infarcts can be present in any region of the brain, they are commonly discovered in the thalamus, medial temporal lobes, posterior cerebral artery, and basal ganglia.

Similar to that for AD, the treatment for VD is prevention through healthy lifestyle choices. The interventions that typically occur after the onset of VD are contingent upon the clinical presentation. In light of the language impairment, speech therapy is likely to be beneficial. Furthermore, occupational therapy can be used to retrain specific body parts to complete tasks. Medications in the form of acetylcholinesterase inhibitors, which slow the progression of cognitive decline, can be prescribed with some success (Black et al., 2003).

## Frontal Lobe Dementia

Similar to the other dementias, frontal lobe dementia (FLD) is characterized by memory impairment. However, FLD is also characterized by significant behavioral disturbances consistent with executive dysfunction. Behavioral dysregulation of disinhibition is often a hallmark feature. For example, people with FLD demonstrate socially inappropriate behaviors and make comments that are out of the ordinary from their previous functioning. Increased apathy, personality changes, vulgar language, and poor judgment often occur as a result of FLD. In fact, these substantial personality changes are the hallmark feature of FLD.

It has been estimated that 40% of individuals who have been diagnosed with FLD have a close relative with a similar disorder. Genetic analysis has suggested that the tau gene on chromosome 17 plays an important role in the presence of the disease. However, several other genetic locations have been identified, and these continue to be explored. A repeat and expansion on a gene on chromosome 9 has been consistently found in people with FLD (Mahoney et al., 2012).

The most common form of FLD is Pick's disease, which typically occurs between ages 50 and 60. It is identified by microscopic imprints of intraneuronal tau inclusions called Pick's bodies. Cell loss is typical in the frontal region of the brain, as well as in the temporal and frontal temporal areas. Typically, the progression of the disease is relatively slow.

Cognitive impairment accompanies FLD, particularly in areas of executive functioning, language, and attention. Individuals who have been diagnosed with FLD will often have progressive aphasia, demonstrate extreme disinhibition, and have severe limitations in problem-solving, planning, and organization. Individuals with FLD will have limited attention, demonstrating difficulty maintaining attention long enough to sustain a meaningful conversation. Ventromedial and prefrontal atrophy have been linked to the decreased insight in patients with the disease (Hornberger et al., 2012).

The treatment for FLD is to focus on providing safety for the individual and support for family and caretakers. Cognitive behavioral therapy is unlikely to be effective, given the limited insight and behavioral regulation. Medications to slow cognitive decline are not proven to be effective. SSRIs, such as citalopram, have proven to decrease symptoms of depression commonly found in FLD (Herrmann et al., 2012). Other medications, such as Clomipramine, can be effective in treating the obsessive–compulsive symptoms commonly found in FLD.

## Dementia of Lewy Body

Dementia of Lewy body (DLB), until recently, has often been misdiagnosed as Parkinson's disease or AD because of its symptom presentation. DLB is characterized by increased likelihood of psychotic symptoms, including hallucinations, delusions, and paranoia. There are periods of fluctuating alertness with delirium episodes. Late onset of these psychotic symptoms is suggestive of a DLB. Furthermore, DLB increases the risk for tremors, which explains the misdiagnosis of Parkinson's disease for many years. DLB is now thought to be the second most common form of dementia (Corey-Bloom & Merdes, 2004). It is characterized by pathological findings of intraneuronal inclusions, which are round bodies of proteins, highly concentrated in the brain stem, cortex, and striatal pathways. They appear similar to neurotic plaques found in AD, helping to explain the misdiagnoses of AD in this population. Up to 85% of individuals with DLB have visual hallucinations, and 45% have auditory hallucinations—remarkably more than in other dementias. Similarly, up to 75% have delusions. Individuals who have been diagnosed with DLB will frequently fall during the early course of the disease, often associated with syncopal events. As noted by Braem, Verguts, Roggeman, and Notebaert (2012), the genetic component of the disease remains unknown. Mutations on the protein building gene glucosadise beta acid known as GBA appear to play a role in the expression of the disease. Dopaminergic dysregulation is common in DLB.

Similar to AD, DLB has been found in imaging studies to have atrophy in the hippocampal and temporal lobes. Also, increased ventricular findings and decreased temporal lobe volume have been found in this population. Functional imaging has demonstrated the role of the frontal lobe and occipital lobe involvement in DLB. fMRI has been able to distinguish AD from DLB by the occipital lobe involvement. There has also been the suggestion that thinning in the dorsal cingulate, orbital frontal, and posterior temporal portions of the brain might distinguish AD from DLB (Lebedeve et al., 2012).

Although it is not diagnostic, a common clinical symptom of DLB is the high level of sensitivity to neuroleptic medications. Adverse events have commonly been reported in DLB patients who are administered these medications to control their symptoms. There are increased mortality rates, increased confusion, and severe parkinsonism with the medication. Neuroleptic use is therefore contraindicated for this diagnosis. DLB has severe cholinergic alterations, so, acetycholinesterase inhibitors are prescribed; they have been found to improve sustained attention, episodic memory, and executive functioning (Wesnes et al., 2002). Often, treatment will also focus on improving sustained healthy sleep. Behavioral interventions are similar to those for the other dementias, and they include support for the family and assistance with determining safety, mood, and psychosis.

## SEIZURE DISORDERS

A seizure can have a variety of manifestations. Though many people associate seizures with involuntary and uncontrolled movements, a seizure can result in alterations in consciousness, perceptual experience, and behavior. There are certain types of seizures that result in staring spells and some that occur only at night. Seizures are very prevalent, with the most common cause of seizure being epilepsy. Epilepsy is not a disease in its own right, but rather a complex number of different disorders that alter brain function. Epilepsy is a broad term used to describe abnormal electrical discharges from the brain, associated with alterations in behaviors. Epilepsy is diagnosed only when a person has recurrent seizures that are not psychogenic (seizures caused by emotional dysregulation) in nature. Temporal lobe epilepsy is the most common form, and it indicates the region of the brain where the electrical discharges are originating. It has been estimated that 50 million people worldwide have epilepsy. There are approximately 2.5 million people in the United States alone who have been diagnosed with epilepsy. The onset of most seizure disorders is at a young age. Thirty percent of seizures occur before age 4. Seventy-five percent of seizure disorders begin before age 20. If there is an onset of seizure after age 35, without the causation of a closed head injury, infectious disease, or nutritional imbalance, there is a high likelihood of a brain lesion. Immediate medical evaluation is necessary. Seizures are also almost three times as common in males as in females.

Individuals who experience seizures have increased risk for emotional problems relating to functional limitations. For example, people who have uncontrolled seizures often have restrictions or are not allowed to drive until they are seizure-free for 6 months. This arrangement can result in limited ability to work or participate in social activity, and it can potentially affect relationships. Seizure patients might feel dependent upon family and friends. A seizure history can impede one's ability to have a normal life. It can cause psychological distress when the individual knows that there was a seizure in front of peers, something that can be particularly emotionally and socially challenging for children and adolescents. Observing a seizure can result in interpersonal changes among peers. It is necessary for the clinician to have a basic understanding of the various types of seizures, as these can be mistaken for or can coexist with psychiatric conditions.

## Seizure Subtypes

There are several different types of seizures. The determination of the type of seizure is based on EEG results. Specifically, the electrical spikes and wave complexes that are measured on the surface of the brain determine the etiology of seizures. That said, the diagnosis of a seizure is based upon behavioral observation, as seizures typically occur in the absence of EEG recording, and approximately 20% of patients with epilepsy will have normal EEGs (Engel, 1992). It is also possible to have seizure-like brain activity yet not experience seizures. Thus, it is not necessary for the mental health clinician to know how to read an EEG, or even what electrical waves are altered; however, it is helpful for the clinician to be aware of the different types of seizures and the subsequent disturbances involved with each type.

Each of the subtypes of seizures results in a different behavioral manifestation. In 2001, the International League Against Epilepsy Commission Report provided a diagnostic schema similar to the *DSM-IV* in terms of five axes. Within that report, descriptors were recommended to focus on the neuroanatomy and pathophysiology to describe seizure type. *Ictal* is a term used to describe an event relating to a seizure episode; its phenomenology is Axis I within the classification system. Mental health providers are typically more interested in Axis II, which is the seizure type. Axis III describes the epileptic syndrome; Axis IV specifies the genetic, medical, or other known etiology; and Axis V provides the level of impairment. Although space is limited in this textbook, several medical books, chapters, and scientific articles are available that address this issue in depth. The mental health practitioner should be familiar with the most common types of seizures, behavioral phenomena, and outcomes. Such familiarity can help guide the clinician in conceptualization and treatment.

Generalized seizures are bilaterally symmetrical, occurring in both the right and left hemispheres of the brain. These types of seizures occur deep within the brain, and they are distributed throughout the brain; they result in loss of consciousness. The term *generalized seizure* reflects their occurrence throughout the brain and not in only one specific (or focal) region. Generalized seizures can be broken into several subtypes, including tonic–clonic seizures, tonic seizures, clonic seizures, myoclinic seizures, atonic seizures, absence seizures, and spasms. Generalized tonic–clonic seizures, formerly referred to as *grand mal seizures*, involve a loss of consciousness, with tonic rigidity followed by clonic jerking. This event can last from 1 to 2 minutes, and it results in complete loss of consciousness, often along with urinary and fecal incontinence and tongue biting. The tonic component of the seizure includes a titanic muscular phase, meaning that there are sustained muscular contractions of several seconds, followed by repetitive jerking of up to 2 minutes, which is the clonic phase. After the seizure, there is post-ictal (after-seizure activity of) confusion, muscle ache, headache, exhaustion, and amnesia of the seizure event. These seizures typically originate in either the thalamus or the cortex, and they are more likely to impair cognitive functioning. Although common belief holds that a finger or other object should be inserted into the patient's mouth during this type of seizure, it is important that nothing be inserted into the mouth. Rather, any loose objects near the person should be removed, the head should be protected, and the person should be provided emotional support when the seizure concludes. Though tonic–clonic seizures can occur as partial seizures (a seizure type with a focal onset), they are very rare.

Another common type of generalized seizure is absence seizure. In absence seizures, formerly referred to as *petit mal*, the behavioral symptoms include a vacant stare with fluttering eyelids. This type of seizure starts as generalized and with loss of consciousness of only a few seconds. Typically, EEG results demonstrate a 3-second wave of bilateral and synchronous spike discharges during an ictal event. There is no motor phase to this type of seizure. Therefore, absence seizures are easily not diagnosed at all, or are often misdiagnosed as attention problems. Making this even more likely, this type of seizure is almost exclusive to children, and there is no decrease of intelligence in this disorder. Typically, absence seizures are resolved after puberty.

Focal seizures (also referred to as *partial seizures*) occur focally or in one specific region of the brain, typically the cerebral cortex. The symptoms typically include focal motor or sensory symptoms. These symptoms can range from a twitching finger to a Jacksonian march. If these types of seizures result in no alterations in consciousness, they are referred to as *simple partial seizures*. When seizures are accompanied by alteration in consciousness such as various levels of confusion or dizziness, they are referred to as *complex partial seizures*. Although often considered temporal lobe seizures, this assumption can be erroneous, as they can also originate in the medial frontal lobes. Thus, there are temporal complex partial and nontemporal lobe complex partial seizures. The localization is based upon sharp waves occurring in either the temporal lobe or some other region of the brain. A complex partial seizure is the most common form of seizure and can result in a wide range of symptoms, including aura (unpleasant taste, odors, and sensations that occur prior to the event), rapid eye blinking, hallucination, déjà vu (new experiences seeming familiar), jamais vu (familiar experiences seeming unusual or novel), and purposeless automatisms, including scratching, blinking, rubbing, kissing, and chewing. Despite these symptoms, the person can engage with the environment, but not in a purposeful or sequential capacity. This seizure typically occurs for 2 minutes and is followed by confusion, amnesia to the event, and severe drowsiness. Seventy percent of people with focal seizures have hippocampal volumetric decline and reduced gray matter in areas connected to the hippocampus. Focal seizures also increase the risk for dementia.

Beyond generalized and focal seizures, there is another classification of seizure disorders labeled *continuous seizure type*, referred to as *status epilepticus*. These occur when two or more seizures are superimposed on each other without recovery of consciousness. This type of seizure can consist of convulsions, tonic–clonic events, or nonconvulsions. This type of seizure is fairly rare, and it is considered a medical emergency.

Mental health practitioners often work with people who are diagnosed with psychogenic seizures. Formerly referred to as *pseudoseizures*, or *nonepileptic seizures*, psychogenic seizures have no clear abnormal EEG results and are thought to be associated with internalized stress of anxiety. Up to 20% of people who are referred to seizure clinics and experience a video-monitored and EEG-recorded seizure event are found to have normal EEG readings. Psychological evaluations of these individuals often reveal a conversion type of presentation of coping with anxiety. Although there is evidence for abnormal fMRI findings in individuals with psychogenic seizures, it is thought that treatment for the anxiety will resolve the seizure-like events. Mental health practitioners often need to assist the patient in understanding the interplay between physical symptoms and stress, while supporting the notion that the patient is not faking seizures. In fact, even those individuals who have EEG-identified seizures have symptoms of anxiety, and psychogenic seizures can often accompany nonpsychogenic seizures.

## Genetics and Seizure Disorder

Although genetic factors play a role in seizure disorders, there are very few genetic disorders that directly result in seizures. The incidence of seizure disorder increases from 1% of the population to upward of 8% in individuals who have a first-degree family member with seizures. Generalized seizures are more likely to have an inherited trait, with absence seizures being the most likely to run in families. If a mother has a seizure disorder, there is twice the likelihood of it occurring in children than if the father has the disorder. As with most disorders, the interplay of genetic factors and environmental influences results in seizures.

## Neurochemistry and Seizure Disorder

During epileptic activity, a hormone is secreted by the pituitary gland, releasing prolactin up to 4 times as much as the baseline level. An elevated prolactin level is the only neurochemical that has been used in the medical evaluation of seizure disorders. The alteration in prolactin occurs within 15 minutes and returns to normal within an hour. Interestingly, elevated prolactin has been found in psychogenic seizures as well, and it cannot be used to differentiate among seizure types.

## Neuroimaging and Seizure Disorder

As mentioned previously, EEGS are the most widely used and important evaluations for seizure disorders. EEGs can pinpoint the type of seizure when the event occurs during EEG monitoring. However, a negative EEG finding does not eliminate seizure disorder as a diagnosis, as it is not uncommon for a person not to experience a seizure during EEG monitoring or for a person to have co-occurring psychogenic seizures. Other neuroimaging techniques with seizure disorders are somewhat limiting. Structural imaging studies using computed tomography (CT) and MRIs have demonstrated focal onset of seizures and can help determine severity of brain alterations as well as possible locations for intervention. Functional imaging techniques demonstrate hypometabolic processes between seizure events and hypermetabolic processes during the course of the seizure.

## Cognition and Seizure Disorder

Cognitive functioning in individuals with seizures is heterogeneous. Whether a person has cognitive difficulties is associated with the severity of the seizure, age of onset, course of the disorder, and etiology of the seizure. For example, generalized tonic–clonic seizures are more likely to result in cognitive impairment. Localized temporal lobe seizures are often accompanied by verbal memory impairment and language deficits. Studies have demonstrated that performance in areas of memory, executive functioning, and visuomotor abilities might be worse after even the first seizure episode (Äikiä, Salmenperä, Partanen, & Kälviäinen, 2001). Cognitive impairment seems fairly stable throughout the course of the illness as well, with improved cognitive performance with seizure remission.

## Treatment for Seizure Disorder

The treatment for seizure disorder is typically medical intervention, including medication, surgery, or potentially deep brain surgery. Antiepileptic medications aim to modify the balance of neuronal excitation and inhibition. Contingent upon the type of seizure activity involved, various medications will be used. Tonic–clonic seizures have traditionally been treated with phenobarbital, valproic acid, and carbamazepine. Absence seizures have been treated with valproic acid and ethosuximide. Several new medications, including gabapentin, toprigate, lamotrigine, and vigabatrin, have been used to treat seizures as well. Though there are various amounts and types of side effects, each of the medications has side effects that can result in noncompliance or the need to treat other symptoms, such as depression.

Another option for the treatment of seizure disorder is neurosurgery. Neurosurgery occurs in patients with medically intractable epilepsy. The goal of the procedure is to locate the site of the focal seizure and remove it without causing any functional damage to the cortex.

## MOVEMENT DISORDERS

### Huntington's Disease

Huntington's disease (HD) is a debilitating genetic disease characterized by choreiform movements, which are rapid, jerky, dancelike movements. The symptoms of HD typically occur between ages 30 and 50, with a mean age of onset at age 40, but individuals with HD can present as young as age 2 and as old as age 90. The progression of the disease is typically more rapid when the onset of symptoms is earlier. Individuals who inherit the disease from their fathers are more likely to have an earlier age of onset (Ridley et al., 1988). Eventually, HD leads to dementia, psychiatric problems, and the notable movement disorder. HD usually causes death within 15 years of the onset of symptoms, if not sooner, due to the emotional turmoil and co-occurring depression. There is often a change in personality, paranoia, and psychotic symptoms that can occur. Individuals who have been diagnosed with HD have an increased risk for suicide, with an estimated rate of approximately 12%. Often, the emotional symptoms and behavioral difficulties precede the chorea and get diagnosed as bipolar disorder, given the severe mood dysregulation. In fact, up to 25% of HD patients are diagnosed with bipolar disorder. Given the overlap of neurobiological pathways and brain regions that are involved with HD and bipolar disorder, this finding is not surprising. Choreiform movements can be misdiagnosed as nervousness or as tics early in the disease process. Other symptoms include difficulty with coordination of limbs, problems with speech (although not with language), and slower initiation.

HD is an inherited genetic disease from an abnormal expansion of the HTT gene, a seratonin transporter protein gene, on chromosome 4. An expansion of the HTT gene on chromosome 4 leads to neuronal loss and dysfunction throughout the brain (Ross & Tabrizi, 2011). An offspring of a parent who has this disease has a 50% chance of inheriting the disease. Genetic testing in at-risk individuals can identify those individuals who will have the disease. This brings up ethical questions for patients and families (see the Ethics box).

The basal ganglia, important for the integration of movement, is grossly affected in HD. Specifically, the striatum, including the caudate nucleus and globus pallidus, undergo dramatic deterioration and cell death. MRI scans demonstrate the deterioration of these areas of the brain, as well as of the limbic system and

> **ETHICS**
>
> **Genetic Testing: To Test or Not to Test?**
>
> We have entered an era where our neuroscientific capabilities have outpaced the consideration of possible emotional effects on the outcome of neuroscience. This state of affairs leads to clear ethical concerns as we try to determine what is in our patients' best interest. One clear area that is a concern is genetic testing. A genetic test provides valuable clinical data about specific genetic conditions. It has been argued that regardless of the outcome of genetic tests, families who have children who are tested for a genetic mutation face potentially serious psychological harm (Wertz, Fanos, & Reilly, 1994). For example, a child might experience survivor's guilt, decreased self-esteem, and harmful child–parent bonds.
>
> As clinicians, we are ethically bound to practice only in the areas in which we are trained. Therefore, it is unlikely that a mental health professional would make the recommendation to genetically test a high-risk child. However, it is possible that families will seek the opinion of a mental health professional before pursuing the endeavor, or that they may address related concerns after testing. It has been estimated that commercially available services for genetic tests for a variety of medical conditions will expand (Burke & Psaty, 2007). The balance of clinical benefits versus emotional and clinical risks is often an individual decision that cannot be determined by a professional. Rather, the individual or individual's parents are the final decision makers in risk–reward analysis. This can be particularly challenging in a disease such as HD. Though genetic testing can be important in identifying early disease for more informed and timely medical treatment, HD can be determined well before the onset of any symptoms. In comparison, other diseases can be treated early, and the early treatment can prevent irreversible cognitive and emotional damage. In the example of HD, a test might limit or alter the course of the future for the child who tests positive. In other genetic disorders, a negative result in the child might still require the child, only after becoming older, to determine whether he or she wants to produce offspring, knowing that he or she is at high risk to pass on a disease.

motor cortex. Considering the clinical symptoms of mood disorder and abnormal movement presentation, it is not surprising that the deep brain matter structures responsible for mood and movement are the ones that deteriorate. Furthermore, white matter deterioration occurs even in individuals who have not yet had onset of symptoms (Dumas et al., 2012). Decreased prefrontal, dorsal cingulated, and temporal cortical metabolism have been found on fMRI studies within HD on task-based fMRI. A recent study has also demonstrated decreased activity in the left frontal area, even during a resting state (Dumas et al., 2013), in both gene-carrying individuals and those in the early stages of HD.

The neurochemical basis of HD is complex and is not fully understood. It is clear that dopamine and glutamate contribute to striatal deterioration. Dysregulation of the chemicals that play a role in memory and movement affects in the behavior and cognitive dysfunction.

Cognitive deficits typically, although not universally, occur before onset of other symptoms. Cognitive decline can occur in a variety of domains, but it often includes attention, visual, spatial and constructional abilities (in light of the motor difficulties, this is not unexpected), memory, and executive functioning. Specifically, memory impairments are often observed in encoding and learning new material, as well as in storage of new information. Executive functioning impairments can occur in a variety of different contexts, but they typically include difficulty with cognitive flexibility, problem solving, and planning. In comparison with AD, individuals with HD typically perform better on recognition components of memory tests. Furthermore, people with HD do not demonstrate language difficulties, with the exception of worsened verbal fluency.

Unfortunately, there is no known treatment that prevents the onset of the disease in those affected. Rather, treatment is often focused on the emotional component of the disease. Psychopharmacological interventions and behavioral strategies can be used for those who suffer from the disease. In 2012, the Guideline Development Subcommittee of the American Academy of Neurology provided a detailed literature review and specific medication recommendations for medical providers (Armstrong & Miyasaki, 2012). The recommendations provided include tetrabenazine, amantadine, and riluzile. During the early portions of the disease, low-dose neuroleptic medications, including haldol, can be prescribed to decrease the involuntary motor movements, but there is not enough evidence to indicate effectiveness. However, this treatment might increase cognitive impairment. Medications aimed to slow down cognitive decline can be prescribed, although the results are unclear. Psychotic symptoms can improve with neuroleptic treatment. Given the overlap of bipolar symptoms, lithium is often prescribed as well. A mental health provider can assist not only in ensuring that these medications and the psychiatric symptoms are treated appropriately, but also in making sure that the individuals and their family receive appropriate support and behavioral strategies to assist with the disorder. Specifically, coping with the challenges of this debilitating disease is difficult, and being involved in social support groups, therapy, and other resources, such as those offered by the Huntington's Disease Society, is recommended. Awareness of the emotional component of the disease and the earlier onset of cognitive problems can assist in developing appropriate strategies when working with families with a member who has been diagnosed with HD. Use of therapy for the verbal-based strengths can help with mood in the form of supportive techniques. Being aware and attempting to prepare the environment for the risk of falling can be beneficial.

## Parkinson's Disease

Parkinson's disease (PD) affects up to 1 million people, with a typical onset between ages 50 and 70. PD is characterized by tremors, muscle rigidity, bradykinesia, and postural instability. Depression often occurs during the course of the disease, and there is approximately a 40% chance of dementia occurring. The individual with PD will demonstrate slower initiation, lack of facial expression, and he or she will often experiences falls due to the combination of motor, physical, and balance symptoms.

PD has been consistently linked to autosomal dominant mutation on chromosome 4. Several causative genes of PD have been identified. The website at www.pdgene.org has up-to-date results of a meta-analysis on the genetic components of PD.

Dopaminergic cell loss in the substantia nigra relates directly to the hallmark motor abnormalities of PD. Dopaminergic neurons are also highly concentrated in the frontal lobes and brainstem, and decreased volume is noted in those brain regions in PD as well. However, there is also a decrease of GABA, serotonin, and norepinephrine. Medications, including levadopa and dopamine agonists, are often administered to patients who have been diagnosed with PD to improve upon these symptoms. If these treatments are not effective, recent advances in deep brain stimulation are particularly effective for treating the motor symptoms. With deep brain stimulation, electrodes are implanted in the brain and impulses occur. Figure 12.2 demonstrates deep brain stimulation.

Cognitive impairment does not occur in all patients with PD. Typical cognitive impairment is often prevalent in the visual cognitive domain. Specifically, cognitive impairment can include impairments in facial recognition, spatial capacities, and visual analysis. Language-based difficulties can occur in areas of fluency and naming. Furthermore, cognitive impairments on measures examining motor abilities are typical, considering the motor limitations.

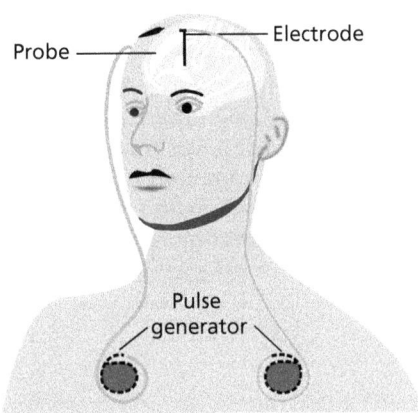

**FIGURE 12.2 Deep brain stimulation location.**
Adapted from the National Institute of Mental Health, National Institutes of Health, Department of Health and Human Services.

Treatment for PD from a mental health provider perspective often focuses on coping with motor abnormalities. Psychotherapy can focus on developing strategies to overcome the limitations of having tremors. Use of heavier utensils, voice recognition technology, and assistance with other motor-focused activities are important. Given the high occurrence of depression, cognitive behavioral therapy is often implemented to assist in dealing with the emotional challenges.

## OTHER MEDICAL EVENTS

### Tumors

Tumors that occur in the central nervous system, particularly in the brain, are common. Brain tumors are among the leading causes of death due to central nervous system dysfunction. Tumors can be either primary or metastatic, and they are often labeled by the location and cell type. Primary tumors include gliomas—whether astrocytomas or gliobastomas, meningiomas, pituitary adenomas, pinealomas, and acoustic neuromas. Though the general mental health clinician may not be expected to be familiar with the differences in these tumors, one should be aware of the importance in psychiatric presentations. In fact, primary tumors are 10 times as likely among patients who have been diagnosed with psychiatric disorders than in individuals without psychiatric disorders. In fact, confusion and behavioral disturbance are often early indicators of a tumor, followed by headaches and seizures.

A tumor can occur in any part of the brain. Focal symptoms often result. For example, frontal lobe tumors will result in changes in personality and behavioral disinhibition. Specific areas of frontal lobe tumors will result in different symptoms. Factors that play a role in symptom presentation include the rate of growth of the tumor and whether it has caused intracranial pressure. Although tumor location is not the most important factor of symptom occurrence, it plays an important role. For example, orbital frontal tumors are likely to result in agitation and irritability, along with poor judgment. In comparison, dorsolateral tumors result in depressive-like symptoms of apathy, slowing, and lack of initiation. Tumors in other lobes of the brain have similarly unique behavioral and emotional changes. Temporal lobe tumors will often result in psychotic symptoms.

Given the high quantity of psychiatric symptoms that occur in patients with tumors, the mental health practitioner should familiarize himself or herself with symptoms in order to determine when to make referrals. Symptoms of possible tumors include recent- or new-onset seizures, new-onset dull headaches, vomiting in conjunction with headaches, and changes in sensory functioning, including blurred vision and hearing loss. If there are symptoms of balance problems or localized weakness, immediate medical attention is warranted (Price, Goetz, & Lovell, 2011).

## Anoxia and Hypoxia

Hypoxia and anoxia have been long identified as contributing to emotional, cognitive, and behavioral changes. Hypoxia is defined as having oxygen intake lower than atmospheric concentration. This can occur either from acute situations, such as those that occur in carbon monoxide poisoning, decreased oxygen in depressurized airplane cabins, or chronic hypoxic events, including sleep apnea, chronic obstructive pulmonary disease (COPD), and lung disease. Anoxia is defined as having a complete loss of oxygen. This occurs in situations such as near-drowning, cardiac arrest, and hepatic encephalopathy. Acute or chronic respiratory insufficiency can result in cognitive, emotional, and behavioral difficulty.

There is research suggesting that genetic variables can play a role in the outcome following anoxia and hypoxia. It has been suggested for more than a decade that anoxic tolerance can be genetically detected (Haddad, Sun, Wyman, & Xu, 1997). There is an interest in discovering the factors that play a role in decreasing the cellular damage associated with oxygen deprivation. Studies have shown that within animals, mutations in genes encoding specific compounds can suppress the long-term effects of anoxic events (LaRue & Padilla, 2011). They found that animals that were fed metformin, which induces dietary restrictions in animals, have an increased survival rate when experiencing an anoxic event. Although in its infancy, further such research will likely discover how genetics either plays a role as a protective factor or increases risk for cognitive impairment subsequent to decreased oxygen-related events in humans.

Oxygen deprivation can have profound effects on cell differentiation, proliferation, and tumor growth. Different cell types in the same brain region can have different levels of tolerance of variable oxygen deprivation. Specific areas in the hippocampus are found to be more sensitive to oxygen deprivation. This sensitivity can result in memory problems and delayed onset of cognitive problems. It is not unusual for slower cognitive processing and processing speed to occur as a result.

Anoxic events can have profound effects on brain structure. Imaging findings typically include infarctions within brain regions that have a higher amount of metabolic demands, including the basal ganglia and cortex. Though important for numerous activities, the basal ganglia is particularly important for the selection and execution of activity, and is actively involved in the motor systems. It has been found that the medial occipital cortex and the hippocampi in the medial temporal lobe are the most affected parts of the brain after anoxic events. Damage to the corpus callosum, hippocampus, and white matter can be delayed subsequent to an anoxic event.

Neuroimaging results of hypoxia are limited in humans. CT scans are often deceptive, as they show little change in the brain after severe hypoxic events. On CT scans, cortical gray matter and subcortical white matter are often more difficult to differentiate. This means that there are likely to be changes occurring at a level that our technology has limited capacity to differentiate in less severe hypoxic events. There is also evidence that white matter lesions occur in specific types of hypoxic events, including those that are the result of pulmonary disease (van Dijk et al., 2004). It has been suggested that the lower the oxygen saturation, the greater the increased risk for severe white matter periventricular lesions.

Several studies have documented the presence of cognitive impairment, even subsequent to mild hypoxia. Up to 40% of patients who have been diagnosed with COPD have demonstrated moderate to severe cognitive decline. Subcortical cognitive decline in areas of attention, memory, cognitive slowing, planning, and visual spatial skills, in the absence of language impairment, is typical. It is not uncommon for psychiatric symptoms to occur after the event as well.

The type of treatment for anoxia or hypoxia is contingent upon the cause of the limited, or lack of, oxygen. For example, treatment for breathing-related sleep disorders should occur in the event of sleep apnea. Immediate medical treatment for pulmonary or cardiac embolism is necessary to help decrease the cognitive, emotional, and behavioral effects of the event. The goal for any breathing-related disorder is to obtain immediate and consistent clear air flow. When a near-drowning occurs there is evidence that use of induced therapeutic hypothermia with a core temperature between 32°C and 34°C after successful resuscitation is neuroprotective and improves long-term cognitive outcome. The challenge is that resuscitation requires increased core body temperature.

## CONCLUSIONS

Medical diseases do not occur exclusively from cognitive, emotional, and behavioral symptoms. The neurobiological aspects of the disease, including the neurochemical changes, structural brain abnormalities, and functional changes in glucose metabolism are evident in these disorders. Furthermore, the emotional outcome resulting from illness is often burdensome. The clinician needs to be prepared to assist the client not only in coping with the illness, but also by explaining the processes that occur with medical illness. The clinician needs to be knowledgeable about the different behavioral, cognitive, and emotional outcomes that accompany these diseases to ensure that proper treatment and care occur.

## SUMMARY POINTS

- Several medical events contribute to cognitive, emotional, and behavioral dysfunction.
- There are several types of dementias that are progressive. Each has different neuropathology but similar yet distinct cognitive decline.
- Though no medications have been proven to stop or prevent dementia, behavioral and cognitive strategies exist that can prevent or slow down cognitive decline.
- There are several different seizure disorders that can be misdiagnosed for other disorders.
- The mental health practitioner plays an important role in treating and caring for patients with medical problems.

CHAPTER 13

# Traumatic Brain Injury

It is estimated that at least 1.7 million people sustain a traumatic brain injury (TBI) in the United States annually. From these injuries, 52,000 people die, 275,000 are hospitalized, and 1.37 million are treated and released from emergency departments (Faul, Xu, Wald, & Coronado, 2010). Furthermore, in any given year there are countless quantities of individuals who sustain a TBI but do not receive medical attention, with estimates ranging between 1 and 4 million people, with many sports-related injuries in particular going unreported. Considering the wide variability of estimates in the frequency of TBIs and the heterogeneity of symptoms following a TBI, there is limited agreement regarding the most frequent causes for TBI. Historically, motor vehicle accidents have been considered the most common reason for TBI (Yudofsky & Hales, 2008), whereas other research suggests that sports-related injuries are far more common (NIH Consensus Developmental Panel, 1999). On the other hand, the Centers for Disease Control recently reported that the main cause of TBI is falls, accounting for 35% of all head injuries (Faul et al., 2010). In its report, the CDC reports that motor vehicle traffic crashes account for 17%, being struck by or striking against an object accounts for 16.5%, and assaults account for 10% of TBIs. The CDC also reports another 21% of TBIs as having unknown or other reasons, such as bicycle accidents. It has also been estimated that approximately 20% of cases of TBI are due to violence (National Institute of Neurological Disorders and Stroke, 2007). There are recent reports that 10% to 20% of military personnel have TBIs, with 66% having long-term concussive symptoms (Jaffee et al., 2009; Lew et al., 2009). The number of TBIs from blast wounds from recent wars in Iraq and Afghanistan far exceeds those from previous wars (Snell & Halter, 2010). Though the incidence rate of this injury in young children is unknown, physical abuse has been reported to be the leading cause of serious TBI and death in children age 2 or younger (Kennan et al., 2003).

Previous estimates indicate that 50,000 children sustain TBI in bicycle accidents annually (U.S. Department of Health and Human Services, 1989), although there has been an 18% reduction of injuries in states where legislation requires that helmets be worn (Lee, Shoeffer, & Koppelman, 2005).

Despite the variability in reports examining the causes of TBI, it is commonly agreed that more than 5.3 million people in the United States live with TBI-related disabilities. As a result, the cost of TBI-related hospitalizations, acute care, and rehabilitation is substantial. TBI is a serious problem that has received increased attention in recent years. This attention is in part owing to the recent recognition that even less severe TBIs result in potentially long-term emotional, behavioral, and psychosocial difficulties. Evidence of long-term effects from military personnel and athletes have heightened awareness of TBI in the United States.

## TBI RISK FACTORS

Risk factors for obtaining a TBI are numerous; they include age, gender, environment, and occupation. It has been consistently reported that males are 1.5 times as likely to have sustained a TBI than are females. Age is also a risk factor, with children ranging between ages 0 and 4 accounting for 18% of all TBI-related emergency department visits. It has been estimated that 22% of TBI-related hospitalizations are for adults age 75 and older. Adolescents between ages 15 and 19 are also at high risk for TBI. It has been reported that TBI declines during middle age, as impulsive behavior declines (Bruns & Hauser, 2003). It stands to reason that people who participate in high-risk pursuits, including working certain jobs such as being in the military, taking part in recreational activities without using helmets, all-terrain vehicle riding, and consuming alcohol socially, are at higher risk of sustaining a TBI than those who do not engage in such activities. For example, it has been estimated that 56% of all motor vehicle accidents involve a legally intoxicated (above .08 blood alcohol level) driver (Kraus, Morgenstern, Fife, Conroy, & Nourjah, 1989). People living in areas with lower average incomes have been found to sustain more TBIs than those with higher income (Wagner, Sasser, Hammond, Wiercisiewski, & Alexander, 2000). The interaction of various factors increases risk as well. For example, older studies also suggest that young, urban-dwelling minority males are at highest risk for TBIs. Also, males between ages 21 and 40 suffer the highest rates of injury (Ip, Hesch, Brandys, Dornan, & Schentag, 2000). Bushnik, Hanks, Kreutzer, and Rosenthal (2003) found that TBI due to violence tended to involve people who had minority status (56% African American, 11% other), were unmarried, unemployed at the time of the injury, and had a premorbid history of illegal substance use and encounters with law enforcement. It is clear that there exists a complex relationship among various controllable and noncontrollable factors with the incidence of TBI.

## TBI DEFINED

The term *traumatic brain injury* (TBI) is often used interchangeably with head injury. However, it is important to define the two terms, as there can be substantial differences between a TBI and a head injury. A head injury occurs as a result of a blow to the head, in which the force of the blow might be absorbed by the skull without necessarily resulting in functional changes within the individual. For example, a simple fall by a child learning to walk that might even result in a bruise on the head does not typically result in a TBI. A TBI is an insult to the brain involving an external force that leads to impairment in cognitive, emotional, behavioral, or physical functioning for at least a brief, and at times a sustained, period of time. Although alterations in alertness or awareness are not necessary to a classification of TBI, loss of consciousness as well as possible amnesia does often occur.

A TBI occurs as a result of either a closed-head or open-head injury. A closed-head injury typically occurs when the brain is either struck by an object or strikes another object. Although not all closed-head injuries result in TBI, it is also unnecessary for the head to strike an object for a closed-head TBI to occur. For example, it is possible for a TBI to occur when there is movement inside of the skull without external or visual trauma to the head. This can occur, for example, as a result of a motor vehicle accident in which the individual is restrained and the head does not strike any part of the vehicle, but for which there are neurocognitive sequelae. In these types of injuries, there can be a whiplash mechanism that results in the brain striking the skull and/or axonal stretching of the neurons. Blast waves as a result of exposure to bombs can also cause closed-head injuries without visible trauma to the head. Further details of different neuromechanisms involved in these and in other TBI injuries are provided later in this chapter.

A less common type of TBI is an open-head injury. This results when there is penetration of an object onto or through the brain's gray matter. As a result, there is exposure of the brain tissue. Bullet wounds are a common example of open-head injury. A famous case example of another type of open-head injury is that of Phineas Gage, described in Chapter 1. Regardless of whether it is a closed-head or open-head injury, a TBI disrupts the normal neuronal or vascular activity of the brain, which can in turn create alterations in consciousness as well as behavioral, cognitive, or emotional symptoms.

## NEUROBIOLOGY OF TBI

TBI can result in a variety of types of brain damage, with resultant medical diagnoses. Table 13.1 lists common medical diagnoses associated with TBI, along with their definitions. Understanding the neuropathology involved with the most common types of nonfatal TBI is particularly important for the mental health practitioner, as the knowledge can lend itself to conceptualization of the patient's emotional problems and appropriate forms of treatment. Specifically, understanding contusions, diffuse axonal injury, and concussion is particularly important for the mental health practitioner, as most of the 5.3 million individuals with TBI-related difficulty have sustained one or several of these types of injuries.

A contusion is a bruise of the brain tissue that is localized—that is, occurring in one location of the brain. Contusions can be multiple, and they often result from coup–contrecoup types of injuries. A coup–contrecoup injury occurs when a moving head stops abruptly. The coup injury is the point in the brain that strikes the inside of the skull upon impact. The bruising that results is a contusion. The contrecoup injury can occur directly opposite the side of the initial impact, as a result of the violent collision of the brain and the skull. For example, when a person falls backward and the head strikes pavement, the initial blow (coup) could occur in the orbital area of the brain. Thereafter, the brain might slam on the opposite side of the skull at the frontal lobe (contrecoup). Figure 13.1 presents an example of a fall and the head hitting concrete. This might explain deficits in executive functioning, attention, and memory even when there is no observable impact to that location of the brain. In fact,

**TABLE 13.1 Common Medical Diagnoses Associated With TBI**

| | |
|---|---|
| Cerebral edema | Accumulation of water within the brain. |
| Concussion | When a blow or jolt to the brain results in intracranial changes and subsequent neurobehavioral consequences, such as headaches, loss of consciousness, and confusion. |
| Contusion | Bruise of brain tissue, typically occurring in the frontal and temporal lobes. |
| Basilar skull fracture | A break in a bone in the head. These can be further classified as linear or depressed. A linear skull fracture is a break in the bone that is a straight line. A depressed skull fracture is when a portion of the skull is crushed in toward the brain. |
| Diffuse axonal injury | Death of neurons in the brain due to rapid acceleration/deceleration within the brain. The tearing of the axons occurs throughout the brain, resulting in global deficits and often loss of consciousness for more than 6 hours, coma, or even death. |
| Encephalitis | Inflammation of the brain, usually due to swelling. |
| Encephalopathy | Any diffusion of the brain that affects the brain in terms of structure, function, or normal capacity. Typically results in altered mental state. |
| Epidural hematoma | Accumulation of blood between the skull and the dura mater, the tough outer layer of the central nervous system. |
| Hydrocephalus | Abnormal accumulation of cerebral spinal flood in the ventricles. Also known as *water on the brain*. |
| Increased intracranial pressure | Rise of pressure inside of the skull that can be caused by or result in traumatic brain injury. This pressure can cause significant damage to the structure and function of the brain. |
| Subarachnoid hemorrhage | A bleed that occurs between the thin tissues covering the brain and the brain itself. Typically results in acute onset of a severe headache that is more severe in the back of the head. |
| Subdural hematoma | A collection of blood on the surface of the brain from the tearing of tiny veins between the dura and the surface of the brain. Can either be acute or chronic. Acute results in a rapid collection of blood on the surface of the brain, resulting in the compression of brain tissue. Chronic results in a slow leak of blood and begins several weeks after initial bleeding. |

contrecoup injuries most often occur in the frontal and temporal areas due to the bony features that surround the frontal and temporal lobes (Gennarelli & Graham, 1998).

Diffuse axonal injuries result from chemical or mechanical changes in the axons of cerebral white matter. This typically occurs throughout several areas of the brain, and it can be widespread. Extensive lesions occur in the white matter tracks of individuals who sustain diffuse axonal injuries. They can result from rotational forces or from deceleration forces that cause the injury to the axons. As a result, axons are stretched, causing cell disruption of molecular processes that can last for several weeks. As a result, metabolic changes occur in which there is an increase in energy needs but a decrease in available energy. This commonly occurs as the result of high-speed motor vehicle accidents. The basal ganglia, corpus callosum,

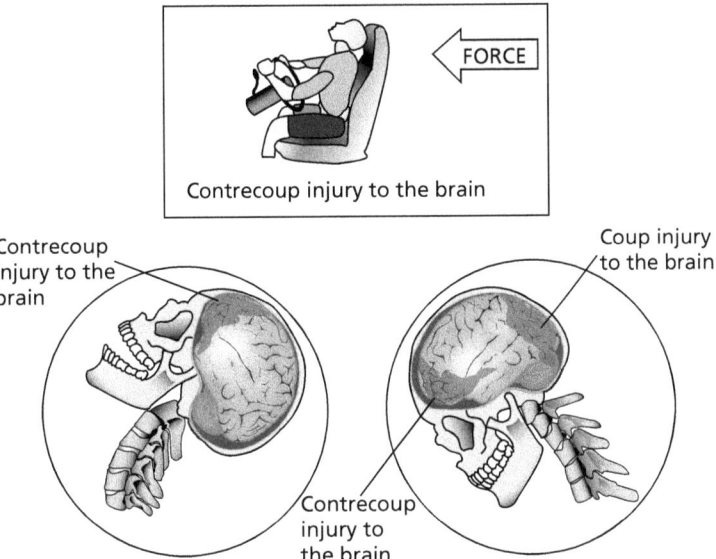

**FIGURE 13.1 Coup–contrecoup injury.**
Adapted from Holtz (2011).

hypothalamus, cerebellum, and limbic system are the brain regions most likely to suffer from diffuse axonal injuries. It is possible that these areas are more susceptible due to the difference in density between those portions of the brain and the rest of the brain (Singh & Stock, 2008).

## NEUROIMAGING AND TBI

There have been many attempts to image TBI with available technology for both clinical and research purposes. Neuroimaging is used to examine the severity and extent of brain damage due to TBI. Contusions, fractures, and hematomas can be documented by computed tomography (CT) and magentic resonance imaging (MRI). Positive imaging findings can occur immediately after the TBI, or might take upward of 3 months after the injury (Cope et al., 1988). From a clinical perspective, it has been suggested that a mild TBI accompanied by positive imaging findings increases the likelihood of slower recovery over a 6-month time frame (Williams, Levin, & Eisenberg, 1990). In terms of research, it is possible to correlate positive imaging results and cognitive, emotional, and behavioral functioning. When findings are positive, consistency in the results are rare. Other than the limitations in available technology, the heterogeneity of TBI and difficulty controlling for co-occurring psychiatric problems or substance abuse problems present unique challenges for imaging findings in this population. The newer experimental neuroimaging techniques have provided increased consistency in discovering structural, functional, or neurochemical abnormalities associated with TBI. Belanger, Vanderploeg, Curtiss, and Wardon (2007) provide a review of studies that have used newer techniques, such as functional magnetic resonance imaging (fMRI), single-photon emission computed tomography (SPECT), and diffusion tensor imaging (DTI) in TBI. Although these techniques are not used for clinical purposes at this point, it is anticipated that advances in technology will lead to more effective imaging practices for TBI patients. At present, however, there are often no positive results on imaging technology. Traditional imaging techniques, such as CT and MRI, have not proven consistently effective in observing brain changes following a TBI. Borg (2007) reported that up to 95% of TBI patients have normal CT findings. Other studies have suggested that approximately half of mild TBI patients have normal MRI findings (Hofman et al., 2001; Hughes et al., 2004). This is likely because of limitations in available technology. Furthermore, the medical diagnosis associated with the head injury might not be indicative of behavioral, emotional, or cognitive short-term or long-term outcome. For example, a severe diffuse axonal injury might in fact yield worse long-term outcome than a linear basilar skull fracture. However, the severity of the TBI has been traditionally thought to be associated with long-term prognosis.

## TBI: SEVERITY CLASSIFICATION

Although there are several possible medical outcomes associated with TBI, classifying the severity of the injury can be challenging. There are several scales used to classify severity of TBI. TBI is traditionally classified into mild, moderate, and severe subtypes. However, there is no agreed-upon standard that experts use in differentiating among severity levels. Bigler (2008) provides a thorough summary of many of the different scales. In general, the main considerations involved with determining the severity of TBI include response to motor, verbal, and visual stimuli, as well as the duration of posttraumatic amnesia (PTA)—that is, the period of loss of memory after the injury. Furthermore, performance on cognitive measures and whether there are positive findings on an MRI or CT of the brain following a TBI have also been used to assess the severity of TBI.

A common, easily applied scale to determine severity immediately following a TBI is the Glasgow Coma Scale (GCS; Teasdale & Jennett, 1974). As demonstrated in Table 13.2, the GCS evaluates visual, motor, and verbal responses to stimuli. However, the GCS is limited in terms of anticipating cognitive, behavioral, or emotional responses. This scale is practical for personnel who are on the scene or in a hospital subsequent to a TBI. However, the GCS has not been found to be associated with outcomes 1 year after the injury (Hiekkanen, Kurki, Brandstack, Kairisto, & Tenovuo, 2009). Rather, it has been argued that the duration of PTA is a better predictor. Table 13.3 provides an example of assessing PTA. The Galveston Orientation

**TABLE 13.2 Glasgow Coma Score Rating and Interpretation.**

| Eye Opening (E) | Verbal Response (V) | Motor Response (M) |
| --- | --- | --- |
| 4 = Spontaneous | 5 = Normal conversation | 6 = Normal |
| 3 = To voice | 4 = Disoriented conversation | 5 = Localizes to pain |
| 2 = To pain | 3 = Words, but not coherent | 4 = Withdraws from pain |
| 1 = None | 2 = No words; only sounds | 3 = Decorticate posture |
| | 1 = None | 2 = Decerebrate |
| | | 1 = None |

Total Score = E + V + M

A coma score of 13 or higher correlates with a mild brain injury, 9 to 12 is a moderate injury, and 8 or less is a severe brain injury.

*Source:* Teasdale and Janette (1974).

TABLE 13.3 Estimates of TBI Severity Based on Duration of Posttraumatic Amnesia (PTA).

| PTA Duration | Severity Classification |
| --- | --- |
| <5 Minutes | Very mild |
| 5–60 minutes | Mild |
| 1–24 hours | Moderate |
| 1–7 days | Severe |
| 1–4 weeks | Very severe |
| >4 weeks | Extremely severe |

Source: Jeanette and Teasdale (1981).

and Amnesia Test (GOAT; Levin, O'Donnell, & Grossman, 1979) and the Westmead PTA scale are also formal tests that attempt to determine the length of PTA. A study completed by Shores et al. (2008) indicated that the Revised Westmead PTA scale can be used in conjunction with the GCS to predict cognitive impairment in patients who have mild TBI. An assessment commonly used for determining function and classifying a TBI during the first few weeks or months after a TBI is the Rancho Los Amigos Scale (Hagan, 1984), which observes the patient's reaction to external stimuli.

Although it is widely agreed that it is futile to attempt to classify the severity of TBI, such attempts continue to occur. Furthermore, severity level is very prevalent in the nomenclature of TBI. Though most TBIs are classified as mild, a report sent to Congress by experts in the field clearly states "it is clear that mild TBI does not result in mild symptoms" (NIH Consensus Development Panel, 1999). Continued classification of TBI occurs, as it can be helpful to treatment providers to differentiate severity of injury, as well as the possible course of recovery and overall outcome. However, research has indicated that severity of injury does not predict rehabilitative gain toward functional independence (Toschlog et al., 2003). Furthermore, many individuals who sustain so-called mild TBI struggle significantly in their functional skills. As a result, mental health practitioners are likely to be involved in some capacity with the treatment of mild TBI patients, even when the TBI is not the presenting problem or initial reason for treatment.

## MILD TBI: DEFINITION AND SYMPTOMS

Despite the uncertainty about what criteria must exist to classify among different levels of TBI, it is commonly agreed upon that mild TBI occurs most frequently. The Head Injury Interdisciplinary Special Interest Group of the American Congress of Rehabilitation Medicine defines mild head injury as "a traumatically induced physiologic disruption of brain function, as manifested by any periods of loss of consciousness, any loss of memory for the events immediately before or after the accident, or any alteration in the mental state at the time of the accident." Furthermore, it reports that focal neurologic deficits might or might not be transient. Vos et al. (2002) point out that all definitions of mild TBI (mTBI) result from the consequences of a blunt impact with sudden acceleration, deceleration, or rotation of the head, with a GSC score between 13 and 15.

The outcome subsequent to mTBI is variable. It is common for individuals to have symptoms of concussion immediately after mTBI. A concussion is the temporary altered mental state following a head trauma. Symptoms of a concussion include headache, sleep difficulty, and feelings of mental fogginess, among other things. The British Columbia Postconcussion Symptom Inventory is a checklist commonly used for determining symptoms of concussion, and such symptoms include headaches, nausea, dizziness, irritability, sadness, nervousness, increased temper, poor concentration, reading difficulty, poor sleep, and memory problems. These symptoms typically resolve within 1 week. Iverson (2007) has noted that the best-controlled studies demonstrate good-to-complete recovery of these symptoms in individuals who sustain an mTBI. Yet, for some, the consequences of mTBI can be devastating. Up to 10% experience long-term cognitive, emotional, or behavioral difficulty (Wood, 2004). If these symptoms persist longer than 1 week, the individual meets the criteria for postconcussion syndrome (PCS). The *DSM-IV-TR* provides criteria for PCS under the section for further research. It lists eight different symptoms of PCS, including (1) becoming easily fatigued, (2) disordered sleep, (3) headache, (4) vertigo or dizziness, (5) irritability, (6) anxiety, depression, or affective lability, (7) changes in personality, and (8) apathy or lack of spontaneity. It suggests that these symptoms must lead to functional impairment in either the home or work environment and not be due to some other possible etiology, such as depression. The *DSM-IV-TR* criteria suggest that the symptoms must last for at least 3 months. The *DSM-5* includes a diagnosis of major or mild neurocognitive disorder due to TBI and warns that the condition can be persistent.

When symptoms persist longer than 6 months, the individual meets the criteria for persistent postconcussion syndrome (PPCS). Individuals who have been diagnosed with PPCS have been shown to have cognitive difficulty in areas of attention, memory, and executive functioning. PPCS has led to retirement from professions, marital difficulties, and long-term emotional problems. Though it is unclear what leads to PPCS, several factors have been identified that increase the likelihood of PPCS. These factors include aspects associated with the injury, such as severity of injury, type of injury, duration of PCA, and the presence of total anosmia—that is, the lack of olfaction functioning. Psychosocial factors have also been demonstrated as being important in influencing outcomes following TBI. Specifically, history of emotional or behavioral functioning, including depression, attention disorders, learning problems, and drug or alcohol abuse, has been found to be particularly important. It has been found that subsequent to a TBI, these emotional difficulties are exacerbated. Individuals with lower premorbid IQ are more likely to have PPCS. The presence of PTSD following TBI also increases the likelihood of PPCS. Lower levels of education, older age, lower income, and lower socioeconomic status are also risk factors for PPCS. The presence of the apolipoprotein E (ApoE) gene has also predicted poorer recovery from TBI, and it is more likely to be present in PPCS (Isoniemi, Kurki, Tenovuo, Kairisto, & Portin, 2006). The ApoE is a protein responsible for the redistribution of cholesterol from cells during synthesis, as well as neurotic extension growth and branching (Graham, Horsburgh, Nicoll, & Teasdale, 1999). However, it appears that the presence of ApoE is not predictive of performance on cognitive measures after TBI (Han et al., 2007). Although this finding was not expected, it is only one of several challenges associated with PPCS. Other challenges involved with diagnosing PPCS include the overlap with other possible explanations of the symptoms, such as depression. For example, Iverson (2006) has provided evidence that PPCS is misdiagnosed when the symptoms raise from depression. Pain following the trauma also increases the likelihood of PPCS (Sheedy, Geffen, Donnelly, & Faux, 2006). Also, individuals can try to fake symptoms for secondary gain. Therefore, consideration of motivation needs to occur in order to rule out malingering.

## SPECIAL TOPICS

### Sports-Related TBI

Evaluation of athletes has become an increasingly important topic over the past few years. It is clear from sports-related injuries that even mild blows to the head can have a devastating effect. Athletes who are involved in contact sports have been found be at an increased risk for significant brain changes. Two medical issues that occur as a result of TBI associated with sports are second impact syndrome (SIS) and chronic traumatic encephalopathy (CTE).

SIS is the devastating effect of a second closed-head injury occurring prior to the resolution of an initial TBI. After the second TBI, the brain swells rapidly, which results in severe impairment or even fatality. A series of metabolic events, including the dilation of blood vessels, which result in dysregulation of cerebral blood flow, are unleashed in the brain after the second TBI. The increased blood flow and resultant brain volume lead to a severe increase of cranial brain pressure, which results in the brain expanding beyond the available space within the skull. In turn, the symptoms include respiratory failure, limited or no eye movement, and unconsciousness. The symptoms of SIS most often occur immediately after the second blow to the head. However, there are documented cases in which the symptoms do not occur for days, or even weeks, after the second blow. SIS occurs only in young adults, adolescents, and children.

CTE has been identified in the brains of late NFL football players John Grimsley, Mike Webster, Andre Waters, Justin Strzelczyk, Terry Long, and Chris Henry. Although CTE has been associated with boxing-related injuries as far back as 1928 (Martland, 1928), the public realization of CTE being associated with professional football players is relatively new. CTE results from multiple blows to the head, which can result in concussions or subconcussive impacts, that are typically associated with contact sports. Postmortem studies have identified an elevated level of tau protein and tau DNA-binding protein 43 (TDP-43) in the spinal cord tissue of these athletes (McKee et al., 2010). As a result, these athletes experience early symptoms of dementia, including memory problems, rage, severe depression, and confusion. Though CTE typically occurs in middle-aged to older adults, in 2010, a 21-year-old college football player who committed suicide showed early stages of CTE.

Whether with children or adults, it is clear that repeated blows to the head, even if they do not lead to full-blown concussion symptoms, can be dangerous. Careful identification, evaluation, treatment, and follow-up care for any athlete who sustains a blow to the head are important for both short-term and long-term health.

## MALINGERING ISSUES IN TBI

Those working with individuals who sustain TBI must consider whether there are potential benefits for the individual to have symptoms following a TBI. Unfortunately, even in the most mundane cases, the possibility of secondary gain needs to be examined, given the high number of lawsuits. This is particularly important with the vast amounts of media attention on TBI, as it has become easy for nonclinicians to learn the clinical presentation of concussion and postconcussive syndrome. Those who are attempting to profit from their injury can fake their symptoms, or malinger. Therapists need to be aware of this possibility and at times consider assessing for this possibility. Physicians and psychologists have developed some strategies to assess for malingering. For example, physicians will explore whether there are unusual or inconsistent symptoms. Similarly, psychologists have developed many tests that examine effort. Self-report behavioral questionnaires are also used to examine unusual or vague symptoms.

## OUTCOME

Outcome can vary considerably, even with each type of brain injury, as there is a wide range of severity for each type. For example, medical treatment immediately after a TBI varies from craniotomy, shunting, steroids, to merely rest, among others. Given the wide range of possible locations on the brain where a TBI can occur, the outcomes vary considerably, from no neurobehavioral changes to death. Assuming survival, the person can have a variety of emotional, behavioral, or cognitive difficulties that might affect functioning. The normal abilities to resume employment responsibilities, carry on relationships with family, regain driving abilities, overcome or learn to accommodate physical limitations, and deal with sensory problems and limitations in terms of abilities needed to succeed in school are some of the areas that can be negatively influenced following TBI.

For children, TBI can result in academic problems. Subsequent to TBI, children have been found to have problems with academic areas such as math and reading. Careful consideration of participation in sports is important. See Box 1 for special consideration of sports-related TBI for both children and adults.

## APPLICATION OF TREATMENT TO MENTAL HEALTH PROVIDERS

There is no one set standard of treatments for TBI. The goal is for patients who suffer TBI to receive immediate, careful medical evaluation and follow-up management. The American Association of Neurological Surgeons has established guidelines for the management of severe TBI patients. Unfortunately, fewer than 50% of patients are managed according to these guidelines, resulting in poorer clinical outcomes. Furthermore, treatment for mild TBI patients remains less clear. It is necessary that mental health providers assist in the identification, treatment, and long-term care of patients who sustain TBI. Mental health providers can be involved in numerous levels of treatment with patients with TBI. Completing a thorough assessment in order to determine whether the emotional–cognitive symptoms that the patient is experiencing could be resulting from ongoing symptoms associated with a TBI—whether it is a remote injury or a recent injury—can be particularly useful in guiding recommendations and treatment.

### Cognitive Rehabilitation

Given the significant increase in the amount of TBI, it is expected that there will be an increased need for mental health workers to be trained in cognitive rehabilitation. Cognitive rehabilitation works to improve upon cognitive dysfunction either by aiming to achieve improved postinjury cognitive status or by developing skills to overcome permanent cognitive impairment. Cognitive rehabilitation has been defined as "a systematic, functionally oriented service of therapeutic activities intended to improve cognitive functioning through reestablishing previously learned patterns of behaviors; establishing new patterns of cognitive activity through compensatory cognitive mechanisms for impaired neurological systems; establishing new patterns of activity through external compensatory mechanisms or environmental support; and/or enabling persons to adapt to their cognitive disability to improve their overall quality of life" (Cicerone et al., 2000). It is highly specific to the individual's needs, and thus there is no one right way to conduct cognitive rehabilitation. It typically includes strengthening the individual's cognitive skill set in order to increase functional abilities. For example, treatment might focus on relearning cognitive abilities that were lost or damaged due to TBI, substituting a new skill for an ability that was damaged or lost, and/or developing skills to control impulses. It has also been suggested that web-based treatment, particularly of the kind that involves the family in a rural community, can improve outcomes following TBI (Wade, Wolfe, Brown, & Pestian, 2005). Studies have shown cognitive rehabilitation to be effective; however, more research needs to be conducted to help specify which treatments are most practical for specific symptoms.

## Psychotherapy

Mental health providers are skilled at providing therapy for a variety of patients who suffer from emotional and behavioral difficulties. A common postconcussive and persistent postconcussive symptom is emotional instabilities, such as irritability, anger, and depression. Cognitive behavioral therapy can assist in developing skills to cope with the anger outburst or depression. Identifying stressors that increase the likelihood of mood dysregulation can be important for patients in an attempt to better manage their environments. It can also help the patient to learn effective strategies to deal with relationships, and with the potential changes that might occur in relationships as a result of TBI. Along those lines, family therapy can also be useful in terms of psychoeducation regarding TBI, as well as how to most effectively cope with an individual's changes.

## Neuropsychological Assessment

It is not uncommon for TBI to lead to cognitive problems in attention, memory, and/or executive functioning. If there are concerns regarding cognitive functioning and the usefulness of therapy, neuropsychological testing can be recommended. Such evaluations can assist in determining the patient's strengths and weaknesses, which can guide treatment strategies. For example, the individual might be found to have verbal learning deficits, which can result in limited learning from traditional psychotherapy techniques. Rather, speech therapist referral would be warranted. It might also be discovered that underlying visual perceptual deficits are occurring, which would then lead to referral to other specialists such as an ophthalmologist or neurologist. The role of the mental health professional is to help identify the patients for whom such recommendations would be beneficial.

## Speech and Language

Difficulty can also occur in terms of language skills. Stuttering, word finding problems, and difficulty expressing oneself can result from TBI. Speech therapists help the patient improve in these areas through repetition, articulation therapy, and other techniques. Pragmatic language skill development might also be necessary, depending on the severity of the injury.

### Sensory Evaluations

Hearing, balance, tinnitus, and auditory and visual difficulties can also result from TBI. Audiology evaluations can assist in determining the severity of hearing-related difficulty following TBI. Symptoms can include earache, muffled hearing, sensitivity to noise, and ringing in the ears. If the TBI is due to a blast injury, damage can occur in the middle ear cavity. Blast-related injuries can result in either short-term or permanent hearing loss.

TBI can also lead to blurred vision, double vision, or sensitivity to light. Studies have found that visual image misalignment can occur in TBI patients. An evaluation by ophthalmology can help correct this difficulty. For example, it has been found that individualized prismatic spectacle lenses can alleviate these symptoms (Doble, Feinberg, Rosner, & Rosner, 2010).

It is not uncommon for patients who have been diagnosed with TBI to have balance problems or vertigo. Some studies have indicated that balance-related problems shortly after a TBI result in slower functional recovery (Greenwald et al., 2001). Occupational training, including vestibular rehabilitation, can help the patient overcome these balance problems.

Given the clear neurobehavioral difficulties that can occur as a result of TBI, ameliorating any sensory disruption will help the clinician pursue higher quality of life for the patient.

### Headache

Headaches are one of the most frequently reported symptom after TBI. The International Headache Society provides criteria for several different types of headaches associated with TBI (Headache Classification Committee of The International Headache Society, 2004). Treatment for the headache should be carefully considered and individually tailored. A careful review of the course, location, and feeling of the headache is important. Over-the-counter medication might result in analgesic rebound, thereby delaying recovery. Thus, behavioral strategies should be implemented to aid recovery. For example, a headache calendar should be used to track changes in the symptoms. Similarly, behavioral strategies of avoiding alcohol and nicotine are recommended. Establishing and/or maintaining a healthy sleep pattern can decrease headaches. Recommendation for massage or acupuncture might be beneficial. Specific medications might be necessary, and a referral to a headache specialist should occur. Also, neurologic reevaluation might be necessary.

### Medication Effects

Much as with patients who have been diagnosed with common psychiatric disorders, medication can be used to help TBI patients' symptoms improve. Although there have been limited studies, it is typical for physicians to prescribe medications aimed at improving emotional and behavioral symptoms. Patients who have been diagnosed with TBI, however, are often more sensitive to the effects of medication. It is common for medication to start at a low dose and be slowly titrated up. Research has shown that numerous neurotransmitters are affected by a TBI, including serotonin, glutamate, dopamine, and acetylcholine. Therefore,

> **ETHICS**
>
> **Confidentiality**
>
> As previously stated, ethical standards vary according to practice. Nevertheless, regardless of the field, confidentiality is of utmost priority for patients and providers. The Ethics Code for Psychologists, Standard 4.01 (American Psychological Association, 2010), states that "psychologists have a primary obligation and take reasonable precautions to protect confidential information obtained through or stored in any medium, recognizing that the extent and limits of confidentiality may be regulated by law or established by institutional rules or professional or scientific relationship." Though such standards would seem to apply in a straightforward manner to most populations, individuals working with patients who have sustained a TBI are in a unique situation. Given the potential for numerous other providers, including physicians, team trainers, occupational and vocational therapists, along with possible involvement of employers (such as in the case of those TBIs that are a resulting from military or sports accidents) or schools, a strong possibility exists that others involved in the care of the patient will be highly invested in the patient's recovery. It is particularly important to discuss the limits of confidentiality before evaluating and/or treating in order to inform the patient of who has the authority to be informed of the outcome. For example, a patient who is evaluated while in the hospital would need to know what other medical providers would be informed of his or her performance and functioning. An athlete should be made aware of the role that the psychologist or neuropsychologist assessment will have in determining whether she or he is capable of playing. Considering the recent advancement of baseline computerized assessment in athletes, conducted and interpreted by team trainers, there is uncertainty as to the limits of confidentiality, as the team trainers do not follow the same set of ethical guidelines. Regardless, psychologists must remain strict in their adherence to ensure that confidentiality regarding functioning and test performance is protected while working with the patient to ensure his or her safety as well as that of the public in certain circumstances.

augmentation of these through medications, such as SSRIs for serotonin, has been used. The role of the mental health professional, if prescription privileges are not granted in the clinician's state, is to help facilitate the patient's evaluation by a psychiatrist or other medical doctor to determine the necessity of medication.

## Diet/Nutrition/Exercise

There are limited controlled studies that have examined specific foods that can be helpful for recovery from TBI. There is some indication that certain types of food should be consumed, whereas others should be avoided when recovering from TBI. Mental health professionals should always defer to medical doctors or dieticians when discussing dietary options. In general, foods that are slow to digest or that increase sugar slowly are thought to be better dietary options. For example, though brown rice is a good option, white bread is thought to be a poor choice. Other foods to avoid include excessive caffeine, alcohol, highly refined sugars, and fruit juices.

When an individual no longer feels PCS symptoms at rest, exercise can be recommended. Yoga, meditation, and slow-paced aerobic exercise should be considered initially. Exercise requiring intense breathing is not suggested until after the patient gradually improves and is symptom-free following lighter workouts. Increasing exercise too quickly can result in prolonged and slower recovery. Such intensifying of exercise should be under the care of a specialist or physician in order to ensure safety.

## CONCLUSIONS

TBI is a very common diagnosis that affects millions of people of all ages. Recognition of the symptoms associated with TBI and persistent PPCS is necessary for a clinician dealing with patients of various populations, ages, and settings. Knowledge of the different outcomes and terminology associated with TBI can help the clinician have an awareness of the severity of injury, and it can help guide treatment planning.

## SUMMARY POINTS

- TBI is highly prevalent and can lead to short-term and long-term emotional, behavioral, cognitive, and social dysfunction.
- There are various degrees of TBI and levels of severity, which can contribute to various levels of functional difficulty.
- The symptoms of TBI can vary and range depending on the severity of the injury, preexisting factors, and approaches to recovery.
- Knowledge about the variability of symptoms can aid the development of treatment plans, which can include several other types of medical providers.
- The long-term consequences of repeated low-impact head injuries are now being considered a factor in the development of CTE, which might influence organized sports and activities.

CHAPTER 14

# Personality Disorders

Personality disorders are maladaptive character traits associated with rigid thinking and unhealthy patterns of behavior. Personality disorders are associated with relationship difficulties, functional problems, and atypical social tendencies. These are thought to be pervasive and enduring. Personality disorders are among the most common, if not the most common, psychiatric disorders, and they are highly associated with disability (Grant et al, 2004; Torgensen, Kringlen, & Cramer, 2001). Many clinicians overlook the fact that personality in itself is a product of multiple factors, not the least of which includes the brain's processing of the environment and experiences that the individual undergoes. Even the conceptualization of personality disorders being listed as an Axis II disorder in the *DSM-IV-TR* gives some people the false belief that personality might be less biologically driven than the Axis I disorders. Interestingly, the other diagnosis on Axis II within the *DSM-IV-TR* along with the personality disorders is mental retardation (now known as intellectual disability), a brain-based disorder characterized by intellectual deficits. This distinction has been eliminated in the *DSM-5* and now, with the elimination of the axis approach, personality disorders are classified along with the other disorders without a specific axis.

To understand the biological underpinnings of personality disorders, the mental health practitioner must conceptualize behaviors and thoughts as a core neurobiological component of personality traits. The challenge lies in personalities being thought to be inherent within an individual and potentially more difficult to alter through treatment than are other conditions. However, treatments have been found to be effective, and behaviors associated with personality disorders can in fact be improved upon.

Throughout the text, the link between behaviors and cognitions with brain-based functioning has been clearly indicated in a variety of symptoms, diseases, and disorders. This last chapter provides evidence of how personality is also, at least partially, explained neurobiologically. It is clear that experience plays a role in developing one's personality, and this chapter is not suggesting otherwise. Rather, the goals of this chapter are to examine the evidence of the neurobiologically underpinnings of personality,

## SPECIAL TOPICS

### Psychopathy, Neuroscience, and the Law

The rise of technology has led to the use of neuroscience in the legal setting. Books have been written describing violent behaviors in personality disorders, and, likewise, expert testimony from various fields of mental health, including psychiatrists and psychologists, has used neuroscience data to explain violent behavior for years. With the advancement of technology, a more recent trend has focused on explaining violent criminal behavior in an attempt to decrease the punishment of the convicted criminal. This has been used in several capacities, but perhaps the most controversial capacity is in homicide cases.

Neuroscientist James Fallon has done studies examining the basis of psychopathic behavior. He has reported that violent psychopaths have consistent characteristics, including the monoamine oxidase (MAOA) gene—also known as the warrior gene—a history of being abused, and decreased orbital frontal activity on neuroimaging. The warrior gene regulates serotonin in the brain, whereas the orbital frontal cortex is important for impulse control and decision making (see Figure 14.1). This combination of factors significantly increases the risk for violent behaviors. These findings have been used to reduce the sentences of criminals within the courts. Further studies have shown that diffuse tensor imaging has been used in over a thousand cases since 2010. These cases attempt to understand how white matter tracks within the brain might be responsible for the behaviors the criminal demonstrates. It has also been clearly demonstrated that overall cortical thinning is linked to psychopathy (Blair, 2012). This begs the question of whether human behavior is hardwired, and, if so, whether violent behaviors should be punished less severely due to the inability of the individual to prevent his or her violent tendencies.

On the other hand, some are concerned that explaining behaviors through merely biological factors could lead

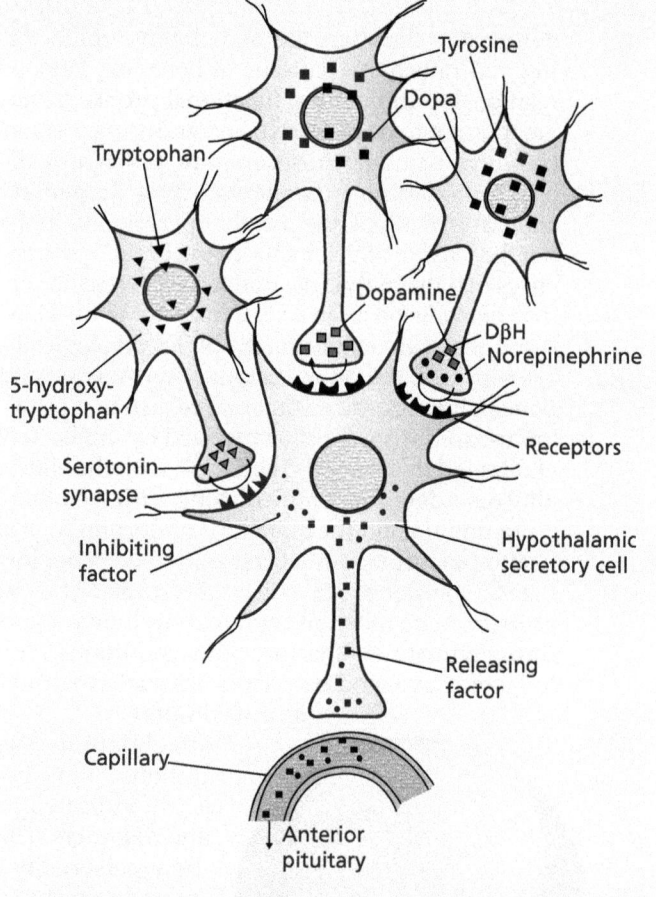

FIGURE 14.1 MAO inhibitor.

*(continued)*

*(continued)*

to a deterministic understanding of human functioning. This possibility eliminates, or at the very least significantly reduces, the role of the concept of *free will*. As humans, we prefer to believe that we have the power and authority to determine right from wrong and to act appropriately in response to this understanding. It can be asked why some individuals who have met all risk factors for committing a violent crime choose not to do so. The common, potentially simplistic answer is *free will*. The concept of *free will* implies that a person chooses to act or not act upon innate internal impulses. Although this might be theoretically correct, it might overlook the role of neurobiology itself. Free will might be another function of the prefrontal cortex that is responsible for higher-order cognitive and emotional functioning. It fact, studies have shown that individuals with frontal lobe damage are more likely to be impulsive, emotionally unstable, and violent. Is it then possible that abnormalities occurring at the neuronal, chemical, electrical, or other microscopic level might in fact account for an inability to decipher and/or act upon knowing right from wrong? If that is so, how should criminal intent be understood, and how should punishment occur? Unfortunately, we are just on the verge of further understanding these behaviors from a scientific perspective. Even if there is a clear 100% certainty that a biological explanation can account for criminal activity, which would be highly unlikely in all scenarios, nobody would go so far to state that criminals should be excused of their misconduct. Rather, remediation, treatment, and the services provided for the individuals who are incarcerated might alter. Also, the length and type of sentencing might be called into question. For example, though an individual who committed homicide might not be punishable by death, after serving a prison term, can the person in fact stop his or her innate biological tendencies to harm again after meeting the conditions of parole? It seems that a treatment-based, rather than time-based, approach to help determine such a possibility is likely to be effective in criminal sentencing.

The interaction of the legal system, neuroscience, and behavior will continue to grow. The burgeoning technological tools to explain behavior adds a new, complex dimension to the legal system such that people have a hard time determining how to use the information. Although how our knowledge of biology of behavior can be used from a legal perspective will be an ongoing debate, it will assuredly continue to be used. The skilled mental health professional can help educate the public, including individuals within the legal system, concerning the underpinnings of behaviors.

help link experiences with personality, and explore the neurobiological evidence concerning two major personality disorders. Examining the evidence, including cognitive studies, imaging studies, and even genetic studies, will help the clinician conceptualize personality disorders as brain-based disorders. When considered in this way, the rationale for treatment and understanding of the effectiveness of treatment becomes clearer. The two specific personality disorders examined in this final chapter are antisocial personality disorder and borderline personality disorder. The overlap of personality disorders, including obsessive–compulsive personality disorder (OCPD), avoidant personality disorder, and schizotypal personality disorders, with other psychiatric disorders, makes examination of these disorders challenging and limited. As a result, the neurobiological aspects and research in these areas, though explored, are not as clear. Furthermore, the research regarding OCPD is intertwined with obsessive–compulsive disorder (OCD). Similarly, schizotypal personality disorder is linked with the psychotic disorders. Avoidant personality disorder is difficult to study because of the inherent characteristics of the individual to avoid situations allowing for further research and study. Regardless, there is no doubt that neurobiological aspects underlie all of the personality disorders, and future studies will continue to research these personality traits.

Although borderline personality disorder and antisocial personality disorders have high comorbidity with other disorders, enough research has been completed to provide a reasonable neurobiological basis for these disorders. Though antisocial personality disorder and borderline personality disorder do not represent all of the personality disorders, they represent a good portion of the most diagnosed personality disorders often treated by the clinician (Grant et al., 2004). Having a biological understanding of these two disorders can help provide a conceptualization that will help the clinician guide treatment. It is difficult for the clinician, other treatment providers, and family to consider personality as being a neurobiological-based construct. Instead, the maladaptive behaviors demonstrated by individuals with personality disorders can be frustrating, as there appears to be a lack of willingness to comply with social norms and expectations. This chapter aims to elaborate on the neurobiological predisposition of seemingly controllable maladaptive behaviors that occur in people with personality disorders. Within this framework, the special topics section of this chapter examines psychopathy, its association with criminal behavior, and an attempt to understand the neurobiology of the behavior.

## ANTISOCIAL PERSONALITY DISORDER

Antisocial personality disorder (ASPD) is a behavior pattern associated with a complete disregard for the rights of other people. Individuals with ASPD will also violate the rights of other people for personal gain. ASPD is associated with unlawful behaviors, manipulation, and deceit—irresponsible and aggressive behaviors. The individual who has been diagnosed with ASPD shows no remorse for his or her behaviors and has little or no empathy for others. These individuals are able to make relationships, but they are unable to sustain relationships for a long time. They do not have the ability to learn through punishment, and they have limited capacity to gain from experiences. ASPD has been associated with lower socioeconomic status, particularly in urban areas, with males, and with substance abuse (Lykken, 1995; Moeller & Dougherty, 2006). As discussed in the childhood disorder chapter (Chapter 5) of this book, childhood maltreatment is a significant risk factor for antisocial behavior.

Criminal behavior is highly associated with ASPD, with some estimates that nearly 50% of male prisoners and approximately 25% of female prisoners meet criteria for ASPD (Fazel & Danish, 2002). In fact, psychopathy and ASPD are often used interchangeably in much published research (Fitzgerald & Demakis, 2007), although there is ongoing debate as to the accuracy of such an approach. Further evaluation regarding the complex interaction of psychopathy, neuroscience, and the legal system is discussed in the Special Topics section of this chapter. Regardless of whether ASPD is associated with psychopathy, it is clear that the identification of ASPD is becoming increasingly important within our society, given the apparent rise of violence and criminal behavior. Mental health clinicians who work in prisons or other legal-based environments are likely to work with individuals who are diagnosed with or who demonstrate characteristics of ASPD.

## Genetics and ASPD

Genetic factors playing a role in ASPD have been examined extensively in recent years. Studies have shown that there appear to be genetic risk factors that play a role in behavioral characteristics associated with the disorder. A recent meta-analytic review determined that 56% of the variance in ASPD is explained by genetic variables, in comparison to 11% due to shared nongenetic influences and 31% due to nongenetic influences (Ferguson, 2010). The combination of alterations in specific dopamine gene receptors responsible for novelty-seeking and impulsivity, as well as alterations in proteins responsible for effective synaptic transmissions, have been found to increase the likelihood of ASPD (Basoglu et al., 2011). A deletion or polymorphism on serotonin transporter receptor gene 5-HTT predicted violent behavior thought to be associated with ASPD (Retz, Retz-Junginger, Supprian, Thome, & Rossler, 2004). Multiple studies have implicated the MAOA gene, which is located on the X-chromosome. Low MAOA gene activity in male genotype and exposure to maltreatment significantly increase behaviors associated with ASPD, in comparison with high MAOA gene activity and exposure to maltreatment (Caspi et al., 2002; Kim-Cohen et al., 2006; Nilsson et al., 2007). It has long been known that the MAOA gene metabolizes norepinephrine, serotonin, and dopamine, neurochemicals associated with mood, aggression, and motor functioning. Aggressive behaviors are known to occur in humans and animals when there is MAOA deficiency (Bruner, Nelen, Breakefield, Ropers, & Van Oost, 1993; Cases et al., 1995).

The results regarding neurobiological activity and MAOA involvement in aggressive behavior have led to implications in the legal system when determining the extent of punishment for violent behavior. The Special Topics section discusses this in more detail. Although there are clearly genetic factors that contribute to ASPD, there is no single gene that determines ASPD. Rather, it is the influence of multiple factors and potentially the interaction of these factors, such as genetic factors and childhood maltreatment, that play a role in determining the presence of ASPD (Ferguson, 2010).

## Neuroimaging and ASPD

A neurobiological explanation of ASPD is not a new concept (Moffitt, 1993; Raine, 1995). These early studies hypothesized an abnormal neurodevelopmental basis to the cluster of symptoms. Neuroimaging techniques have supported these findings. Numerous studies have since demonstrated limbic system abnormalities. Within the limbic system, also known as the emotional regulation system, the amygdala, thalamus, hippocampus, and hypothalamus have all been demonstrated to have abnormal structure and function in ASPD (George et al., 2004) For example, atypical hippocampal asymmetry has been found in individuals who have either successfully or unsuccessfully committed violent crimes (Raine et al., 2003; Raine et al, 2004). It is thought this abnormality might account for hippocampal-prefrontal circuitry disruption. As a result, there is dysregulation of affect, limited or no learning from fear-based contextual conditioning, and insensitivity to identifying social cues. As a result, the individual is more likely to act upon violent impulsive urges with no remorse. Similarly, ASPD has been found to have attenuated thalamic-striatal activity during threatening situations. Interestingly, this result is the direct opposite finding of increased activity in the same area of the brain in violent patients diagnosed with schizophrenia (Kumari et al., 2009). Though violence is a link, the underlying neuropathology appears to involve the same pathways but in different capacities.

The anterior cingulate system, important for cognitive flexibility, decision-making, and emotional processing, among other behavioral and cognitive functions, has been found to be implicated in ASPD as well (Davidson, Putnam, & Larson, 2000; Raine & Yang, 2006). Thus, the individual with ASPD might have limited capacity to feel emotions, make correct decisions, and learn from previous experiences. Furthermore, abnormalities have been documented in the orbitofrontal cortex, which is important for adapting to reinforcement and maintaining reactive emotional responses (De Oliveira-Souza et al., 2008). Because of these things, individuals with ASPD have a difficult time learning from social experiences and implementing more appropriate and adaptive social functioning. Brain abnormalities in the orbitofrontal cortex might result in reactive aggression and social avoidance.

## Cognition and ASPD

It is clear that, by definition, individuals who have been diagnosed with ASPD make poor decisions. For example, repeated lying, failure to plan ahead, consistent irresponsibility, and unlawful behaviors are all core features of ASPD. Cognitive-based research with ASPD indicates that decision making is one of several areas of cognitive limitations consistently identified. Cognitive studies are consistent with the imaging studies in terms of identified areas of abnormalities. One example of a limitation that has been found in ASPD is poorer performance on the Stroop task (Mercer, Selby, & McClung, 2005). The Stroop task requires the inhibition of overlearned behavior in favor of a novel response. The individual performing the Stroop task must inhibit his or her initial desire to read the name of a color (such as red) printed in different color ink (such as blue) and instead read the color of the ink. Problems have also been demonstrated on other measures used to assess executive functioning, including the Wisconsin Card Sorting Task, a problem-solving and flexibility measure; the Trail Making Test, another measure of cognitive flexibility; and the Category Test, still another problem-solving and cognitive flexibility task (Blair et al., 2006; Dinn & Harris, 2000). Selective attention has also been demonstrated to be impaired in ASPD (Hiatt, Schmitt, & Newman, 2004). Individuals who have been diagnosed with ASPD have also been demonstrated to have impairment on a computerized measure of planning and set-shifting (Dolan & Park, 2002). These cognitive limitations are consistent not only with the behavioral manifestation of the disorder (not learning from social cues, poor capacity to resolve social problems) but also with previously described imaging findings. Individuals who have been diagnosed with ASPD have clear cognitive limitations that are implicated in their poor decisions, lack of ability to learn from social norms, and more effective problem-solving strategies. These cognitive limitations can lead to challenges for the treating clinician.

## Treatment for ASPD

Not surprisingly, individuals who have been diagnosed with ASPD are not likely to seek treatment unless they are under court order or there is high involvement from family members. People with ASPD typically do not admit or believe that they have a need for help. Though treatment for these individuals can be challenging due to this complicating factor, often there might be a co-occurring disorder accompanying the ASPD. Treatment might focus on improving the symptoms from the other disorder, including mood or substance use. Medications aimed at decreasing aggression have been found to be effective in some individuals who have been diagnosed with ASPD. Mood stabilizers can be used to decrease impulsivity. However, it is strongly recommended that medications that have abusive characteristics, such as psychostimulants, be avoided. It is also important to point out that no medications have been found efficacious at treating ASPD.

Though individual cognitive behavioral treatment strategies can be beneficial, some patients benefit more from group therapy. However, evidence remains unclear concerning the clinical utility of psychotherapy of any sort for ASPD. The U.K. National Institute for Health and Clinical Excellence (NICE) provides a thorough examination of treatment for ASPD. The thorough examination of the literature conducted by NICE indicated that three review studies conducted between 2003 and 2007 failed to demonstrate any high-quality evidence for the effectiveness of treatment for ASPD. In comparison, treatment for criminal offenders has been moderately effective, although it remains unclear whether similar treatment approaches would in fact be effective for ASPD. In fact, counter to intuitive belief, approaches to treating specific symptoms such as anger management do not appear to be an effective therapeutic strategy for ASPD (Duggan, Mason, Banerjee, & Milton, 2007). This might be due to the underlying neurobiological predisposition of rejecting societal assistance and rules.

Overall, psychotherapeutic interventions and medication management for ASPD have proven limited. Rather, prevention strategies and early intervention and treatment of conduct disorder are likely to be the most effective form of treatment. It is possible that future treatments must become more invasive, such as deep brain stimulation, for effective strategies to use with ASPD to be developed. However, there is no literature at the present examining this possibility.

## BORDERLINE PERSONALITY DISORDER

Borderline personality disorder (BPD) is characterized by unstable personal relationships, self-identity issues, impulsivity, and affective lability. Interpersonal features of feelings of abandonment and dependency characterize the disorder. The individual who has been diagnosed with BPD demonstrates impulsive tendencies and demonstrates recurrent suicidal threats or gestures, such as self-mutilation. BPD patients have difficulty maintaining appropriate boundaries (see the Ethics box).

Although BPD can start as early as childhood, typically the symptoms are not present until late adolescence or early adulthood (Gunderson, 2009; Zelkowitz, Paris, Guzder, & Feldman, 2001). Approximately 1.6% of adults have BPD at any given time in the United States (Lenzenweger, Lang, Loranger, Kessler, 2007). BPD is associated with clinical instability over time, more than the other personality disorders are. There is a comorbidity of up to 85% with an Axis I disorder, with a mood disorder and posttraumatic stress disorder (PTSD) being the most likely co-occurring conditions (Lenzenweger et al., 2007). The mortality rate from suicide is 10%, 50 times that of the general population (Oldham, 2006). Individuals who have been diagnosed with BPD have a very difficult time maintaining healthy, consistent relationships. Their behavior and difficulty in developing healthy relationships contribute to interpersonal and psychosocial conflicts.

Risk factors for BPD include inherited and environmental influences. Although it is commonly viewed as being more prevalent in women, there is no evidence that BPD actually occurs more commonly in women (Skodol et al., 2005). It is possible that, instead, women are more likely to seek treatment and to therefore be diagnosed with BPD. Environmental risk factors do include unstable family environment,

# ETHICS

## Dual-Role and Relationships with BPD

Considering the potential for misinterpretation of relationships, individuals who have been diagnosed with personality disorders might be likely to attempt to engage a mental health provider in a capacity that is not appropriate for treatment. Inherent symptoms of BPD are social relationship difficulties, pushing boundaries, and splitting relationships. Thus, the clinician might have to cope with the ethical conflict of a dual relationship. Although the psychology trainee and advanced professional might scoff at the notion of breaking this ethical boundary by saying, "I would never do that," dual relationships can occur in a variety of manners, and they are inconsistently addressed in the existing ethical guidelines. The BPD client might misinterpret the meaning of therapy beyond the client–therapist goals of treatment. The patient might view an empathetic, caring therapist either as a friend or as somebody with whom to potentially develop a romantic relationship, considering that the therapist typically has unconditional positive regard for the client, despite the client's flawed and misdirected behavior. At times, the therapist experiences a reciprocal positive emotion with his or her patients. In light of the inherent position of power and authority, regardless of the attempts of the clinician to equalize the power in the relationship, the therapist must exert good judgment and restraint to develop a relationship outside of therapy. Therefore, guidelines have been established to help the clinician.

One area of a dual relationship that would seem to be a boundary violation would be engaging in sexual behavior with a past patient. However, even that situation is not clear cut across disciplines. For example, though the ethics code for the American Psychiatric Association does not have a time frame during which psychiatrists should refrain from participating in romantic relationships with their former patients, the American Psychological Association code of ethics clearly states that at least 2 years must pass from the termination of a therapeutic relationship and that only under the "most unusual circumstances" should psychologists engage in sexual behavior with former clients even after that point (American Psychological Association, 2010).

Although the conduct of a mental health worker who engages in a relationship with a current patient, student, or supervisee seems to be an obvious violation of trust and moral behavior, the ethical guideline is listed for a reason. Specifically, it is a code that has not been followed and that continues to be broken all too frequently. This code is frequently identified as one of the more likely ethical breaches of confidence. The relationships that are built in therapy can become emotionally intimate, and the secrets that are shared in a safe therapeutic environment might create the misunderstanding from the patient's perspective that the therapist is more than just a therapist. It is up to the mental health professional to maintain appropriate professional boundaries and to use good judgment not to exploit an individual who has likely been exploited in the past. The therapist must demonstrate good role-modeling behavior in being able to develop a good relationship that does not cross boundaries beyond the therapeutic relationship.

Although there is no current neuroscience research used to identify those patients (or therapists) who might be at risk for attempting to breach the code of dual relationship, advancements in technology and brain understanding might mean that such neuroscience means might not be far off. If so, more efforts might be made to ensure that the patient is provided safe and appropriate treatment with a skilled therapist.

limited social support, and early childhood abuse or neglect. The National Institute of Mental Health (NIMH) reports that adults who have been diagnosed with BPD are at a higher risk of being victims of violent crimes, including rape.

## Genetics and BPD

There is clear evidence that genetic factors contribute to the expression of BPD. A review conducted as recently as 2007 reported that although the study of genes in BPD is in its early stages, the symptoms of impulsiveness and aggression have been shown to be highly heritable (Lis, Greenfield, Henry, Guilé, & Dougherty, 2007). Furthermore, older twin studies examining BPD indicated that there is a strong genetic influence (Coolidge, Thede, & Jang, 2001; Torgersen, Kringlen, & Cramer, 2001). Serotonin transporter genes have been demonstrated to play a crucial role in the expression of BPD. In fact, more recent genetic studies have reported the presence of a short allele on a serotonin receptor (5-HT) that has been linked to the development of impulsivity in significant life events (Wagner et al., 2009). This genetic abnormality has previously been linked with impulsive and self-destructive behavior in monkeys when they are removed from their mothers and raised in nonnurturing environments (Gabbard, 2005). In humans, studies examining eating disorders have linked the short allele on the 5-HT with individuals who have co-occurring BPD. Furthermore, this short allele has also been demonstrated to be correlated with limited therapeutic effectiveness of Prozac for resolution of depressive symptoms.

Much as with ASPD, the MAOA gene has been found to potentially play a role in BPD. Studies have indicated that the MAOA gene might be protective against a negative outcome when a patient had been subject to childhood abuse. This gene has a role in aggression and impulsivity. Studies have demonstrated that dopamine dysfunction and the genes that interact to help metabolize dopamine play a role in BPD as well. Dopamine is important for emotional processing and for impulsivity and motor control. Compromise of dopamine can influence limbic system dysregulation associated with social situations. If a gene that influences dopamine is altered, that would increase the likelihood for difficulty modulating emotional response. The literature indicates that individuals with BPD have specific genetic risk factors that play a role in impulsivity and aggression.

## Neuroimaging and BPD

Studies examining BPD have been inconsistent in terms of findings. Early CT studies did not reveal statistically significant differences between BPD and nonpsychiatric volunteer groups. It is likely that the amygdala has a role in the behavior manifestation of the BPD, yet the findings are inconsistent. The amygdala is known to be important in mediating social and emotional responses. The dysregulation of this structure could result in difficulty navigating the complexities in social interaction and handling the emotional challenges involved in relationships. Many studies have revealed reduced volumetric gray matter density of the amygdala (Driessen et al., 2000; Tebartz van Elst, Hesslinger, & Thiel, 2003). However, a more recent MRI study revealed that the amygdala can in fact be larger (Minzenberg, Fan, New, Tang, & Siever, 2008). The abnormal amygdala volume can be associated with emotional instability and self-identity problems. Minsenberg et al. (2008) also found that the cingulate system has a smaller amount of gray matter and decreased activation on functional magnetic resonance imaging (fMRI) studies in this population. The hippocampus and left orbital frontal cortex have also been shown to have a reduction of gray volume and decreased activation in functional studies. Furthermore, functional imaging studies support dysregulation of the frontal and temporal regions as well as of the amygdala (Donegan et al., 2003; Driessen et al., 2004). fMRI studies have revealed that individuals who have been diagnosed with BPD have decreased activation in the amygdala bilaterally, suggestive of decreased activation of emotional response to social situations.

There has been suggestion of changes in the prefrontal cortex accounting for personality issues. Positron emission tomography (PET) studies have documented changes in glucose metabolism in the frontal regions. Furthermore, orbital frontal cortex has been demonstrated to be more activated in BPD patients than in normal controls who observe aggressive behavior. Additionally, increased amygdala activation is the most consistent fMRI finding when viewing an aversive, emotion-inducing picture (Herpertz et al., 2001). As discussed previously, the amygdala is important for processing emotions and anxiety. This likely suggests that emotional control dysfunction is occurring in BPD. The anterior cingulate has also been shown to have less activation in BPD patients asked to use CBT strategies to cope with a negative emotional situation. This might account for the difficulty that BPD patients experience when implementing strategies to change their negative emotional responses. The inability to be flexible in emotional processing makes for the long-standing emotional behavioral symptoms. Although the hypothalamic–pituitary–adrenal axis was initially thought to play a role in the expression of the disorder, recent studies have more clearly indicated that the involvement of the HPA axis is limited to comorbid PTSD. Taken together, there is a clear frontal–limbic–anterior cingulated circuitry that plays a role in the expression of the disorder.

The imaging data support the Jacksonian biopsychosocial model of BPD (Meares, Stevenson, & Gordon, 1999). This model asserts that neurobiological dysregulation in the connections between the prefrontal cortex and limbic structures cause identity disturbance, affective instability, dissociation, and somatic symptoms that occur in BPD. Thus, executive functioning impairment would also be explained by this dysregulation. Although the Jacksonian model provides a sound theoretical understanding of BPD, other factors, including genetic and cognitive functioning, also contribute to the expression of the disorder.

## Neurocognition and BPD

There have been several studies examining the cognitive functioning in individuals with BPD. Early studies examining BPD revealed intact performance on intelligence measures (Carr, Goldstein, Hunt, & Kernberg, 1979). However, some theorists postulated that people diagnosed with BPD experienced problems separating the essential visual information they were experiencing from the extraneous information presented to them (O'Leary, Brouwers, Gardner, & Cowdry, 1991). These researchers documented evidence that on objective measures examining complex visual memory and auditory learning, individuals who had been diagnosed with BPD demonstrated significant deficits in comparison to a healthy group of individuals. Interestingly, in the same study on simple visual memory and auditory memory tasks, there were no differences. These results might account for the patient's difficulty remembering past experiences, including content of previous therapy sessions. More recent results have revealed cognitive declines in the areas of attention and memory, with intact, mostly language and visuospatial abilities (Seres, Unoka, Bodi, Aspán, & Keri, 2009). Others have replicated the findings that BPD patients having impaired performance on nonverbal memory tests, visuospatial tasks, and on tasks that measure visual decision-making (Bazanis et al., 2002). A meta-analysis conducted on 10 studies documented that individuals who have been diagnosed with BPD perform more poorly on a variety of cognitive measures, including learning, memory, visuospatial abilities, and planning (Ruocco, 2005).

Neurocognitive deficits can also be associated with suicide attempts in BPD. For example, a review of literature indicated that executive functioning deficits that include visual inhibition, visual decision making, visual organization, and verbal fluency, as well as memory, were linked with suicide attempts. In comparison, performance on measures examining working memory abilities, IQ, and planning were not as affected in BPD, and were less linked with suicidality. Impairment on the Wisconsin Card Sorting Test and Stroop Color Word Test were most frequently found in BPD (LeGris & van Reekum, 2006). These cognitive abilities are thought to be modulated by the dorsolateral prefrontal cortex and orbital frontal pathways. As such, impairment in these areas on cognitive tests can assist the clinician working with BPD in identifying who is at higher risk for self-harm.

## Treatment for BPD

Individuals with BPD are much more likely to seek treatment for their difficulty than are people with ASPD. Thus, research and treatment for the disorder have substantially improved, and there is robust evidence that much can be accomplished in treatment to reduce behavioral difficulty and decrease stress. Psychological treatment for BPD has proven to be much more promising than has psychopharmacological treatment for BPD (Silk, 2008). In fact, there have been more randomized clinical trials for psychotherapeutic treatment for BPD than for any other disorder (Duggan et al.). Various forms of cognitive behavior therapy have proven to be an effective treatment for BPD in several studies. The reader is encouraged to examine other texts for specific therapeutic interventions and strategies for working with BPD. Briefly, one evidence-based approach and technique found particularly effective for BPD is Marsha Linehan's dialectical behavior therapy (DBT). DBT aims to increase mindfulness and attention to the current situation in an attempt to control emotions and improve relationship abilities. It uses both individual patient therapy and group therapy to identify difficulties, practice more effective strategies, and improve social skill functioning. The therapist aims to help the patient identify areas of stress and develop new skills to cope with stress. While simultaneously validating the client's experiences and emotions, the therapist also aims to assist in developing positive change in the patient's life. Specific strategies are implemented to reach the goal of helping the client recognize the connection among various events, emotions, and behaviors in order to help allow for positive change. Although DBT is not the only means of effectively treating BPD, it has consistently proven to be effective, so much so that it is now being used to treat other disorders as well.

Numerous articles have documented the effectiveness of DBT, CBT, and other forms of treatment for BPD. Several books and book chapters have given guidance to professionals on how to administer treatment for BPD. Similarly, books have been written to assist families with loved ones who have BPD, as the behaviors can result in challenging relationships with loved ones. The clinician is encouraged to become familiar with available resources, particularly if he or she will be working with adult populations, considering the likelihood of working with an individual with BPD and the importance of behavioral strategies.

No medications have been approved by the U.S. Food and Drug Administration for treating BPD. However, psychotropic medications are often prescribed for this population in an attempt to control the symptoms associated with the disorder or co-occurring diagnosis. It is not uncommon for antidepressants to be prescribed in an attempt to improve mood, decrease impulsivity, and control aggression. Antianxiety and antipsychotic medications can also be prescribed in an attempt to decrease symptoms. A recent meta-analysis indicated that mood stabilizers, with the exception of Depakote, demonstrated large-effect size for decreasing anger in BPD, whereas antipsychotic medications had moderate effect sizes in decreasing anger (Mercer, Douglass, & Links, 2009). Clozapine has been found to be effective for decreasing self-mutilation in dual diagnosis of psychosis and BPD (Chengappa, Ebling, Kang, Levine, & Parepally, 1999). However, replication of studies is difficult due to comorbidity, length and dose of treatment, and variability in outcome measures used.

## CONCLUSIONS

Personality disorders are no longer characterized in the *DSM-5* on a separate axis distinct from other psychiatric disorders. Rather, its biological roots are as likely to contribute to behavioral functioning as much as to other disorders. The two best-understood personality disorders from a biological perspective are ASPD and BPD. Interestingly, these disorders share similar clinical and biological features. Treatment aimed to improve certain areas of interpersonal functioning is necessary.

## SUMMARY POINTS

- Personality disorders are highly prevalent psychiatric disorders that are now recognized to be in the same vein as the other disorders listed in the *DSM-5*.
- There are several personality disorders, but the two best-understood from a biological perspective are BPD and ASPD.
- A variety of therapeutic techniques have been attempted as treatment for people who have been diagnosed with personality disorder, although no medication has been consistently proven to be effective.
- The mental health clinician should conceptualize the behaviors from people with personality disorders as being similar to other disorders as they relate to the underlying neurobiological components playing a role.

# Glossary

**ablation**—one of the oldest scientific techniques used to examine brain-based behavior, where in a portion of the brain is destroyed or removed and subsequent behavior is observed.

**absence seizure**—type of silent seizure associated with a form of epilepsy, often found in children, characterized by staring spells, and by brief, altered state of consciousness. Also referred to as *petit mal seizure*.

**acalculia**—reduction in the ability to perform mathematics/arithmetic.

**ACC**—see **anterior cingulate cortex (ACC)**.

**acetylcholine**—neurochemical important for learning and memory.

**acoustic nerve**—eighth cranial nerve; it assists in the transmission of auditory information, as well as sense of balance, to the brain.

**ACT**—acceptance and commitment therapy.

**ACTH**—see **adrenocorticotropic hormone (ACTH)**.

**action potential**—abrupt change in polarization when a stimulus exceeds the threshold of an axon. Occurs when sodium enters the neuron.

**addiction**—use of drugs despite adverse effects and withdrawal symptoms.

**Addison's disease**—condition resulting from a deficiency in secretion of adrenal hormones, the symptoms of which include weight loss, eating disorder, weakness, and fatigue.

**adenosine triphosphate (ATP)**—extracellular molecule that plays a role in numerous functions of the sensory system.

**ADHD**—see **attention deficit hyperactivity (ADHD)**.

**adrenal gland**—endocrine gland that is located on top of the kidneys, and that is responsible for releasing hormones during stressful situations.

**adrenaline**—neurochemical that is produced during stressful situations.

**adrenocorticotropic hormone (ACTH)**—neurochemical that stimulates the adrenal gland. Plays a role in the production of cortisol.

**afferent nerve cells**—neuronal pathways that transmit information toward the cerebral cortex from the peripheral nervous system.

**agenesis**—complete or partial absence.

**agnosia**—inability to recognize sensory stimuli.

**agoraphobia**—anxiety or intense fear of being in situations from which it is hard to escape.

**agraphia**—a condition of deficiency in, or lack of, writing ability.

**aha moments**—events of great relevance that transform a science.

**akathisia**—extreme motor restlessness. This is accompanied by subjective feelings of anxiety and restlessness.

**akinesia**—decreased motor activity.

**alcohol abuse**—condition in which an individual has increased tolerance, withdrawal symptoms, and the need to increase the use of alcohol to obtain a physiological response to exposure to alcohol.

**alexia**—the inability to read.

**allele**—alternative form of the same gene.

**alpha waves**—electrical brain activity and neuronal discharge, frequently associated with a relaxed state.

**Alzheimer's disease**—progressive dementia characterized by neurofibrillary tangle and senile plaque resulting in continued cognitive decline that is a decrease from previous abilities and that leads to functional impairment.

**amenorrhea**—absence of menstrual bleeding.

**amnesia**—partial or total impairment of memory functions. It can be for memory immediately before or after a trauma.

**amphetamine**—stimulant medication aimed to increase arousal of inhibitory neurons.

**amygdala**—limbic system structure that is important for the regulation of mood and for the recognition of others' emotion and that is involved with emotional memory.

**amyloid plaques**—protein deposits that occur in Alzheimer's disease.

**amyotrophic lateral sclerosis (ALS)**—deterioration of motor neurons of the central nervous system that results in muscle weakness and lack of motor ability. Also known as *Lou Gehrig's disease*.

**aneurysm**—weakening of a vein or artery wall that dilates and fills with blood, resulting in destruction of surrounding neural tissue.

**Angelman syndrome**—genetic disorder that is characterized by severe language impairment, developmental delays, small head, distinctive facial features, ataxia, and intellectual deterioration.

**anomia**—language impairment in which the patient has difficulty finding correct words or names.

**anorexia nervosa**—eating disorder characterized by body image distortion, significant low body mass index, and restrictive behaviors, including purging, laxative use, and excessive exercise.

**anosmia**—the inability to smell.

**anosognosia**—unawareness of existing deficits despite objective evidence that deficits exist.

**anoxia**—absence of oxygen.

**Antabuse**—medication used to treat alcohol-related problems, causing a person to experience unpleasant effects with the consumption of alcohol.

**anterior cerebral artery**—artery that originates from the internal carotid artery and that is found in the frontal lobes and corpus callosum and in the olfactory and optic tracts.

**anterior cingulate cortex (ACC)**—subcortical brain system that is involved in numerous cognitive and emotional functions, including cognitive flexibility. Disruption of the ACC has been found in many psychiatric disorders, including OCD, anorexia, and hoarding.

**anterior communicating artery**—artery that originates from the anterior cerebral artery and that is found in the caudate nucleus and in the anterior part of the Circle of Willis.

**anterograde amnesia**—inability to form new memories, often occurring subsequent to brain trauma.

**antianxialegic medication**—classification of drugs aimed to rapidly decrease symptoms of anxiety. Long-term use can lead to cognitive dullness and withdrawal.

**anticholinergic medications**—classification of drugs that interfere with passage of nerve impulses through the parasympathetic nerves, aimed at controlling cholinergic regulation.

**antidepressant medications**—a classification of drugs that aim to regulate serotonin; first line of medication treatment for depression and anxiety.

**antipsychotic medications**—classification of drugs that aim to control dopamine, among other neurotransmitters; used to treat psychotic symptoms, including hallucinations and delusion, and referred to as narcoleptics.

**antisocial personality disorder**—psychiatric disorder in which a person has long-standing characteristics of manipulating, exploiting, and violating other people's rights.

**anxiety**—state of increased stress that results in cognitive, physical, and emotional symptoms.

**apathy**—presentation of indifference, lack of concern, or lack of emotional responsiveness.

**aphasia**—broad term used to describe the lack of ability to use certain aspects of language. It can be either an expressive or a receptive language impairment.

**ApoE gene**—apolipoprotein, which is mapped on chromosome 19. Multiple ApoE genes increase the risk for Alzheimer's disease.

**apraxia**—broad term used to describe impaired ability to perform previously learned skills or behavior.

**aprosody**—condition in which the coloring, rhythm, melody, cadence, intonation, or emphasis of speech is impaired.

**aqueduct of Sylvius**—narrow canal that connects the third and fourth ventricles.

**arachnoid layer**—middle layer of the meninges, which protects the brain from the skull.

**arachnoid space**—space around the arachnoid layer that is filled with fibrous tissue and that acts as a conduit for cerebrospinal fluid.

**articulation therapy**—language-based treatment to improve upon articulation by teaching sounds in isolation.

**Asperger's disorder**—previously considered a pervasive developmental disorder; a type of autism spectrum disorder that does not result in the decline of language abilities, or intellectual abilities, but that does include social skills deficits.

**astrocytoma**—type of neoplasm that develops from astrocyte cells. These tumors are typically intracerebral.

**ataxia**—loss or failure of muscular coordination in the absence of motor impairment. Movement is clumsy. Difficulty with walking increases greatly when the patient is asked to walk with eyes closed.

**atomoxetine**—selective norepinephrine inhibitor that is used to treat ADHD; brand name is Strattera.

**atonia**—complete lack of muscle tone.

**atonic seizures**—also known as *drop attacks*; there is a brief loss of muscle tone.

**ATP**—see **adenosine triphosphate (ATP)**.

**atrophy**—shrinkage of (brain) tissue due to loss of neuronal processes.

**attention**—capacity of an individual to screen out certain aspects of the environment to perceive and process other aspects.

**attention deficit hyperactivity disorder (ADHD)**—characterized by inattention and executive dysfunction.

**auditory nerve**—sometimes called the *vestibular* or *acoustic nerve*. It is the eighth cranial nerve, and it transmits auditory information and is also involved in the sense of equilibrium. (See also **acoustic nerve**.)

**aura**—sensory phenomenon or feeling that can precede a seizure.

**autism spectrum disorder (ASD)**—collection of neurodevelopmental disorders that result in social skill limitations, delays in language, and other cognitive difficulties.

**autoimmune disorders**—impairment of bodily processes such that immune functions are affected.

**autonomic nervous system**—that part of the nervous system concerned with visceral and involuntary functions.

**avoidant personality disorder (APD)**—psychiatric disorder in which a person is socially inept or avoids social situations out of fear of being humiliated or disliked.

**axial**—view of the brain from the top, looking down; created from MRI scans.

**axon**—portion of a neuron that transmits energy from the cell body to the receptors of other neurons.

**axon terminal**—last portion of a neuron to receive an electrical impulse, as well as the area where the impulse is changed to a neurochemical signal.

**Babinski response**—extension (instead of flexion) of the toes on stimulation of the sole of the foot, occurring in persons with lesions of the pyramidal tract.

**basal ganglia**—system of the brain, including the striatum, globus palladus, substantia nigra, and nucleus accumbens, that is involved with a number of abilities, including voluntary movement.

**benzodiazepines**—psychiatric medications that affect GABA, often inducing a relaxed or sedative state.

**beta waves**—electrical brain activity and neuronal activity, most active during periods of wakefulness and alertness.

**binding proteins**—proteins that assist in connecting proteins together.

**binge eating disorder**—psychiatric disorder in which an individual excessively overeats in a short period of time, accompanied by a fear of loss of control.

**bipolar disorder I**—formerly referred to as *manic depressive disorder*, in which an individual experiences extreme fluctuations in mood state, causing social, occupational, or academic impairment. Mania is a hallmark feature.

**bipolar disorder II**—similar to bipolar disorder I; however, hypomania is the hallmark feature; it occurs for a shorter duration, and impairment is not as debilitating.

**blast injury**—type of closed-head injury that occurs from direct or indirect exposure to an explosion.

**blastocyst**—collection of cells that form early in embryonic development.

**blood–brain barrier**—separation of blood from the brain in the central nervous system, allowing only some material from the blood to enter the brain.

**body mass index (BMI)**—number (ration) calculated from an individual's weight and height.

**borderline personality disorder**—psychiatric condition in which there is a long-standing instability of mood, pervasive relationship difficulties, impulsive behavior, and poor self-image, among other symptoms.

**BPAP**—bilevel positive airway pressure.

**bradykinesis**—motor disorder, frequently seen in Parkinson's disease, that results from rigidity of muscles. It is manifested by slow finger movements and loss of fine motor skills, such as writing.

**brainstem**—located at the base of the brain connecting the cerebrum with spinal cord.

**British Columbia Postconcussion Symptom Inventory (BC-PSI)**—behavior questionnaire examining symptoms following traumatic brain injury.

**Broca's area**—area of the left-frontal lobe involved in the production of speech.

**bulimia nervosa**—psychiatric disorder in which a person binges on food, then uses compensatory techniques, such as laxatives or vomiting, to prevent weight gain.

**buspirone**—nonbenzodiazepine anxiolytic psychotropic medication; the trade name is Buspar.

**CBT**—see **cognitive behavior therapy (CBT)**.

**caffeine**—alkaloid that functions as a fast-acting stimulant drug.

**cataplexy**—brief, sudden weakness of muscles, often occurring during intense emotional states.

**catecholamines**—set of neurochemicals, including dopamine, norepinephrine, and epinephrine, made by the adrenal glands and excessively released during stressful states.

**caudate nucleus**—part of the brain located in the basal ganglia, important for learning and memory, among other functions.

**CDC**—see **Centers for Disease Control and Prevention (CDC)**.

**central nervous system (CNS)**—collection of neurons that coordinate the activity of all parts of the human body; it comprises the spinal cord and the brain.

**central sleep apnea**—sleep disorder in which an individual has diminished breathing as well as periodic episodes of complete lack of breathing.

**Centers for Disease Control and Prevention (CDC)**—an agency of the U.S. government that serves as the national public health institute. It is in Druid Hills, Georgia.

**cerebellum**—also known as the *little brain*, it is a portion of the back of the brain, under the occipital lobes, with functions including balance, coordination, equilibrium, as well as components of memory and emotion.

**cerebral anoxia**—condition in which the cells of the brain do not receive sufficient oxygen to perform their normal functions.

**cerebral blood flow**—blood flow to the brain at any given time.

**cerebral cortex**—outer part or surface of the brain, referred to as gray matter.

**cerebrovascular accident**—ischemic disorder produced by a disruption of blood flow in the brain caused by dysregulation of a portion of the vascular system. Also known as *CVA*.

**ChEIs**—see **cholinesterase inhibitors (ChEIs)**.

**cholesterol**—fat produced by the liver.

**choline**—essential, water-soluble, alkaline nutrient; a precursor of acetylcholine.

**cholinesterase inhibitors (ChEIs)**—psychotropic medications that delay the breakdown of acetylcholine. Most effective in the early treatment of Alzheimer's disease.

**chorea**—sudden involuntary asymmetric movement that serves no purpose. These movements are known as choreiform, and they are brief in duration.

**chromosome**—DNA molecule in the nucleus of each cell.

**chronic traumatic encephalopathy**—degenerative progressive dementia caused by repetitive low-impact, single high-impact, or episodic blows to the head, strong enough to cause acceleration–deceleration force.

**cingulate**—brain system or set of structures found deep in the brain, which is important for many functions, including cognitive flexibility.

**cingulate gyrus**—portion of the brain involved in regulation of emotion and pain.

**circadian rhythms**—24-hour cycle of cognitive, behavioral, and physical alterations that respond to light and dark.

**clinical case report**—method of learning about brain-based behavior by examining an individual patient.

**clonic seizure**—spasmodic alteration of contraction and relaxation, occurring as a form of epilepsy.

**closed-head injury**—blow to the brain that disrupts normal cognitive, emotional, and behavioral functioning. No brain matter is exposed, and it might or might not cause loss of consciousness.

**clozapine**—atypical antipsychotic psychotropic medication used to treat psychotic symptoms; brand name is Clozaril.

**CNS**—see **central nervous system (CNS)**.

**cocaine**—addictive drug obtained from the leaves of the coca plant. When used the drug results in feelings of euphoria, excitement, and strength, followed by depression, paranoia, and anxiety.

**codeine**—narcotic pain medication, typically used to treat mild to moderate levels of pain.

**cognitive behavior therapy (CBT)**—type of psychotherapeutic technique used to alter thought processes and behaviors in an attempt to improve emotional and psychosocial functioning.

**cognitive flexibility**—ability to shift between various tasks or thought processes.

**cognitive rehabilitation**—type of therapeutic technique used to relearn memory skills and executive functioning and to improve functional abilities.

**cognitive remediation therapy**—form of behavioral therapy often used to treat people who have schizophrenia. It practices specific behaviors and cognitive strategies aimed at improving social functioning.

**cognitive testing**—objective evaluation of one's mental abilities in areas of attention, memory, visual special skills, language, executive functioning, etc.

**coma**—condition of severe stupor or unconsciousness.

**comorbidity**—co-occurring condition.

**complex partial seizure**—typically a temporal lobe–type of epileptic symptom altering consciousness in which a person cannot interact with his or her environment. Typically begins with a blank stare.

**computed tomography (CT)**—a type of imaging technique that utilizes x-ray technology that can examine inside the brain.

**concussion**—form of closed-head injury resulting from a blow or violent shaking of the brain. Might or might not include a period of unconsciousness or amnesia.

**confabulation**—apparently false or made-up story used to fill in gaps in memory with plausible facts when asked questions. It is a symptom of Korsakoff's syndrome.

**confidentiality**—maintenance of information obtained from clients and that is not to be shared with others.

**constructional apraxia**—see **apraxia**.

**constructional dyspraxia**—difficulty drawing simple geometric shapes and designs. See **apraxia**.

**contracoup**—damage in a closed-head injury, characterized by destruction of brain tissue opposite the site of impact because of the brain's bouncing off the walls of the cranium.

**contralateral**—referring to the opposite side of the body or brain.

**contusion**—bruise, swelling, or hematoma of tissue. Often related to closed-head injury.

**coronal**—view of the brain facing the front of the head; created using imaging.

**corpus callosum**—large set of white matter tracts that connect the right and left hemispheres.

**cortex**—outer layer of brain tissue, comprised of sulci and gyri.

**cortisol**—glucocorticoid that is known as a stress hormone and that is produced by the adrenal gland.

**CPAP**—continuous positive airway pressure. Ventilation treatment for sleep apnea.

**cranial nerves**—twelve pairs of nerves that originate in the brain and that carry sensory and motor signals to and from the periphery of the central nervous system.

**cranial pressure**—pressure inside the skull.

**craniotomy**—brain surgery in which a bone is removed temporarily from the skull to gain access to the brain.

**creatine**—acid naturally produced to supply energy to muscles and to other cells of the body.

**CSF**—cerebral spinal fluid.

**CT**—see **computed topography**.

**cyst**—fluid sac usually associated with an infectious disorder.

**cytokines**—proteins that act as messengers between cells.

**d-amino acid oxidase**—enzyme used in a variety of applications.

**deep brain stimulation (DBS)**—surgical procedure in which a medical device is implanted into the brain to send electrical impulses to specific brain structures or regions.

**déjà vu**—experience in which new experiences seem familiar and relived. Feelings of déjà vu are common with complex partial seizures.

**delirium**—acute, global impairment of cognitive functioning resulting in confusion. Delirium is usually reversible, and it is often due to metabolic disturbances of brain functioning.

**delta waves**—electrical brain activity, or neuronal activity, associated with stages 3 and 4 of sleep.

**delusion**—false belief held despite contradictory evidence that should dispel it.

**dementia**—typically progressive condition of cognitive decline that occurs in the areas of memory and other cognitive abilities.

**dementia of Lewy body**—progressive dementia that is characterized by Lewy body occlusions. Typical symptoms include memory loss, cognitive decline, and psychotic symptoms, particularly recurrent visual hallucinations.

**dendrites**—part of the neuron that conducts electrochemical stimulation received from other neurons.

**Depakote**—anticonvulsant medication also used to stabilize mood in bipolar disorder.

**depression**—mood state characterized by sadness, loneliness, apathy, and feelings of worthlessness, helplessness, and hopelessness, among other vegetative symptoms.

**determinism**—philosophical belief that everything that happens in the world is predetermined, including human behavior.

**dexmethylphenidate**—psychostimulant medication aimed to treat attention disorders; brand name is Focalin.

**dextroamphetamine**—psychostimulant medication aimed to treat attention disorders; brand name is Dexedrine.

**dialectical behavior therapy**—psychotherapeutic technique used to improve behaviors and cognitions that involves both individual and group therapy. Often used for borderline personality disorder.

**diencephalon**—posterior region of the brain that consists of the thalamus and hypothalamus, and which connects the midbrain with the cerebral hemispheres.

**diffuse axonal injury**—traumatic brain injury globally effecting brain cells and regions. Its effects are widespread, and they result in a variety of behavioral, cognitive, and emotional symptoms.

**diffuse tensor imaging (DTI)**—type of imaging technique in which white matter tracts are observed.

**diplopia**—double vision.

**disorder of written expression**—learning disorder characterized by writing impairment in the absence of motor impairment.

**donepezil**—psychotropic medication that is an acetylcholinesterase inhibitor, aimed to slow down memory decline; brand name is Aricept.

**dopamine**—neurotransmitter that plays an important role in pleasure and reward circuits of the brain. Also helps to regulate movement and emotional responses; disruption can result in a variety of disorders, including schizophrenia.

**dorsal medial prefrontal cortex**—portion of the frontal cortex that is important for planning, determining good from bad, and moderating social behavior.

**dorsolateral prefrontal cortex**—portion of the frontal lobe of the brain that is important for making decisions and working memory, among other functions.

**Down syndrome**—condition in which there is an extra copy of chromosome 21, which results in physical, cognitive, and developmental delays. Also known as *Trisomy 21*.

**DSM-IV-TR**—*Diagnostic and Statistical Manual of Mental Disorders*, 4th edition, text revision. A psychiatric classification system published by the American Psychiatric Association in 1994, before text revision in 2000.

**DSM-5**—*Diagnostic and Statistical Manual of Mental Disorders*, 5th edition. A psychiatric classification system published by the American Psychiatric Association in 2013.

**DTI**—see **diffuse tensor imaging (DTI)**.

**dura layer**—tough, outermost membrane of the meninges that protects the brain from the skull.

**dysarthria**—acquired impairment in motor aspects of speech, including slurred speech and raspy sound, resulting in difficulty in being understood.

**dysbindin**—gene that encodes neuronal proteins and that is linked to schizophrenia.

**dyscalculia**—aphasic symptom characterized by impaired ability to perform arithmetic calculations.

**dysfluency**—disturbance of the fluency of speech.

**dysgnosia**—partial loss of the symbolic significance of information reaching the brain.

**dyslexia**—impairment in reading, including letter and word reversal, fluency issues, and comprehension problems, despite intact intelligence.

**dysnomia**—see **anomia**.

**dysphagia**—difficulty in swallowing.

**dyspnea**—labored breathing.

**dyspraxia**—see **apraxia**.

**dysthymia**—symptoms of low mood for at least 2 years, that do not meet criteria for a major depressive disorder.

**dystonia**—involuntary, slow movements that typically contort a part of the body.

**ECT**—see **electroconvulsive therapy (ECT)**.

**ectoderm**—outermost of the three germ layers of the embryo.

**edema**—swelling of the brain subsequent to cerebral insult or injury. It results from the accumulation of fluid in intercellular tissue.

**EEG**—see **electroencepholographic recording (EEG)**.

**efferent nerve cells**—motor neurons that receive information from other neurons and send it to other cells.

**electroconvulsive therapy (ECT)**—type of treatment for depression in which the brain is exposed to electrical charges to induce a mild seizure during an anesthetized state, in an attempt to improve mood state. Formerly known as shock treatment.

**electroencepholographic recording (EEG)**—examination of electrical activity brain waves in regions of the brain, through the use of electrodes, which measure underlying neuron activity.

**embolism**—process in which a foreign object, such as a blood clot, lodged in a vessel or artery, causes an occlusion of blood flow.

**embryogenesis**—process in which cell division occurs, forming an embryo that develops into a fetus.

**EMDR**—eye movement desensitization and reprocessing.

**encephalitis**—inflammation of the brain.

**encopresis**—passage of stool in a child who is toilet-trained that results in soiling of clothing, bed sheets, or other objects.

**endocrine system**—gland system that secretes hormones directly into the blood and influences almost all organs of the human body.

**endoderm**—innermost germ layer of the three germ layers of the embryo.

**endorphin**—neurotransmitter that is produced by the pituitary gland and hypothalamus, and that serves as a endogenous morphine.

**enuresis**—inability to control urination in individuals who are old enough to be toilet trained. Commonly referred to as bedwetting when it occurs at night.

**enzyme**—complex proteins that influence changes of chemicals in the body.

**EPAP**—expiratory positive airway pressure that is used as a treatment for sleep apnea.

**epidemiology**—study of distribution of health-related events and diseases.

**epigenetics**—study of changes in the expression of genes influenced by the environment.

**epilepsy**—condition of abnormal electrical discharges from the brain, associated with a temporary alteration in behavior.

**epinephrine**—neurotransmitter that regulates cardiovascular flow, heart rate, and other important functions. Also known as adrenaline.

**EPSP**—see **excitatory postsynaptic potential (EPSP)**.

**ERP**—see **event-related potentials recording (ERP)**.

**erotomanic**—type of delusion in which a person believes that a stranger, usually higher in status or famous, is in love with her or him.

**estrogen**—steroid compounds necessary for development of and functioning of female reproductive cycle.

**etiology**—cause(s) or origination of a disease or disorder.

**euthymia**—period of clinical stability.

**event-related potentials recording (ERP)**—examines brief changes of EEG signals in response to discrete sensory signals.

**excitatory postsynaptic potential (EPSP)**—brief depolarization of postsynaptic membrane caused by the flow of positively charged ions into the postsynaptic cleft.

**executive functioning**—broad term used to describe higher-order cognitive skills, including problem-solving, planning, inhibition, and organization.

**exocrine glands**—glands that secrete their products into ducts that lead to the external environment.

**exocytosis**—process of cellular excretion or secretion in which neurons are released to extracellular space through fusion.

**facial nerve**—seventh cranial nerve. It is involved in the sense of taste, and it contains a few other somatic sensory afferent fibers. It is also involved in facial expression.

**fetal alcohol syndrome (FAS)**—physical, cognitive, and behavioral syndrome that occurs when a fetus is exposed to alcohol.

**fMRI**—see **functional magnetic resonance imaging (fMRI)**.

**fissure**—any deep fold in the cerebral cortex.

**fornix**—portion of the brain found at the base, important for transmitting information from the hippocampus to the hypothalamus.

**fragile X syndrome**—genetic disorder that causes a range of neurocognitive impairments, including intellectual disability. Physical features include long narrow face, large ears, prominent jaw, large forehead, large feet, and enlarged testicles in males.

**frontal lobes**—first one-third portion of the brain that is responsible for executive functioning abilities, among other skills.

**frontal lobe dementia**—progressive cognitive decline that results in severe executive dysfunction, behavioral disinhibition, and memory problems owing to the neurophysiological changes that initially occur in the frontal lobes.

**functional magnetic resonance imaging (fMRI)**—magnetic resonance scanning technique that permits measurements of regional metabolism of the brain through the examination of high and low use of oxygen. See also **magnetic resonance imagining (MRI)**.

**fusiform gyrus**—portion of the brain located in the occipital and temporal lobe that is thought to play a vital role in facial recognition.

**GABA**—see **gama aminobutyric acid (GABA)**.

**gait**—particular manner in which a person moves while walking.

**galantamine**—cholinesterase inhibitor used to treat memory decline from Alzheimer's disease; brand name is Razadyne.

**Galveston Orientation and Amnesia Test (GOAT)**—behavior-rating scale used during recovery stages of TBI to assess for orientation, retrograde, and anterograde amnesia.

**gama aminobutyric acid (GABA)**—inhibitory transmitter that plays a role in reducing anxiety.

**GCS**—see **Glasgow Coma Scale (GCS)**.

**generalized anxiety disorder**—psychiatric disorder in which an individual has debilitating levels of worry, dread, and fear in a variety of contexts.

**gene clusters**—multiple genes that cluster, based upon proximity and function.

**genes**—unit of DNA that contains information that determines hereditable traits.

**genetic studies**—examination of genetic analysis. Previously commonly referred to as examination of traits commonly found in families.

**genotype**—inherited, internally coded information that serves as a blueprint for individuals.

**Glasgow Coma Scale (GCS)**—behavior rating scale used shortly after a brain injury to measure eye, verbal, and motor control.

**glial cells**—connective tissue of the brain.

**glioma**—any neoplasm arising from glial cells.

**globus pallidus**—portion of the basal ganglia system of the brain that regulates voluntary movement.

**glossopharyngeal nerve**—ninth cranial nerve, consisting largely of sensory afferent fibers. Responsible for gag reflex, sense of taste, and general sensation of the lower third of the tongue.

**glucocorticoid**—hormone produced by the adrenal cortex, which regulates the metabolism of carbohydrates, proteins, and lipids.

**glucose**—simple sugar, a carbohydrate, used as a primary source of energy in cells.

**glutamate**—neurotransmitter classified as an amino acid that is found abundantly in the central nervous system and that is excitatory. Considered the most prominent neurotransmitter and dysregulation, glutamate is linked with multiple psychiatric and medical conditions that result in cognitive and behavioral disorders.

**GOAT**—see **Galveston Orientation and Amnesia Test (GOAT)**.

**Golgi body**—part of the cell that stores, packages, and passes along lipids and proteins.

**grand mal seizure**—generalized tonic–clonic seizure that is characterized by unconsciousness, muscle rigidity, and convulsive behaviors.

**grandiosity**—feeling of being on top of the world, or of greatness.

**gray matter**—portion of the brain dominated by gray cell bodies and that is found on the cerebral cortex, cerebrum, and cerebellum and that is important for numerous functions, including memory, emotions, and sensory abilities.

**gyrus**—convolution on the surface of the brain.

**haldol**—generic name for Haloperidol, an antipsychotic psychotropic medication.

**hallucinations**—false perceptions of reality that are typically visual, auditory, or tactile experiences that occur without external stimulus.

**hallucinogens**—drugs used to create deleterious mental experiences and profound distortions in reality.

**heart hypothesis**—ancient Greek theory that the heart controlled cognitive functioning, mostly because it is the warmest part of the body.

**hematoma**—collection of blood outside of a blood vessel.

**hemianopsia**—loss of vision in half of a visual field.

**hemorrhage**—bleeding.

**heredity**—transmission of physical, mental, and behavioral traits to offspring.

**heterogeneity**—variation.

**hippocampus**—portion of the temporal lobe that is primarily responsible for memory, among other cognitive abilities.

**histamine**—neurotransmitter that plays an important role in numerous human functions, particularly sleep, digestion, and immune system repair.

**H.M.**—short for *Henry Molaison*. He underwent neurosurgery and removal of his temporal lobe. His case is important for the attendant discovery of the significance of the hippocampus for memory.

**homeostasis**—stability or state of equilibrium.

**homogeneity**—state of being the same or uniform.

**hormone**—chemical messenger that plays an important role in emotion, behavior, and personality.

**HPA axis**—see **hypothalamic pituitary adrenal (HPA) axis**.

**Huntington's Disease (Chorea)**—genetic abnormality in gene IT-15, chromosome 4 that results in chorieform movements, depression, and eventually progressive dementia.

**hydrocephalus**—abnormal accumulation of cerebrospinal fluid within the brain, resulting in enlarged ventricles and destruction of neural tissue.

**hypercortisolemia**—high levels of cortisol production.

**hypersomnia**—excessive sleepiness.

**hypertension**—high blood pressure.

**hypnic myoclonia**—rapid contraction of muscle movement, or jerks, that often occur in the early stages of sleep.

**hypocretin**—neurotransmitter that regulates arousal and wakefulness. It also plays a role in appetite.

**hypoglossal**—twelfth cranial nerve. It is important for tongue and motor functions.

**hypomania**—symptoms of elevated mood state that do not meet full criteria for a manic episode.

**hypometabolism**—slow functioning, or abnormal decrease, of metabolism that plays a role in weight gain.

**hypothalamic–pituitary–adrenal (HPA) axis**—interaction and feedback system of the limbic system; plays an important role in the body's response to stress and fear.

**hypothalamus**—structure dorsal to the thalamus that regulates sleeping, sexual activity, eating, emotions, and other behaviors.

**hypothyroid**—slow functioning, or abnormal decrease, of thyroid functioning. Can result in numerous physical and mental symptoms, including depressive symptoms.

**hypoxia**—limited or decrease of oxygen supply to the brain.

**hypoxic-ischemic encephalopathy**—condition in which the brain does not receive enough oxygen, causing brain damage.

**ICD**—international classification of disease used by the World Health Organization.

**ictal**—behavior related to a seizure (epileptic) episode. For example, cursing is an ictal behavior associated with some forms of temporal lobe epilepsy.

**idiopathic**—condition of an unknown or uncertain cause.

**infarct**—region of dead brain tissue associated with an occlusion of the vasculature.

**insomnia**—inability to sleep or obtain healthy sleep.

**insula**—refers to the insular cortex, located beneath the frontal and temporal lobes.

**in vivo examination**—evaluation of an individual's brain while he or she is alive.

**interictal**—period between seizure episodes in an epileptic individual.

**intrinsic**—existing within the brain itself.

**ipsilateral**—on the same side.

**IPSP**—inhibitory postsynaptic potential.

**ischemic stroke**—any local and temporary deficiency of blood.

**jamais vu**—experience in which familiar surroundings and experiences seem unusual or unreal.

**k-complex**—neuronal discharge during stage 2 of sleep as recorded on an EEG, and that consists of brief, high-voltage waves followed by low-voltage waves, which are thought to decrease arousal and aid memory consolidation.

**ketamine**—fast-acting anesthetic drug that has become known as a date rape drug, as it can result in limited speech and movement, as well as amnesia.

**Korsakoff's syndrome**—deterioration of the brain and cognitive abilities (particularly memory) caused by chronic and severe alcohol abuse and resulting thiamine deficiency.

**lateralization**—concept that the left side of the brain is responsible for functions on the right side of the body, and the right side of the brain is responsible for functions on the left side of the body.

**learning disorders**—classification of psychiatric disorders that is characterized by difficulties in academic areas due to delayed or limited capabilities to retain or apply concepts in the specific area.

**lesion**—abnormal change in brain tissue, typically as a result of disease or injury.

**levadopa**—neurochemical precursor to dopamine.

**libido**—sexual drive or desire.

**limbic system**—collection of brain structures and pathways that include the amygdala, hippocampus, and thalamus, among other structures, and that are responsible for human emotion regulation.

**linear basilar skull fracture**—break in the skull bone that is straight and involves no displacement.

**locus coeruleus**—widespread noradrenergic system in the brain that releases norepinephrine throughout the CNS.

**longitudinal studies**—repeated examination of a person (or persons) or event over time.

**LSD**—Lysergic acid diethylamine; most common and potent hallucinogen.

**Lunesta (eszopiclone)**—sedative prescribed to treat insomnia.

**lysosome**—cellular organelle that contains enzymes that eliminate waste materials and cellular debris.

**magnetic resonance imaging (MRI)**—imaging method by which a large magnet and a radio frequency pulse of a certain resonance generate a signal from the brain to produce an image.

**major depressive disorder (MDD)**—psychiatric disorder characterized by at least a 2-week period of extreme sadness, irritability, insomnia, feelings of helplessness, hopelessness, and other symptoms of depression. See also **depression**.

**malingering**—intentionally feigning or exaggerating symptoms in an attempt for gain or external reward.

**mania**—feeling of extreme energy, lack of need of sleep, grandiosity, and other extreme feelings of euphoria that are hallmark features of bipolar disorder.

**MAOA gene**—monoamine oxidase A, also known as *the warrior gene*. Plays a role in aggression, antisocial behavior, mood, and anxiety.

**MAOI inhibitors**—monoamine oxidase inhibitors, the first generation of antidepressants, which function by altering the enzyme responsible for metabolizing epinephrine and norepinphrine and serotonin and dopamine.

**marijuana**—cannabis, one of the most common illegal drugs, used to relax people and decrease pain. Used for medicinal purposes in some states.

**mathematics disorder**—psychiatric disorder characterized by difficulty in learning or retaining basic and complex arithmetic skills at an age-appropriate rate.

**MDD**—see **major depressive disorder (MDD)**.

**MDMA**—methylenedioxy-N-methylamphetamine; also known as *ecstasy*.

**medial prefrontal cortex**—middle portion of the front part of the frontal cortex, which is thought to be important in the HPA axis.

**meditation**—mindful activity to relax and calm oneself down.

**medulla oblongata**—lowest portion of the brain stem.

**MEG**—magnocephalography, which is an imaging technique of recording magnetic fields generated by electrical fields.

**melatonin**—neurochemical important for sleep.

**memantine**—medication aimed to alter glutamate receptors in order to slow the progression of Alzheimer's disease during the moderate stages of the illness; brand name is Namenda.

**memory**—retention and recall of learned information or experiences.

**meninges**—three membranes that protect the brain, comprising the dura mater, the pia mater, and the arachnoid layer.

**meningioma**—neoplasm arising in the meninges.

**meningitis**—inflammation of the meninges, especially of the pia mater and the arachnoid.

**mental retardation**—older term used to describe *intellectual disability*. Characterized by an IQ score of less than 70, functional impairment in at least two domains, and onset before age 18.

**mesencephalon**—midbrain, which is important for numerous functions, including hearing, vision, arousal, and motor control.

**mesoderm**—middle of the three germ layers.

**mesolimbic system**—subsystem of the limbic system; involved in the release and dispersion of dopamine in the brain.

**metabolism**—all processes in the body that use energy.

**methadone**—synthetic opioid that is used to reduce the effects of opioid dependency.

**methamphetamine**—psychotropic medication that has stimulating effects on the central nervous system.

**methylphenidate**—psychotropic drug approved by the FDA for the treatment of attention deficit hyperactivity disorder (ADHD); brand name is Ritalin.

**metencephalon**—anterior portion of the developing brain, which includes the cerebellum and pons.

**microglial**—type of glial cell that acts as the main immunological cells wound within the CNS.

**mild cognitive impairment**—constellation of cognitive and behavioral symptoms that are used to describe the cognitive problems before onset of a progressive dementia.

**mitochondria**—organelles that convert energy into useable forms within a cell.

**mitosis**—division of cells into two identical sets of chromosomes.

**monoamine oxidase**—enzyme in the cells of most tissues that catalyzes the oxidation of neurotransmitters, such as serotonin and norepinephrine.

**mood disorder**—type of psychiatric disorder characterized by either extreme highs or lows that negatively affect functional abilities.

**mood stabilizers**—classification of psychotropic medications that result in stable mood; typically prescribed for bipolar disorder.

**morphine**—opiate analgesic; severe pain–reducing medication.

**morphology**—study of the form and structure of the central nervous system.

**motor cortex**—portion of the frontal cortex that is responsible for the execution of movement.

**MRI**—see **magnetic resonance imaging (MRI)**.

**MRS**—magnetic resonance spectroscopy. Varied frequency of radio waves allows for measurements of nonwater molecules within the brain. For example, allows for examination of proteins, phospholipids, and cell membranes in the brain.

**Mr. Tan**—individual case that resulted in the identification of the role of Broca's area.

**multiple sclerosis**—disease resulting from degeneration of myelin, characterized by the development of multiple plaques throughout the brain and spinal cord.

**muscle atonia**—lack of strength within muscle.

**mutism**—condition of not speaking.

**myasthenia gravis**—autoimmune neuromuscular disease the symptoms of which include fluctuating muscle weakness and fatigue.

**myelin sheath**—fatty axon-enwrapped coat that serves to increase the speed of axon transmission. Created from Schwann cells and oligodendrytes.

**myelopathy**—disintegration of the myelin sheath. Not to be confused with *myopathy*.

**mylencephalon**—subregion of the brain that is used to describe areas including the medulla oblongata.

**myoclonic seizure**—abrupt contraction followed by relaxation of musculature and characterized by jerking motion.

**myopathy**—degeneration of muscle fiber. Not to be confused with *myelopathy*.

**myotonia**—delayed relaxation of the muscles. Myotonic dystrophy appears in adulthood and is characterized by an inability of the patient to release a grasp or quickly undo any motor contraction.

**N-acetyl**—acetylcysteine. A drug used to replenish cysteine.

**naltrexone**—opioid receptor antagonist that is used to treat opioid and alcohol dependence.

**narcolepsy**—inability to sustain wakefulness; abrupt periods of falling asleep during daytime hours.

**negative symptoms**—absence or decrease of emotional and cognitive responses and functioning associated with schizophrenia.

**neoplasm**—term referring to a brain tumor. Literally means *new growth*.

**neuregulin**—protein that promotes the development of the nervous system.

**neuroblast**—cell that divides and subsequently turns into a neuron during cell migration.

**neurocognitive disorder**—psychiatric or medical condition that causes impaired or diminished functioning of intellectual skills, including in areas of attention, memory, visual spatial, language, motor, and executive functioning.

**neurofibrillary tangles**—abnormal collection of twisted threads that occur inside a nerve cell: a neuropathological feature of Alzheimer's disease.

**neurogenesis**—process by which neurons are created.

**neuron**—basic building blocks of the central nervous system made of electrically excitable cells.

**neuropathology**—study or etiology of nervous system diseases.

**neuropeptides**—microscopic molecules used to communicate between neurons.

**neuroplasticity**—brain's ability to reorganize neuropathways to regain or maintain important cognitive, emotional, motor, and behavioral functions.

**neuropsychology evaluation**—objective examination of cognitive capabilities, often includes a clinical interview, objective cognitive testing, evaluation of emotional functioning, and clinical report of findings.

**neurosurgery**—brain surgery, including removal of portion of brain, dissection of tumor, or other cranial operations.

**neurotransmitter**—chemicals that transmit signals from a neuron to a cell through a synapse.

**nicotine**—alkaloid that is found in certain plants and used in tobacco; causes a mild euphoria.

**night terrors**—parasomnia characterized by apparent high level of alertness and wakeful behavior during which the individual is, in fact, asleep and cannot be aroused. Common in children and associated with anxiety.

**nocturnal seizure**—abnormal movement disorder that occurs at night. See **Rolandic seizure**.

**nodes of Ranvier**—tiny gaps formed between myelin sheaths of a cell.

**nonverbal learning disorder**—type of learning disorder characterized by impaired measures of visual spatial and constructional abilities, as well as decreased social skill functioning.

**norepinephrine**—neurotransmitter classified as a catecholamine that underlies fight or flight sensation.

**nucleus accumbens**—brain structure that plays an important role in reward circuitry and pleasure sensation.

**nystagmus**—spasmodic movement of the eyes, either rotary or side to side.

**obsessive–compulsive disorder (OCD)**—psychiatric disorder characterized by feelings of anxious intrusive thoughts and repetitive behaviors resulting in functional difficulty.

**obstructive sleep apnea**—sleep disorder condition in which there is pause of regular airflow during sleep.

**occipital lobe**—smallest lobe of the brain located in the back of the cranium that is responsible for vision and visual processes.

**OCD**—see **obsessive–compulsive disorder (OCD)**.

**oculomotor nerve**—third cranial nerve, which has efferent fibers to the eye muscles. It plays a role in pupil dilation, among other motor activities.

**olanzapine**—psychotropic medication classified as an atypical antipsychotic; brand name is Zyprexa.

**olfactory nerve**—first cranial nerve, which facilitates the sense of smell. Often not evaluated in a neurological evaluation of the cranial nerves, but can be severed or damaged as a result of a traumatic brain injury.

**open-head injury**—injury to the brain that results in exposure of brain matter. Most commonly occurs in gunshot wounds or other penetration-type injuries.

**ophthalmologist**—medical doctor who specializes in visual processes and the structural components of the eye.

**opioid**—chemical that bonds to opioid receptors, primarily in the CNS, resulting in decreased pain sensation.

**orbital frontal cortex**—portion of the prefrontal cortex that plays a role in decision making and appropriate social behavior.

**orlistat**—medication aimed to treat obesity; brand name is Xenical.

**optic nerve**—second cranial nerve, serving the sense of sight.

**osteoporosis**—disease in which bones become frail and are more likely to be fractured.

**panic attack**—feeling of abrupt fear or discomfort. Symptoms include increased heart rate, sweaty hands, shortness of breath, and racing thoughts. Often confused with a heart attack.

**paranoia**—irrational fear or delusion that you are a target of another person or that somebody is out to harm you or is following you.

**paraphasia**—disturbance in the verbal output of a patient; can be verbal or semantic in nature.

**parasomnia**—classification of sleep disorders in which an individual has periods of uncontrolled, wakeful behavior during the sleep cycle.

**parasympathetic nervous system**—one of two main divisions of the autonomic nervous system, responsible for maintaining homeostasis.

**parathesia**—abnormalities of sensation, especially tactile and somatic sensation.

**parietal lobe**—one of four lobes of the brain that is near the top portion of the cranium and responsible for somatosensory functions.

**Parkinson's disease**—disorder that affects primarily the motor functions of the cerebellum; characterized by tremors and gait disturbances.

**paroxetine**—SSRI antidepressant medication; brand name is Paxil.

**pathognomonic sign**—any sign or symptom that is characteristic of a disease or pathological condition and that does not occur in the absence of pathology.

**PDD**—see **pervasive developmental disorder (PDD)**.

**peripheral nervous system (PNS)**—nervous system that resides outside of the brain and spinal cord.

**persistent postconcussion disorder**—ongoing symptoms of altered behavioral, emotional, or cognitive functioning subsequent to traumatic brain injury for greater than approximately a 6-month period (the exact time frame is not universally agreed upon).

**pervasive developmental disorder (PDD)**—a developmental disorder characterized by social skills deficits and cognitive impairments. See **autism spectrum disorder (ASD)**.

**PET scan**—see **positron emission tomography (PET) scan**.

**phencyclidine**—synthetic general aesthetic medication.

**phenobarbital**—barbiturate that has sedative features and that is used as an anticonvulsant for all types of seizures.

**Phineas Gage**—subject of a well-known individual case study; sustained an open-head injury and subsequent personality changes.

**phrenology**—study of brain functions by examination of bumps and grooves found on the skull.

**pia mater**—the innermost layer of the meninges, which serves to protect the brain from the skull.

**Pick's disease**—form of dementia that affects the frontal and temporal lobes and that is characterized by early loss of social skills and inhibition.

**pineal gland**—tiny endocrine gland that secretes melatonin.

**pituitary gland**—pea-size endocrine gland located near the brain that releases neurochemicals involved in multiple cognitive, emotional, and behavioral functions.

**PNS**—see **peripheral nervous system (PNS)**.

**polymorphism**—Involves multiple variants of a particular DNA sequence.

**positive symptoms**—psychotic symptoms found in schizophrenia that include active hallucinations and delusions.

**positron emission tomography (PET) scan**—measures metabolic brain activity through the use of an injection of water with unstable radioactive molecules to examine high and low blood flow during a cognitive activity.

**postconcussion disorder**—neurobehavioral symptoms that occur after a closed-head injury and that are sustained over time, typically thought of as over 3 months' time.

**posterior cingulate**—back portion of the cingulated gyrus, or cingulated system.

**postpartum depression**—decreased level of mood subsequent to pregnancy and birth of a child for women.

**posttraumatic amnesia**—memory difficulty following the time of a closed-head injury. Difficulty recalling and forming new memories.

**Prader–Willi syndrome**—rare genetic disorder that occurs on chromosome 15 in which the characteristic features include obesity, insatiable appetite, low muscle tone, incomplete sexual development, and cognitive delays, including intellectual impairment.

**premorbid**—preexisting or prior functioning and abilities.

**prefrontal cortex**—frontal portion of the frontal lobes important for executive functioning, along with other abilities.

**premotor cortex**—portion of the frontal lobe that is responsible for planning for the execution of movement.

**prolactin**—hormone produced by the pituitary gland, which stimulates breast development and milk production.

**prosopagnosia**—acquired inability to recognize familiar faces.

**provigil**—psychotropic medication used to treat sleep apnea or sleepiness caused by narcolepsy.

**Prozac (fluoxetine)**—SSRI psychotropic medication used to treat depression.

**pseudodementia**—older term used to describe any form of apparent cognitive impairment that is not global and that mimics dementia. A common form is pseudodementia secondary to depression.

**pseudoseizure**—older term used to describe psychogenic seizure.

**psychogenic seizure**—seizure-like events that are not linked to abnormally recorded EEG findings and that are thought to be activated by emotional functioning.

**psychopathy**—personality characterized or defined by lack of remorse, antisocial behavior, and limited behavioral control.

**psychopharmacology**—study of medicines.

**psychosis**—loss of reality, or abnormal condition of the mind, that impairs functioning due to inability to think clearly.

**psychostimulant**—classification of medications aimed at improving attention and increasing concentration by promoting alertness.

**Purkinje cells**—class of GABAergic cells located at the base of the brain, characterized by a large number of dendrites.

**putamen**—structure of the basal ganglia system of the brain that is involved with movement, learning, and memory.

**quetiapine**—atypical antipsychotic medication; brand name is Seroquel.

**Ranchos Los Amigos scale**—rating scale used during recovery phases of traumatic brain injury.

**reading disorder**—psychiatric disorder characterized by deficient or impaired reading skills, including problems with speed, decoding, blending of sounds, and other difficulties.

**receptor**—molecule found on the surface of the cell that receives chemical signals from outside of the cell.

**REM**—stage of sleep characterized by rapid eye movement and increased physiological responses.

**retina**—light-sensitive layer found in the back of the eye that covers a significant portion of the interior surface.

**retrograde amnesia**—inability to recall previous memories or events.

**Rhett's syndrome**—genetic disorder that occurs almost exclusively in females and that causes severe physical and cognitive impairment.

**risperidone**—atypical antipsychotic medication used to treat psychosis; brand name is Risperdal.

**rivastigimine tartrate**—acetycholinesterase inhibitor used to treat memory problems found in Alzheimer's disease; brand name is Exelon.

**rolandic seizure**—benign childhood seizure disorder that is characterized by twitching, numbness, and tingling of tongue and face.

**sagittal view**—image of the brain from the side.

**sawtooth theta activity**—reading on an EEG during sleep, resembling a blade, which occurs during REM stage of sleep.

**schizophrenia**—psychiatric disorder characterized by hallucinations, delusions, paranoia, cognitive dullness, and negative symptoms.

**Schwann cell**—supporting cell of the PNS that aids in the recovery of the peripheral nerves after an injury.

**seasonal affective disorder (SAD)**—decreased mood symptoms that occur during the months of the year in which days have fewer daylight hours.

**second impact syndrome**—devastating effect of a second closed-head injury prior to recovery from an initial closed-head injury.

**selective serotonin reuptake inhibitor (SSRI)**—classification of antidepressant psychotropic medications that aim to regulate serotonin. Used for depression and anxiety as well as for other psychiatric disorders.

**serotonin**—monoamine neurotransmitter that plays an important role in the regulation of mood.

**serotonin norepinephrine reuptake inhibitor (SNRI)**—classification of antidepressant psychotropic medications that aim to regulate serotonin and norepinephrine for the treatment of depression and anxiety.

**sertraline**—SSRI used to treat depression, anxiety, OCD, and panic disorder; brand name is Zoloft.

**single-photon emissions computed topography**—see **SPECT**.

**sleep bruxism**—nocturnal teeth grinding.

**sleep spindles**—occasional burst of brain activity that is recorded on an EEG during stage II sleep.

**SNRI**—see **serotonin norepinephrine reuptake inhibitor (SNRI)**.

**social anxiety disorder**—psychiatric disorder characterized by extreme fear of being in public situations without an easy escape.

**sodium potassium pump**—active transport mechanism that creates an electrical potential between the outside and the inside of a cell.

**soma**—muscle relaxer medication that blocks pain sensations from the brain to the nerve.

**somatic nervous system (SNS)**—portion of the PNS that is associated with the voluntary control of body movements.

**somatization**—bodily sensation of stress and anxiety experienced as pain and presented as recurrent complaints of physical symptoms.

**spasm**—involuntary contraction of a muscle group associated with anxiety, fear, or a neurological disorder, including seizure.

**spasticity**—abnormal increases in muscle tone.

**SPECT**—single-photon emissions computed topography. A type of imaging that allows for examination of specific proteins and other neurochemicals.

**spinal accessory nerve**—eleventh cranial nerve. It has effluent fibers for the branchiomeric musculature. A lesion here might result in paralysis of the trapezium.

**SSRI**—see **selective serotonin reuptake inhibitor (SSRI)**.

**Strattera**—nonstimulant medication aimed at improving attention while decreasing anxiety in attention disorders; generic name is atomoxetine.

**stress inoculation therapy**—form of psychotherapy intended to help prepare the patient to deal with stressful situations; used as CBT for PTSD.

**striatum**—portion of the basal ganglia that consists of the caudate, putamen, and globus pallidus.

**stroke**—general term to describe those disorders of the brain that are characterized by disruption of blood flow.

**Stroop task**—cognitive task of efficiency and inhibition.

**subdural hematoma**—lesion that results from bleeding into the subdural space.

**substantia nigra**—brain structure of the basal ganglia that plays a role in addiction, reward, and movement.

**sulci**—divot or fissure on the surface of the brain.

**suppression**—any failure to perceive a stimulus on one side of the body with bilateral simultaneous stimulation. Suppressions can exist with auditory, visual, or tactile stimulation. Also known as *extinction*.

**suprachiasmatic nucleus**—small region of the brain located above the optic chiasm and important for controlling circadian rhythm.

**sympathetic nervous system**—portion of the ANS that prepares the body to respond to stress or emergency situations.

**synapse**—space between the terminal end of an axon and another cell body. Neurotransmitters are released in the synapse and carry signals from one nerve cell to another.

**synaptic pruning**—elimination of unused neuronal connections to increase the efficiency of neuronal connections that see frequent use.

**synaptic transmission**—movement of cellular information from presynaptic cells to postsynaptic cells.

**tardive dyskinesia**—repetitive involuntary body movements of the tongue, lips, face, and extremities.

**tau protein**—substance found in neurons responsible for keeping nerves functioning effectively.

**telencephalon**—anterior subdivision of the embryonic forebrain.

**temporal lobe**—one of four lobes of the brain responsible for many functions, including memory, learning, and speech.

**tentorium**—structure that divides the cerebrum from the cerebellum.

**testosterone**—steroid hormone responsible for male reproductive tissues and that promotes muscle and well-being.

**thalamus**—portion of the brain responsible for relaying motor and sensory information to the cortex. Commonly thought of as the relay center of the basal ganglia.

**thiamine**—also called vitamin B, it is found in many foods; it promotes central nervous system functioning.

**thorazine**—antipsychotic psychotropic medication aimed at controlling symptoms of psychosis.

**thrombus**—blood clot that forms in an artery or vessel, creating an occlusion.

**thyroid**—one of the largest endocrine glands; it is found in the neck and plays a role in how quickly the human body uses energy.

**thyroxin**—main hormone produced by the thyroid, which plays an important role in digestive, vascular, and muscular functioning.

**TIA**—see **transient ischemic attack (TIA)**.

**tic**—stereotyped movements that can be simple or complex. They are most commonly found in the muscles of the face, and they are sensitive to changes in the level of subjective tension.

**TMS**—see **transcranial magnetic stimulation (TMS)**.

**tinnitus**—ringing in the ears.

**Tofranil**—first tricyclic antidepressant medication.

**tonic seizure**—type of seizure, also known as *grand mal seizure*, that affects the entire brain and body.

**topiramate**—anticonvulsant medication, typically used for seizure disorder, but that has also been found to decrease weight.

**trail-making test**—neuropsychological test that involves visual scanning, sequencing, and cognitive flexibility.

**transcranial magnetic stimulation (TMS)**—relatively new, noninvasive type of treatment used to treat a variety of disorders, including depression in which there is depolarization or hyperpolarization of neurons.

**transient ischemic attack (TIA)**—brief episode of insufficient blood supply to selected portions of the brain.

**traumatic brain injury (TBI)**—intracranial injury, usually as the result of the head striking or being struck by an object.

**tremor**—shaking motion, typically of the extremities.

**tricyclic antidepressant**—classification of medication that aims to treat depression and that blocks the reuptake of serotonin and norepinephrine.

**trigeminal nerve**—fifth cranial nerve, which is important for pain and for tactile and thermal sensation. It is responsible for the corneal reflex, the tearing reflex, and sneezing.

**trochlear nerve**—the fourth cranial nerve. It has efferent fibers innervating skeletal muscles.

**tryptophan**—amino acid necessary for growth. It also assists in the states of relaxation, sleepiness, and calmness.

**unconsciousness**—loss of alertness or wakeful state.

**vagus**—tenth cranial nerve. It has an inhibitory effect on heart rate and is important for speech.

**vascular dementia**—form of dementia characterized by lesions and the disruption of blood supply to the brain. Also known as *multi-infarct dementia*.

**venlafaxine**—psychotropic SNRI used to treat depression and anxiety; brand name is Effexor.

**ventricles**—spaces within the brain through which cerebrospinal fluid circulates.

**vertigo**—sensation of spinning, or the perception that external objects are revolving around the individual.

**vestibular rehabilitation**—exercise-based treatment strategy for improving central nervous system function and ameliorating inner ear dysfunction.

**voxel**—unit of measurement used in MRI analysis.

**Wernicke's syndrome**—inability to verbally communicate due to impairment of receptive abilities.

**white matter**—portion of the brain that consists of myelinated axons and glial cells and that transmit signals to various areas of the cerebrum.

**withdrawal**—behavioral symptoms associated with the abrupt discontinuation of certain drugs that have central nervous system effects.

**Wisconsin Card Sorting Task**—cognitive measure used to assess problem-solving, cognitive flexibility, and concentration.

**zonisamide**—anticonvulsant medication.

# References

Abby, F. J., Costa, R., Hagighi, F., Logue, M., Knowles, J., Weismann, M. M., . . . Hamilton, S. P. (2012). Linkage analysis of alternative anxiety phenotype in multiply affected panic disorder families. *Psychiatric Genetics, 22*(3), 123–129.

Abi-Dargham, A., Gil, R., Krystal, J., Baldwin, R. M., Seibyl, J. P., Bowers, M., & Laruelle, M. (1998). Increased striatal dopamine transmission in schizophrenia: Confirmation in a second cohort. *American Journal of Psychiatry, 155*(6), 761–767.

Abi-Dargham, A., Rodenhiser, J., Printz, D., Zea-Ponce, Y., Gil, R., Kegeles, L. S., . . . Laruelle, M. (2000). Increased baseline occupancy of D2 receptors by dopamine in schizophrenia. *Proceedings of the National Academy of Sciences, 97*(14), 8104–8109.

Ackard, D. M., Fulkerson, J. A., & Neumark-Sztainer, D. (2007). Prevalence and utility of DSM-IV eating disorder diagnostic criteria among youth. *International Journal of Eating Disorders, 40*(5), 409–417.

Adler, C. M., & Cerullo, M. A. (2012). Brain imaging techniques and their application to bipolar disorder. In S. M. Strakowski (Ed.), *The bipolar brain: Integrating neuroimaging and genetics* (pp. 3–16). New York, NY: Oxford University Press.

Äikiä, M., Salmenperä, T., Partanen, K., & Kälviäinen, R. (2001). Verbal memory in newly diagnosed patients and patients with chronic left temporal lobe epilepsy. *Epilepsy & Behavior, 2*(1), 20–27.

Aleman, A., Kahn, R. S., & Selten, J. P. (2003). Sex differences in the risk of schizophrenia: evidence from meta-analysis. *Archives of General Psychiatry, 60*(6), 565.

Alóe, F. (2009). Sleep bruxism neurobiology. *Sleep, 2*(1), 40–41.

American Psychiatric Association. (1994). *Diagnostic and statistical manual of mental disorders* (4th ed.). Washington, DC: Author.

American Psychiatric Association. (2013). *Diagnostic and statistical manual of mental disorders* (5th ed.). Washington, DC: Author.

American Psychological Association. (2010). *Ethical principals of psychologists and code of conduct* (with 2010 amendments). Washington, DC: Author.

Anderson, C. M., Teicher, M. H., Polcari, A., & Renshaw, P. F. (2002). Abnormal T2 relaxation time in the cerebellar vermis of adults sexually abused in childhood: Potential role of the vermis in stress-enhanced risk for drug abuse. *Psychoneuroendocrinology, 27*(1), 231–244.

Andreasen, N. C., Arndt, S., Alliger, R., Miller, D., & Flaum, M. (1995). Symptoms of schizophrenia: methods, meanings, and mechanisms. *Archives of General Psychiatry, 52*(5), 341.

Andreasen, N. C., Nopoulos, P., O'Leary, D. S., Miller, D. D., Wassink, T., & Flaum, M. (1999). Defining the phenotype of schizophrenia: Cognitive dysmetria and its neural mechanism. *Biological Psychiatry, 46*, 908–920.

Angermeyer, M. C., & Kühn, L. (1988). Gender differences in age at onset of schizophrenia. *European Archives of Psychiatry and Neurological Sciences, 237*(6), 351–364.

Angst, J. (1988). European long-term follow-up studies of schizophrenia. *Schizophrenia Bulletin, 14*(4), 501–513.

Annas, G. J. (1994). Informed consent, cancer, and truth in prognosis. *New England Journal of Medicine, 330*(3), 223–225.

Armstrong, M. J., & Miyasaki, J. M. (2012). Evidence based guideline: Pharmacologic treatment of chorea in Huntington's Disease. Report of the guideline development subcommittee of the American Academy of Neurology. *American Academy of Neurology, 79*, 597–603.

Arnold, S. E., Franz, B. R., Gur, R. C., Gur, R. E., Shapiro, R. M., Moberg, P. J., & Trojanowski, J. Q. (1995). Smaller neuron size in schizophrenia in hippocampal subfields that mediate cortical–hippocampal interactions. *American Journal of Psychiatry, 152*(5), 738–748.

Artman, H., Grau, H., Adelmann, M., & Schlieffer, R. (1985). Reversible and non-reversible enlargement of cerebrospinal fluid spaces in anorexia nervosa. *Neuroradiology, 27*(4), 302–312.

Atmaca, M., Yildirim, H., Gürkan Gürok, M., & Akyol, M. (2012). Orbito-frontal cortex volumes in panic disorder. *Department of Psychiatry, Firat University School of Medicine, 9*, 408–412.

Attia, E., Haiman, C., Walsh, B. T., & Flater, S. R. (1998). Does fluoxetine augment the inpatient treatment of anorexia nervosa? *American Journal of Psychiatry, 155*(4), 548–551.

Avena, N. M., & Bocarsly, M. E. (2012). Dysregulation of brain reward systems in eating disorders: neurochemical information from animal models of binge eating, bulimia nervosa, and anorexia nervosa. *Neuropharmacology, 63*(1), 87–96.

Avena, N. M., Rada, P., & Hoebel, B. G. (2008). Evidence for sugar addiction: Behaivoral and chemical effects of intermittent sugar intake. *Neuroscience Behavioral Review, 32*(1), 20–39.

Azevedo, F. A., Carvalho, L. R., Grinberg, L. T., Farfel, J. M., Ferretti, R. E., Leite, R. E., . . . Herculano-Houzel, S. (2009). Equal numbers of neuronal and nonneuronal cells make the human brain an isometrically scaled-up primate brain. *Journal of Comparative Neurology, 513*(5), 532–541.

Baddeley, A. (1992). Working memory: The interface between memory and cognition. *Journal of Cognitive Neuroscience, 4*(3), 281–288.

Badner, J. A., & Gershon, E. S. (2002). Meta-analysis of whole-genome linkage scans of bipolar disorder and schizophrenia. *Molecular Psychiatry, 7*(4), 405–411.

Baiano, M., David, A., Versace, A., Churchill, R., Balestrieri, M., & Brambilla, P. (2007). Anterior cingulate volumes in schizophrenia: a systematic review and a meta-analysis of MRI studies. *Schizophrenia Research, 93*(1–3), 1.

Baio, J. (2012). Prevalence of autism spectrum disorders: Autism and Developmental Disabilities Monitoring Network, 14 Sites, United States, 2008. *Morbidity and Mortality Weekly Report. Surveillance Summaries. 61*(3). Atlanta, GA: Centers for Disease Control and Prevention.

Balanzá-Martínez, V., Rubio, C., Selva-Vera, G., Martinez-Aran, A., Sánchez-Moreno, J., Salazar-Fraile, J., . . . Tabarés-Seisdedos, R. (2008). Neurocognitive endophenotypes (endophenocognitypes) from studies of relatives of bipolar disorder subjects: a systematic review. *Neuroscience and Biobehavioral Reviews, 32*(8), 1426–1438.

Baldessarini, R., Henk, H., Sklar, A., Chang, J., & Leahy, L. (2008). Psychotropic medications for patients with bipolar disorder in the United States: Polytherapy and adherence. *Psychiatric Services, 59*(10), 1175–1183.

Balkin, T. J., Braun, A. R., Wesensten, N. J., Jeffries, K., Varga, M., Baldwin, P., & Herscovitch, P. (2002). The process of awakening: A PET study of regional brain activity patterns mediating the reestablishment of alertness and consciousness. *Brain, 125*(10), 2308–2319.

Bammer, R., Augustin, M., Strasser-Fuchs, S., Seifert, T., Kapeller, P., Stollberger, R., . . . Fazekas, F. (2000). Magnetic resonance diffusion tensor imaging for characterizing diffuse and focal white matter abnormalities in multiple sclerosis. *Magnetic Resonance in Medicine, 44*, 4, 583–591.

Banasiak, S. J., Paxton, S. J., & Hay, P. (2005). Guided self-help for bulimia nervosa in primary care: A randomized controlled trial. *Psychological Medicine, 35*(9), 1283–1294.

Barch, D. M., Carter, C. S., Braver, T. S., Sabb, F. W., MacDonald III, A., Noll, D. C., & Cohen, J. D. (2001). Selective deficits in prefrontal cortex function in medication-naive patients with schizophrenia. *Archives of General Psychiatry, 58*(3), 280–288.

Barkley, R. A. (1997). Behavioral inhibition, sustained attention, and executive functions: Constructing a unifying theory of ADHD. *Psychological Bulletin, 121*(1), 65–94.

Barta, P. E., Pearlson, G. D., Powers, R. E., Richards, S. S., & Tune, L. E. (1990). Reduced volume of superior temporal gyrus in schizophrenia: Relationship to auditory hallucinations. *American Journal of Psychiatry, 147*, 1457–1462.

Basoglu, C., Oner, O., Ates, A., Algul, A., Bez, Y., Cetin, M., . . . Munir, K. M. (2011). Synaptosomal-associated protein 25 gene polymorphisms and antisocial personality disorder: Association with temperament and psychopathy. *Canadian Journal of Psychiatry. Revue Canadienne de Psychiatrie, 56*(6), 341–347.

Basta, M., Chrousos, G. P., Vela-Bueno, A., & Vgontzas, A. N. (2007). Chronic insomnia and the stress system. *Sleep Medicine Clinics, 2*(2), 279–291.

Bates, M. E., Labouvie, E. W., & Voelbel, G. T. (2002, March). Individual differences in latent neuropsychological abilities at addictions treatment entry. *Psychology of Addictive Behaviors, 16*(1), 35–46.

Bava, S., & Tapert, S. F. (2010). Adolescent brain development and the risk for alcohol and other drug problems. *Neuropsychology Review, 20*(4), 398–413.

Bazanis, E., Rogers, R. D., Dowson, J. H., Taylor, P., Meux, C., Staley, C., . . . Sahakian, B. J. (2002). Neurocognitive deficits in decision-making and planning of patients with DSM-III-R borderline personality disorder. *Psychological Medicine, 32*(8), 1395–1405.

Bearden, C. E., Hoffman, K. M., & Cannon, T. D. (2001). The neuropsychology and neuroanatomy of bipolar affective disorder: A critical review. *Bipolar Disorders, 3*(3), 106–150.

Beck, A. T. (2008). The evolution of the cognitive model of depression and its neurobiological correlates. *The American Journal of Psychiatry, 165*(8), 969–977.

Beers, S. R., & De Bellis, M. D. (2002). Neuropsychological function in children with maltreatment-related posttraumatic stress disorder. *American Journal of Psychiatry, 159*(3), 483–486.

Belanger, H. G., Vanderploeg, R. D., Curtiss, G., & Warden, D. L. (2007). Recent neuroimaging techniques in mild traumatic brain injury. *The Journal of Neuropsychiatry and Clinical Neuroscience, 19*(1), 5–20.

Bellebaum, C., & Daum, I. (2010). Memory and cognition in narcolepsy. In M. Goswami, S. R. Pandi-Perumal, & M. J. Thorpy (Eds.), *Narcolepsy* (pp. 23–229). New York, NY: Springer.

Benes, F. M., McSparren, J., Bird, E. D., San Giovanni, J. P., & Vincent, S. L. (1991). Deficits in small interneurons in prefrontal and cingulate cortices of schizophrenic and schizoaffective patients. *Archives of General Psychiatry, 48*(11), 996.

Bennett, C. M., Baird, A. A., Miller, M. B., & Wolford, G. L. (2010). Neural correlates of interspecies perspective taking in the postmortem Atlantic salmon: An argument for proper multiple comparisons correction. *Journal of Serendipitous and Unexpected Results, 1*, 1–5.

Bennett, C. M., Wolford, G. L., & Miller, M. B. (2009). The principled control of false positives in neuroimaging. *Social Cognitive and Affective Neuroscience, 4*(4), 417–422.

Beyer, J. L., & Krishnan, K. R. R. (2009). Late-onset bipolar disorder. In C. A. Zarate Jr. & H. K. Manji (Eds.), *Bipolar depression: Molecular neurobiology, clinical diagnosis and pharmacotherapy* (pp. 213–239). Cambridge, MA: Birkhäuser.

Bienvenu, O. J., Davydow, D. S., & Kendler, K. S. (2011). Psychiatric "diseases" versus behavioral disorders and degree of genetic influence. *Psychological Medicine, 41*(1), 33.

Bifulco, A. A., & Brown, G. W. (1996). Cognitive coping response to crises and onset of depression. *Social Psychiatry and Psychiatric Epidemiology, 31*(3–4), 163–172.

Bigler, E. D. (2008). Neuropsychology and clinical neuroscience of persistent post-concussive syndrome. *Journal of the International Neuropsychological Society, 14*(1), 1–22.

Bigler, E. D. (2009). Traumatic brain injury. In Weiner, M. F. & Lipton, A. M. (Eds.), *The American Psychiatric Publishing textbook of Alzhieimer disease and other dementias* (pp. 229–246). Washington, DC: American Psychiatric Association.

Bischof, M., & Bassetti, C. L. (2004). Total dream loss: A distinct neuropsychological dysfunction after bilateral PCA stroke. *Annals of Neurology, 56*(4), 583–586.

Black, B. S., Kasper, J., Brandt, J., Shore, A. D., German, P., Burton, L., . . . Rabins, P. V. (2003). Identifying dementia in high-risk community samples: The Memory and Medical Care Study. *Alzheimer Disease and Associated Disorders, 17*(1), 9–18.

Black , D. N., Taber, K. H., & Hurley, R. A. (2003). Metachromatic leukodystrophy: A model for the study of psychosis. *Journal of Neuropsychiatry and Clinical Neuroscience, 15*, 289–293.

Black, J., Guilleminault, C., Bogan, R., Feldman, N., Hagaman, M., Hertz, G., . . . Zammit, G. (2003). US Xyrem Multicenter Study Group US Sleep: *Journal of Sleep and Sleep Disorders Research, 26*, 31–35.

Blair, K. S., Newman, C. C., Mitchell, D. G. V., Richell, R. A., Leonard, A., Morton, J., & Blair, R. J. R. (2006). Differentiating among prefrontal substrates in psychopathy: Neuropsychological test findings. *Neuropsychology, 20*, 153–165.

Blair, R. J. R. (2003). Neurobiological basis of psychopathy. *British Journal of Psychiatry, 182*(1), 5–7.

Blair, R. J. R. (2012). Cortical thinning and functional connectivity in psychopathy. *American Journal of Psychiatry, 169*(7), 684–687.

Boeka, A. G., & Lokken, K. L. (2011). Prefrontal systems involvement in binge eating. *Eating and Weight Disorders: EWD, 16*(2), e121.

Bogerts, B., Ashtari, M., Degreef, G., Alvir, J. M. J., Bilder, R. M., & Lieberman, J. A. (1990). Reduced temporal limbic structure volumes on magnetic resonance images in first episode schizophrenia. *Psychiatry Research: Neuroimaging, 35*(1), 1–13.

Boggs, C. D., Morey, L. C., Skodol, A. E., Shea, M. T., Sanislow, C. A., & Grilo, C. M. (2005). Differential impairment as an indicator of sex bias in DSM-IV criteria for four personality disorders. *Psychological Assessment, 17*, 492–496.

Boghi, A., Sterpone, S., Sales, S., D'Agata, F., Bradac, G. B., Zullo, G., & Munno, D. (2011). In vivo evidence of global and focal brain alterations in anorexia nervosa. *Psychiatry Research: Neuroimaging, 192*(3), 154–159.

Bohon, C., & Stice, E. (2011). Reward abnormalities among women with full and subthreshold bulimia nervosa: A functional magnetic resonance imaging study. *International Journal of Eating Disorders, 44*(7), 585–595.

Bombin, I., Arango, C., & Buchanan, R. W. (2005). Significance and meaning of neurological signs in schizophrenia: Two decades later. *Schizophrenia Bulletin, 31*(4), 962–977.

Boos, H., Aleman, A., Cahn, W., Pol, H. H., & Kahn, R. S. (2007). Brain volumes in relatives of patients with schizophrenia: A meta-analysis. *Archives of General Psychiatry, 64*(3), 297.

Bora, E., Yucel, M., Fornito, A., Pantelis, C., Harrison, B. J., Cocchi, L., . . . Lubman, D. (2012). White matter microstructure in opiate addiction. *Addiction Biology, 17*(1), 141–148.

Bora, E., Yucel, M., & Pantelis, C. (2009). Cognitive endophenotypes of bipolar disorder: A meta-analysis of neuropsychological deficits in euthymic patients and their first-degree relatives. *Journal of Affective Disorders, 113*(1–2), 1–20.

Bora, E., Yucel, M., & Pantelis, C. (2009). Cognitive functioning in schizophrenia, schizoaffective disorder and affective psychoses: Meta-analytic study. *British Journal of Psychiatry, 195*(6), 475–482.

Borbely, A. A., & Achermann, P. (1992). Concepts and models of sleep regulation: An overview. *Journal of Sleep Research, 1*(2), 63–79.

Borg, J. (2007). *Molecular imaging of the serotonin system in human behaviour.* Solna, Sweden: Institutionen för Kliniskt Neurovetenskap/Department of Clinical Neuroscience.

Bostrom, N., & Sandberg, A. (2009). Cognitive enhancement: methods, ethics, regulatory challenges. *Science and Engineering Ethics, 15*(3), 311–341.

Bostwick, J. M., & Pankratz, V. S. (2000). Affective disorders and suicide risk: A reexamination. *American Journal of Psychiatry, 157,* 1925–1932.

Bottmer, C., Bachmann, S., Pantel, J., Essig, M., Amann, M., Schad, L. R., . . . Schröder, J. (2005). Reduced cerebellar volume and neurological soft signs in first-episode schizophrenia. *Psychiatry Research: Neuroimaging, 140*(3), 239–250.

Braem, S., Verguts, T., Roggeman, C., & Notebaert, W. (2012). Reward modulates adaptations to conflict. *Cognition 125,* 324–332.

Braskie, M. N., Toga, A. W., & Thompson, P. M. (2012). Recent advances in imaging Alzheimer's disease. *Journal of Alzheimer's Disease, 30,* 1–15.

Braun, A. R., Balkin, T. J., Wesensten, N. J., Gwadry, F., Carson, R. E., Varga, M., & Herscovitch, P. (1998). Dissociated pattern of activity in visual cortices and their projections during human rapid eye movement sleep. *Science, 279*(5347), 91–95.

Bremner, J. D. (2003). Long-term effects of childhood abuse on brain and neurobiology. *Child and Adolescent Psychiatric Clinics of North America, 12*(2), 271–292.

Brenneis, C., Brandauer, E., Frauscher, B., Schocke, M., Trieb, T., Poewe, W., & Högl, B. (2005). Voxel-based morphometry in narcolepsy. *Sleep Medicine, 6*(6), 531–536.

Brent, D. A., & Mann, J. J. (2005). Family genetic studies, suicide, and suicidal behaviors. *American Journal of Medical Genetics, 133C*(1), 13–24.

Brent, D. A., Oquendo, M., Birmaher, B., Greenhill, L., Kolko, D., Stanley, B., . . . Mann, J. (2002). Familial pathways to early-onset suicide attempt: Risk for suicidal behavior in offspring of mood-disordered suicide attempters. *Archives of General Psychiatry, 59*(9), 801–807.

Brownley, K. A., Berkman, N. D., Sedway, J. A., Lohr, K. N., & Bulik, C. M. (2007). Binge eating disorder treatment: A systematic review of randomized controlled trials. *International Journal of Eating Disorders, 40*(4), 337–348.

Brunner, H. G., Nelen, M., Breakefield, X. O., Ropers, H. H., & Van Oost, B. A. (1993). Abnormal behavior associated with a point mutation in the structural gene for monoamine oxidase A. *Science, 262*(5133), 578–578.

Bruns, J., & Hauser, A. W. (2003). The epidemiology of traumatic brain injury: A review. *Epilepsia, 44*(10), 2–10.

Bryant, R. A., Moulds, M. L., Guthrie, R. M., Dang, S. T., Mastrodomenico, J., Nixon, R. D. V., . . . Creamer, M. (2008, August). A randomized controlled trial of exposure therapy and cognitive restructuring for posttraumatic stress disorder. *Journal of Consulting and Clinical Psychology, 76*(4), 695–703.

Buchanan, R. W., Breier, A., Kirkpatrick, B., Elkashef, A., Munson, R. C., Gellad, F., & Carpenter Jr., W. T. (1993). Structural abnormalities in deficit and nondeficit schizophrenia. *American Journal of Psychiatry, 150,* 59–59.

Buka, S. L., Tsuang, M. T., & Lipsitt, L. P. (1993). Pregnancy/delivery complications and psychiatric diagnosis: A prospective study. *Archives General Psychiatry 50*(2), 151–156.

Burke, W., & Psaty, B. M. (2007). Personalized medicine in the era of genomics. *JAMA: The Journal of the American Medical Association, 298*(14), 1682–1684.

Bush, G. (2008). Neuroimaging of attention deficit hyperactivity disorder: Can new imaging findings be integrated in clinical practice? *Child and Adolescent Psychiatric Clinics of North America, 17*(2), 385–404.

Bushnik, T., Hanks, R. A., Kreutzer, J., & Rosenthal, M. (2003). Etiology of traumatic brain injury: Characterization of differential outcomes up to 1 year postinjury. *Archives of Physical Medicine and Rehabilitation, 84*(2), 255–262.

Butters, M. A., Bhalia, R. K., Andreescu, C., Wetherell, J. L., Mantella, R., Begley, A. E., & Lenze, E. J. (2011). Changes in neuropsychological functioning following treatment for late-life generalised anxiety disorder. *The British Journal of Psychiatry,* (199), 211–218.

Caldji, C., Diorio, J., & Meaney, M. J. (2000). Variations in maternal care in infancy regulate the development of stress reactivity. *Biological Psychiatry, 48*(12), 1164–1174.

Calignon, A., Polydoro, M., Suarez-Calvet, M., William, C., Adamowicz, D., H., Kopeikina, J. K., . . . Hyman, B., T. (2012). Propagation of tau pathology in a model of early Alzheimer's disease. *Neuro, 73*(4), 685–697.

Callaghan, R. C., Cunningham, J. K., Sykes, J., & Kish, S. J. (2012). Increased risk of Parkinson's disease in individuals hospitalized with conditions related to the use of methamphetamine or other amphetamine-type drugs. *Drug and Alcohol Dependence, 120*(1–3), 35–40.

Callicott, J. H., Bertolino, A., Mattay, V. S., Langheim, F. J., Duyn, J., Coppola, R., . . . Weinberger, D. R. (2000). Physiological dysfunction of the dorsolateral prefrontal cortex in schizophrenia revisited. *Cerebral Cortex, 10*(11), 1078–1092.

Campbell, I. C., Mill, J., Uher, R., & Schmidt, U. (2011). Eating disorders, gene–environment interactions and epigenetics. *Neuroscience & Biobehavioral Reviews, 35*(3), 784–793.

Campbell, K. J., & Hesketh, K. D. (2007). Strategies which aim to positively impact on weight, physical activity, diet, and sedentary behaviours in children from zero to five years. A systematic review of the literature. *Obesity Reviews, 8*(4), 327–338.

Camus, V., Gonthier, R., Dubos, G., Schwed, P., & Simeone, I. (2000). Etiologic and outcome profiles in hypoactive and hyperactive subtypes of delirium. *Journal of Geriatric Psychiatry and Neurology, 13*(1), 38–42.

Cannon, M., Caspi, A., Moffitt, T. E., Harrington, H., Taylor, A., Murray, R. M., & Poulton, R. (2002). Evidence for early-childhood, pan-developmental impairment specific to schizophreniform disorder: Results from a longitudinal birth cohort. *Archives of General Psychiatry, 59*(5), 449.

Cardno, A. G., & Gottesman, I. I. (2000). Twin studies of schizophrenia: From bow-and-arrow concordances to Star Wars Mx and functional genomics. *American Journal of Medical Genetics, 97*(1), 12–17.

Carlsson, A., & Lindqvist, M. (1963). Effect of chlorpromazine or haloperidol on formation of 3-methoxytyramine and normetanephrine in mouse brain. *Acta Pharmacologica et Toxicologica, 20*(2), 140–144.

Carpenter, W. T., Heinrichs, D. W., & Wagman, A. M. (1988). Deficit and nondeficit forms of schizophrenia: The concept. *The American Journal of Psychiatry, 145*(5), 578–583.

Carpenter Jr., W. T., Gold, J. M., Lahti, A. C., Queern, C. A., Conley, R. R., Bartko, J. J., . . . Appelbaum, P. S. (2000). Decisional capacity for informed consent in schizophrenia research. *Archives of General Psychiatry, 57*(6), 533–538.

Carr, A. C., Goldstein, E. G., Hunt, H. F., & Kernberg, O. F. (1979). Psychological tests and borderline patients. *Journal of Personality Assessment, 43*(6), 582–590.

Carrion, V. G., Weems, C. F., Watson, C., Eliez, S., Menon, V., & Reiss, A. L. (2009). Converging evidence for abnormalities of the prefrontal cortex and evaluation of midsagittal structures in pediatric posttraumatic stress disorder: An MRI study. *Psychiatry Research: Neuroimaging, 172*(3), 226–234.

Caseras, X., Giampietro, V., Lamas, A., Brammer, M., Vilarroya, O., Carmon, S., & Mataix-Cols, D. (2010). The functional neuroanatomy of blood-injection-injury phobia: A comparison with spider phobics and healthy controls. *Psychological Medicine, 40*(1), 125–134.

Cases, O., Seif, I., Grimsby, J., Gaspar, P., Chen, K., Pournin, S., . . . De Maeyer, E. (1995). Aggressive behavior and altered amounts of brain serotonin and norepinephrine in mice lacking MAOA. *Science, 268*(5218), 1763–1766.

Caspi, A., McClay, J., Moffitt, T. E., Mill, J., Martin, J., Craig, I. W., . . . Poulton, R. (2002). Role of genotype in the cycle of violence in maltreated children. *Science, 297*(5582), 851–854.

Caspi, A., Sugden, K., Moffitt, T. E., Taylor, A., Craig, I. W., Harrington, H., . . . Poulton, R. (2003). Influence of life stress on depression: Moderation by a polymorphism in the 5-HTT gene. *American Association for the Advancement of Science, 301*, 386–389.

Cassin, S. E., & von Ranson, K. M. (2005). Personality and eating disorders: A decade in review. *Clinical Psychology Review, 25*(7), 895–916.

Celone, K. A., Thompson-Brenner, H., Ross, R. S., Pratt, E. M., & Stern, C. E. (2011). An fMRI investigation of the fronto-striatal learning system in women who exhibit eating disorder behaviors. *NeuroImage, 56*(3), 1749–1757.

Cendron, M. (1999). Primary nocturnal enuresis: Current concepts. *American Family Physician, 59*, 1205–1224.

Centers for Disease Control and Prevention. (2011). Current depression among adults—United States 2006 and 2008. *Morbidity and Mortality Weekly Reports (2010), 59*(38), 1229–1235.

Chen, M. C., Hamilton, J. P., & Gotlib, I. H. (2010). Decreased hippocampal volume in healthy girls at risk of depression. *Archives of General Psychiatry, 67*(3), 270–276.

Chen, Z., Li, M., Deng, W., He, Z., Wang, Q., Jiang, L., . . . Li, T. (2012). Volume increases in putamen associated with positive symptom reduction in previously drug-naive schizophrenia after 6 weeks antipsychotic treatment. *Psychological Medicine, 42*(7), 1475–1483.

Chengappa, K. N., Ebeling, T., Kang, J. S., Levine, J., & Parepally, H. (1999). Clozapine reduces severe self-mutilation and aggression in psychotic patients with borderline personality disorder. *The Journal of Clinical Psychiatry, 60*(7), 477–484.

Chey, J., Lee, J., Kim, Y. S., Kwon, S. M., & Shin, Y. M. (2002). Spatial working memory span, delayed response, and executive function in schizophrenia. *Psychiatry Research, 110*(3), 259–271.

Chien, K. L., Chen, P. C., Hsu, H. C., Su, T. C., Sung, F. C., Chen, M. F., & Lee, Y. T. (2010). Habitual sleep duration and insomnia and the risk of cardiovascular events and all-cause death: Report from a community-based cohort. *Sleep, 33*(2), 177.

Choi, J., Jeong, B., Rohan, M. L., Polcari, A. M., & Teicher, M. H. (2009). Preliminary evidence for white matter tract abnormalities in young adults exposed to parental verbal abuse. *Biological Psychiatry, 65*(3), 227–234.

Choo, I. H., Lee, D. Y., Oh, J. S., Lee, J. S., Lee, D. S., Song, I. C., . . . Woo, J. I. (2010). Posterior cingulate cortex atrophy and regional cingulum disruption in mild cognitive impairment and Alzheimer's disease. *Neurobiology of Aging, 31*(5), 772–779.

Choy, Y., Peselow, E. D., Case, B. G., Pressman, M. A., Luff, J. A., Laje, G., & Guardino, M. T. (2007). Three-year medication prophylaxis in panic disorder: To continue or discontinue? A naturalistic study. *Comprehensive Psychiatry, 48*(5), 419–425.

Cicchetti, D., Rogosch, F. A., & Thibodeau, E. L. (2012). The effects of child maltreatment on early signs of antisocial behavior: Genetic moderation by tryptophan hydroxylase, serotonin transporter, and monoamine oxidase A genes. *Development and Psychopathology, 24*(3), 907.

Cicerone, K. D., Dahlberg, C., Kalmar, K., Langenbahn, D. M., Malec, J. F., Bergquist, T. F., . . . Morse, P. A. (2000). Evidence-based cognitive rehabilitation: Recommendations for clinical practice. *Archives of Physical Medicine and Rehabilitation, 81*(12), 1596–1615.

Coffey, C., Weiner, R. D., Djang, W. T., Figiel, G. S., Soady, S. R., Patterson, L. J., . . . Wilkinson, W. E. (1991). Brain anatomic effects of electroconvulsive therapy: A prospective magnetic resonance imaging study. *Archives of General Psychiatry, 48*(11), 1013–1021.

Cohen, J. D., & Servan-Schreiber, D. (1992). Context, cortex, and dopamine: A connectionist approach to behavior and biology in schizophrenia. *Psychological Review, 99*(1), 45–77.

Cohen, M. E., Dembling, B., & Schorling, J. B. (2002). The association between schizophrenia and cancer: A population-based mortality study. *Schizophrenia Research, 57*(2), 139–146.

Combs, D. R., Adams, S. D., Wood, T. D., Basso, M. R., & Gouvier, W. D. (2005). Informed consent in schizophrenia: The use of cues in the assessment of understanding. *Schizophrenia Research, 77*(1), 59–63.

Comings, D. E., Gade-Andavolu, R., Gonzalez, N., Wu, S., Muhleman, D., Blake, H., . . . MacMurray, J. P. (2000). Comparison of the role of dopamine, serotonin, and noradrenaline genes in ADHD, ODD and conduct disorder: multivariate regression analysis of 20 genes. *Clinical Genetics 57*(3), 178–196.

Commowick, O., Fillard, P., Clatz, O., & Warfield, S. (2008). Detection of DTI white matter abnormalities in multiple sclerosis patients. *Medical Image Computing and Computer-Assisted Intervention, 11*, 975–982.

Coolidge, F. L., Thede, L. L., & Jang, K. L. (2001). Heritability of childhood personality disorders: A preliminary study. *Journal of Personality Disorders, 15*, 33–40.

Cope, M., Delpy, D. T., Reynolds, E. O. R., Wray, S., Wyatt, J., & Van der Zee, P. (1988). Methods of quantitating cerebral near infrared spectroscopy data. In M. Mochizuki, C. R. Honig, T. Koyama, T. K. Goldstick, & D. F. Burley (Eds.), *Oxygen transport to Tissue X* (pp. 183–189). New York, NY: Springer.

Corey-Bloom, J., & Merdes, A. R. (2004). Influence of Alzheimer pathology on clinical diagnostic accuracy in dementia with Lewy bodies. *Neurology, 62*, 160.

Cortese, S., Kelly, C., Chabernaud, C., Proal, E., Di Martino, A., Milham, M. P., & Castellanos, F. X. (2012). Toward systems neuroscience of ADHD: A meta-analysis of 55 fMRI studies. *American Journal of Psychiatry, 169*(10), 1038–1055.

Courchesne, E., Mouton, P. R., Calhoun, M. E., Semendeferi, K., Ahrens-Barbeau, C., Hallet, M. J., . . . Pierce, K. (2011). Neuron number and size in prefrontal cortex of children with autism. *JAMA: The Journal of the American Medical Association, 306*(18), 2001–2010.

Court, A., Mulder, C., Hetrick, S. E., Purcell, R., & McGorry, P. D. (2008). What is the scientific evidence for the use of antipsychotic medication in anorexia nervosa? *Eating Disorders, 16*(3), 217–223.

Cousins, D. A., Butts, K., & Young, A. H. (2009). The role of dopamine in bipolar disorder. *Bipolar Disorders, 11*(8), 787–806.

Cowdrey, F. A., Filippini, N., Park, R. J., Smith, S. M., & McCabe, C. (2012). Increased resting state functional connectivity in the default mode network in recovered anorexia nervosa. *Human Brain Mapping*. doi: 10.1002/hbm.22202

Cramer, V., Torgersen, S., & Kringlen, E. (2004). Quality of life in a city: The effect of population density. *Social Indicators Research, 69*(1), 103–116.

Crow, S. J., Mitchell, J. E., Roerig, J. D., & Steffen, K. (2009). What potential role is there for medication treatment in anorexia nervosa? *International Journal of Eating Disorders, 42*(1), 1–8.

Crow, T. (2007). How and why genetic linkage has not solved the problem of psychosis: Review and hypothesis. *American Journal of Psychiatry, 164*(1), 13–21.

Croy, I., Schellong, J., Gerber, J., Joraschky, P., Iannilli, E., & Hummel, T. (2010). Women with a history of childhood maltreatment exhibit more activation in association areas following non-traumatic olfactory stimuli: An fMRI study. *PloS One, 5*(2), e9362.

Currier, D., & Mann, J. J. (2008). Stress, genes, and the biology of suicidal behavior. *Psychiatric Clinics of North America, 31*(2), 247–269.

Cyprien, F., Courtet, P., Malafosse, A., Maller, J., Meslin, C., Bonafé, A., . . . Artero, S. (2011). Suicidal behavior is associated with reduced corpus callosum area. *Society of Biological Psychiatry, 70*, 320–326.

D'Alessandro, V., Mason II, T., Pallone, M. N., Patano, J., & Marcus, C. L. (2005). Late-onset hypoventilation without PHOX2B mutation or hypothalamic abnormalities. *Journal of Clinical Sleep Medicine: JCSM, 1*(2), 169.

Davidson, K., Halford, J., Kirkwood, L., Newton-Howes, G., Sharp, M., & Tata, P. (2010). CBT for violent men with antisocial personality disorder: Reflections on the experience of carrying out therapy in MASCOT, a pilot randomized controlled trial. *Personality and Mental Health, 4*(2), 86–95.

Davidson, L., & McGlashan, T. (1997). The varied outcomes of schizophrenia. *Canadian Journal of Psychiatry, 42*(1), 34.

Davidson, R. J., Putnam, K. M., & Larson, C. L. (2000). Dysfunction in the neural circuitry of emotion regulation—a possible prelude to violence. *Science, 289*(5479), 591–594.

Davies, H., & Tchanturia, K. (2005). Cognitive remediation therapy as an intervention for acute anorexia nervosa: A case report. *European Eating Disorders Review, 13*(5), 311–316.

Daviglus, M. L., Bell, C. C., Berrettini, W., Bowen, P. E., Connolly, E. S., Cox, N. J., . . . Trevisan, M. (2010, April 26–28). National Institutes of Health State-of-the-Science Conference Statement: Preventing Alzheimer's Disease and Cognitive Decline. *NIH Consensus State of Science Statements, 27*(4), 1–30.

Davis, C., Patte, K., Curtis, C., & Reid, C. (2010). Immediate pleasures and future consequences: A neuropsychological study of binge eating and obesity. *Appetite, 54*(1), 208–213.

Dawson, J. L., & Conduit, R. (2011). The substrate that dreams are made on: An evaluation of current neurobiological theories of dreaming. In D. Cvetkovic & I. Cosic (Eds.), *States of Consciousness* (pp. 133–156). Berlin: Springer.

DeBellis, M. D. (2002). Developmental traumatology: A contributory mechanism for alcohol and substance use disorders. *Psychoneuroendocrinology, 27*(1–2), 155–170.

DeBellis, M. D., Clark, D. B., Beers, S. R., Soloff, P. H., Boring, A. M., Hall, J., . . . Keshavan, M. S. (2000). Hippocampal volume in adolescent-onset alcohol use disorders. *American Journal of Psychiatry, 157*(5), 737–744.

DeBellis, M. D., Hooper, S. R., Spratt, E. G., & Woolley, D. P. (2009). Neuropsychological findings in childhood neglect and their relationships to pediatric PTSD. *Journal of the International Neuropsychological Society, 15*(6), 868–878.

DeBellis, M. D., Keshavan, M. S., Clark, D. B., Casey, B. J., Giedd, J. N., Boring, A. M., . . . Ryan, N. D. (1999). Developmental traumatology part II: Brain development. *Biological Psychiatry, 45*(10), 1271–1284.

DeBellis, M. D., & Kuchibhatla, M. (2006). Cerebellar volumes in pediatric maltreatment-related posttraumatic stress disorder. *Biological Psychiatry, 60*(7), 697–703.

Deckersbach, T., Moshier, S. J., Tuschen-Caffier, B., & Otto, M. W. (2011). Memory dysfunction in panic disorder: An investigation of the role of chronic benzodiazepine use. *Depression and Anxiety, 28*(11), 999–1007.

DeLeon, J., & Diaz, F. J. (2005). A meta-analysis of worldwide studies demonstrates an association between schizophrenia and tobacco smoking behaviors. *Schizophrenia Research, 76*(2), 135–157.

De Oliveira-Souza, R., Hare, R. D., Bramati, I. E., Garrido, G. J., Ignácio, F. A., Tovar-Moll, F., & Moll, J. (2008). Psychopathy as a disorder of the moral brain: Fronto-temporo-limic grey matter reductions demonstrated by voxel-based morphometry. *NeuroImage, 40*(3), 1202–1213.

De Oliveira-Souza, R., Moll, J., Ignácio, F. A., & Hare, R. D. (2008). Psychopathy in a civil psychiatric outpatient sample. *Criminal Justice and Behavior, 35*(4), 427–437.

Depp, C. A., Lebowitz, B. D., Patterson, T. L., Lacro, J. P., & Jeste, D. V. (2007). Medication adherence skills training for middle-aged and elderly adults with bipolar disorder: Development and pilot study. *Bipolar Disorders, 9*(6), 636–645.

DePrince, A. P., Weinzierl, K. M., & Combs, M. D. (2009). Executive function performance and trauma exposure in a community sample of children. *Child Abuse & Neglect, 33*(6), 353–361.

De Rubeis, R. J., Gelfand, L. A., Tang, T. Z., & Simons, A. D. (1999). Medications versus cognitive behavior therapy for severely depressed outpatients: Meta-analysis of four randomized comparisons. *The American Journal of Psychiatry, 156*(7), 1007–1013.

De Santis, A., Noar, S. M., & Webb, E. M. (2009). Nonmedical ADHD stimulant use in fraternities. *Journal of Studies on Alcohol and Drugs, 70*(6), 952–954.

Deshmukh, A., Rosenbloom, M. J., De Rosa, E., Sullivan, E. V., & Pfefferbaum, A. (2005). Regional striatal volume abnormalities in schizophrenia: Effects of comorbidity for alcoholism, recency of alcoholic drinking, and antipsychotic medication type. *Schizophrenia Research, 79*(2), 189–200.

Di Chiara, G., Bassaero, V., Fenu, S., Deluca, M. A., Spina, L., Cardoni, C., . . . Lecca, D. (2004). Dopamine and drug addiction: The nucleus accumbens shell connection. *Neuropharmacology, 47*(1), 227–241.

DiMario, F. J., & Emery, E. (1987). The natural history of night terrors. *Clinical Pediatrics, 26*, 505–511.

Dinn, W. M., & Harris, C. L. (2000). Neurocognitive function in antisocial personality disorder. *Psychiatry Research, 97*(2), 173–190.

Disner, S. G., Beevers, C. G., Haigh, E. A. P., & Beck, T. A. (2011). Neural mechanisms of the cognitive model of depression. *Nature Reviews Neuroscience, 12*, 467–477.

Doble, J. E., Feinberg, D. L., Rosner, M. S., & Rosner, A. J. (2010). Identification of binocular vision dysfunction (vertical heterophoria) in traumatic brain injury patients and effects of individualized prismatic spectacle lenses in the treatment of postconcussive symptoms: A retrospective analysis. *PM&R, 2*(4), 244–253.

Dolan, M., & Park, I. (2002). The neuropsychology of antisocial personality disorder. *Psychological Medicine, 32*(3), 417–427.

Dolan, R. J., Mitchell, J., & Wakeling, A. (1988). Structural brain changes in patients with anorexia nervosa. *Psychological Medicine, 18*(2), 349–353.

Domschke, K., Stevens, S., Pfleiderer, B., & Gerlach, A. (2010, February). Interoceptive sensitivity in anxiety and anxiety disorders: An overview and integration of neurobiological findings. *Clinical Psychology Review, 30*(1), 1–11.

Donegan, N. H., Sanislow, C. A., Blumberg, H. P., Fulbright, R. K., Lacadie, C., Skudlarski, P., . . . Wexler, B. E. (2003). Amygdala hyperreactivity in borderline personality disorder: Implications for emotional dysregulation. *Biological Psychiatry, 54*(11), 1284–1293.

Downar, J., Sankar, A., Giacobbe, P., Woodside, B., & Colton, P. (2012). Unanticipated rapid remission of refractory bulimia nervosa, during high-dose repetitive transcranial magnetic stimulation of the dorsomedial prefrontal cortex: A case report. *Frontiers in Psychiatry, 3*, 1–5.

Draganski, B., Geisler, P., Hajak, G., Schuierer, G., Bogdahn, U., Winkler, J., & May, A. (2002). Hypothalamic gray matter changes in narcoleptic patients. *Nature Medicine, 8*(11), 1186–1188.

Driesen, N. R., Leung, H. C., Calhoun, V. D., Constable, R. T., Gueorguieva, R., Hoffman, R., . . . Krystal, J. H. (2008). Impairment of working memory maintenance and response in schizophrenia: Functional magnetic resonance imaging evidence. *Biological Psychiatry, 64*(12), 1026–1034.

Driessen, M., Beblo, T., Mertens, M., Piefke, M., Rullkoetter, N., Silva-Saavedra, A., . . . Woermann, F. G. (2004). Posttraumatic stress disorder and fMRI activation patterns of traumatic memory in patients with borderline personality disorder. *Biological Psychiatry, 55*(6), 603–611.

Driessen, M., Herrmann, J., Stahl, K., Zwaan, M., Meier, S., Hill, A., . . . Petersen, D. (2000). Magnetic resonance imaging volumes of the hippocampus and the amygdala in women with borderline personality disorder and early traumatization. *Archives of General Psychiatry, 57*(12), 1115–1122.

Drummond, S. P., Brown, G. G., & Salamat, J. S. (2003). Brain regions involved in simple and complex grammatical transformations. *NeuroReport, 14*(8), 1117–1122.

Duchesne, M., Mattos, P., Appolinário, J. C., de Freitas, S., Coutinho, G., Santos, C., & Coutinho, W. (2010). Assessment of executive functions in obese individuals with binge eating disorder. *Revista Brasileira de Psiquiatria, 32*(4), 381–387.

Duggan, C., Huband, N., Smailagic, N., Ferriter, M., & Adams, C. (2008). The use of pharmacological treatments for people with personality disorder: A systematic review of randomized controlled trials. *Personality and Mental Health, 2*(3), 119–170.

Duggan, C., Mason, L., Banerjee, P., & Milton, J. (2007). Value of standard personality assessments in informing clinical decision-making in a medium secure unit. *The British Journal of Psychiatry, 190*(49), s15–s19.

Dumas, E. M., van den Bogaard, S. J., Ruber, M. E., Reilmann, R., Stout, J. C., Craufurd, D., . . . Roos, R. A. (2012). Early changes in white matter pathways of the sensorimotor cortex in premanifest Huntington's disease. *Human Brain Mapping, 33*(1), 203–212.

Dumas, R., Baumstarck, K., Michel, P., Lançon, C., Auquier, P., & Boyer, L. (2013). Systematic review reveals heterogeneity in the use of the Scale to Assess Unawareness of Mental Disorder (SAUMD). *Current Psychiatry Reports, 15*(6), 1–15.

Duong, T. T., Englander, J., Wright, J., Cifu, D. X., Greenwald, B. D., & Brown, A. W. (2004). Relationship between strength, balance, and swallowing deficits and outcome after traumatic brain injury: A multicenter analysis. *Archives of Physical Medicine and Rehabilitation, 85*(8), 1291–1297.

Durmer, J. S., & Dinges, D. F. (2005). Neurocognitive consequences of sleep deprivation. *Seminars in Neurology, 25*(1), 117–129.

Durston, S., Pol, H. E. H., Schnack, H. G., Buitelaar, J. K., Steenhuis, M. P., Minderaa, R. B., & Kahn, R. S. (2004). Magnetic resonance imaging of boys with attention-deficit/hyperactivity disorder and their unaffected siblings. *Journal of the American Academy of Child & Adolescent Psychiatry, 43*(3), 332–340.

Duyckaerts, C., & Dickson, D. (2011). Neuropathology of Alzheimer's disease and its variants. In D. W. Dickson and R. O. Weller (Eds.), *Neurodegeneration: The molecular pathology of dementia and movement disorders* (2nd ed., p. 62). Oxford, UK: Wiley-Blackwell.

Ebdrup, B. H., Glenthøj, B., Rasmussen, H., Aggernaes, B., Langkilde, A. R., Paulson, O. B., & Baaré, W. (2010). Hippocampal and caudate volume reductions in antipsychotic-naive first-episode schizophrenia. *Journal of Psychiatry & Neuroscience: JPN, 35*(2), 95.

Ebdrup, B. H., Skimminge, A., Rasmussen, H., Aggernaes, B., Oranje, B., Lublin, & Glenthøj, B. (2011). Progressive striatal and hippocampal volume loss in initially antipsychotic-naive, first-episode schizophrenia patients treated with quetiapine: Relationship to dose and symptoms. *The International Journal of Neuropsychopharmacology, 14*(1), 69–82.

Eckert, M. (2004). Neuroanatomical markers for dyslexia: A review of dyslexia structural imaging studies. *The Neuroscientist, 10*(4), 362–371.

Eisenberg, D. P., & Berman, K. F. (2009). Executive function, neural circuitry, and genetic mechanisms in schizophrenia. *Neuropsychopharmacology, 35*(1), 258–277.

Ellman, L. M., Huttunen, M., Lonnqvist, J., & Cannon, T. D. (2007). The effects of genetic liability for schizophrenia and maternal smoking during pregnancy on obstetric complications. *Schizophrenia Research, 93*(1–3), 229–236.

Ellman, L. M., Yolken, R. H., Buka, S. L., Torrey, E. F., & Cannon, T. D. (2009). Cognitive functioning prior to the onset of psychosis: The role of fetal exposure to serologically determined influenza infection. *Biological Psychiatry, 65*(12), 1040–1047.

Engel, A. K., & Singer, W. (2001). Temporal binding and the neural correlates of sensory awareness. *Trends in Cognitive Sciences, 5*(1), 16–25.

Engel, J. (1993). Update on surgical treatment of the epilepsies: summary of the Second International Palm Desert Conference on the Surgical Treatment of the Epilepsies (1992). *Neurology, 43*(8), 1612–1617.

Engel, K., Bandelow, B., Gruber, O., & Wedekind, D. (2008). Neuroimaging in anxiety disorders. *Journal of Neural Transmission, 116*, 703–716.

Erhart, S. M., Young, A. S., Marder, S. R., & Mintz, J. (2005). Clinical utility of magnetic resonance imaging radiographs for suspected organic syndromes in adult psychiatry. *The Journal of Clinical Psychiatry, 66*(8), 968–973.

Ettinger, U., Schmechtig, A., Toulopoulou, T., Borg, C., Orrells, C., Owens, S., & Picchioni, M. (2012). Prefrontal and striatal volumes in monozygotic twins concordant and discordant for schizophrenia. *Schizophrenia Bulletin, 38*(1), 192–203.

Everson, C. A., & Nowak, T. S. (2001). Transcriptional responses related to thyrotropin-releasing hormone regulation during suppression of the thyroid hormone axis by sleep deprivation in rats. *Sleep, 24*, A75–A76.

Fair, D. A., Cohen, A. L., Dosenbach, N. U., Church, J. A., Miezin, F. M., Barch, D. M., . . . Schlaggar, B. L. (2008). The maturing architecture of the brain's default network. *Proceedings of the National Academy of Sciences, 105*(10), 4028–4032.

Fairburn, C. G., Agras, W. S., Walsh, B. T., Wilson, G. T., & Stice, E. (2004). Prediction of outcome in bulimia nervosa by early change in treatment. *American Journal of Psychiatry, 161*(12), 2322–2324.

Fairburn, C. G., Cooper, Z., & Shafran, R. (2003). Cognitive behaviour therapy for eating disorders: A "transdiagnostic" theory and treatment. *Behaviour Research and Therapy, 41*(5), 509–528.

Fairburn, C. G., Shafran, R., & Cooper, Z. (1999). A cognitive behavioural theory of anorexia nervosa. *Behaviour Research and Therapy, 37*(1), 1–13.

Farah, M. J., Illes, J., Cook-Deegan, R., Gardner, H., Kandel, E., King, P., & Parens, E. (2004). Neurocognitive enhancement: What can we do and what should we do? *Nature Reviews Neuroscience, 5*, 421–425.

Faraone, S. V., Biederman, J., & Mick, E. (2006). The age-dependent decline of attention deficit hyperactivity disorder: A meta-analysis of follow-up studies. *Psychological Medicine, 36*(2), 159–166.

Farrin, L., Hull, L., Unwin, C., Wykes, T., & David, A. (2003). Effects of depressed mood on objective and subjective measures of attention. *The Journal of Neuropsychiatry & Clinical Neurosciences, 15*, 98–104.

Faul, M., Xu, L., Wald, M. M., & Coronado, V. G. (2010). Traumatic brain injury in the United States: Emergency department visits, hospitalizations, and deaths. Atlanta, GA: Centers for Disease Control and Prevention, National Center for Injury Prevention and Control.

Fazel, S., & Danesh, J. (2002). Serious mental disorder in 23,000 prisoners: A systematic review of 62 surveys. *Lancet, 359*(9306), 545–550.

Fazeli, P. K., Calder, G. L., Miller, K. K., Misra, M., Lawson, E. A., Meenaghan, E., . . . Klibanski, A. (2012). Psychotropic medication use in anorexia nervosa between 1997 and 2009. *International Journal of Eating Disorders, 45*(8), 970–976.

Ferguson, C. J. (2010). Genetic contributions to antisocial personality and behavior: A meta-analytic review from an evolutionary perspective. *The Journal of Social Psychology, 150*(2), 160–180.

Ferraro, F. R., Wonderlich, S., & Jocic, Z. (1997). Performance variability as a new theoretical mechanism regarding eating disorders and cognitive processing. *Journal of Clinical Psychology, 53*(2), 117–121.

Fitzgerald, K. L., & Demakis, G. J. (2007). The neuropsychology of antisocial personality disorder. *Disease-a-Month: DM, 53*(3), 177–183.

Fitzsimmons, J., Kubicki, M., Smith, K., Bushell, G., Estepar, R., Westin, C. F., . . . Shenton, M. E. (2009). Diffusion tractography of the fornix in schizophrenia. *Schizophrenia Research, 107*(1), 39–46.

Fleck, D. E., Eliassen, J. C., Durling, M., Lamy, M., Adler, C. M., DelBello, M. P., . . . Strakowski, S. M. (2010). Functional MRI of sustained attention in bipolar mania. *Molecular Psychiatry, 17*(3), 325–336.

Fleisher, A. S., Sun, S., Taylor, C., Ward, C. P., Gamst, A. C., Petersen, R. C., . . . Thal, L. J. (2008). Volumetric MRI vs clinical predictors of Alzheimer disease in mild cognitive impairment. *Neurology, 70*(3), 191–199.

Flesher, S. (1990). Cognitive habilitation in schizophrenia: A theoretical review and model of treatment. *Neuropsychology Review, 1*(3), 223–246.

Fogel, S. M., Nader, R., Cote, K. A., & Smith, C. T. (2007). Sleep spindles and learning potential. *Behavioral Neuroscience, 121*(1), 1–10.

Foley, D. J., Vitiello, M. V., Bliwise, D. L., Ancoli-Israel, S., Monjan, A. A., & Walsh, J. K. (2007). Frequent napping is associated with excessive daytime sleepiness, depression, pain, and nocturia in older adults. *American Journal of Geriatric Psychiatry, 15*, 344–350.

Fombonne, E. (2009). A wrinkle in time: From early signs to a diagnosis of autism. *Journal of the American Academy of Child and Adolescent Psychiatry, 48*(5), 463–464.

Fossati, P., Ergis, A. M., & Allilaire, J. M. (2002). Executive functioning in unipolar depression: A review. *Encephale, 28*(2), 97–107.

Frank, G. K., Bailer, U. F., Henry, S. E., Drevets, W., Meltzer, C. C., Price, J. C., . . . Kaye, W. H. (2005). Increased dopamine D2/D3 receptor binding after recovery from anorexia nervosa measured by Positron Emission Tomography and [$^{11}$C] Raclopride. *Biological Psychiatry, 58*(11), 908–912.

Frank, G. K., Bailer, U. F., Meltzer, C. C., Price, J. C., Mathis, C. A., Wagner, A., . . . Kaye, W. H. (2007). Regional cerebral blood flow after recovery from anorexia or bulimia nervosa. *International Journal of Eating Disorders, 40*(6), 488–492.

Freitag, C. M. (2006). The genetics of autistic disorders and its clinical relevance: A review of the literature. *Molecular Psychiatry, 12*(1), 2–22.

Fries, A. B. W., & Pollak, S. D. (2004). Emotion understanding in postinstitutionalized Eastern European children. *Development and Psychopathology, 16*(2), 355–370.

Fujii, D. E., Wylie, A. M., & Nathan, J. H. (2004). Neurocognition and long-term prediction of quality of life in outpatients with severe and persistent mental illness. *Schizophrenia Research, 69*(1), 67–73.

Fulda, S., & Schulz, H. (2001). Cognitive dysfunction in sleep disorders. *Sleep Medicine Reviews, 5*(6), 423–445.

Funke, B., Finn, C. T., Plocik, A. M., Lake, S., DeRosse, P., Kane, J. M., . . . Malhotra, A. K. (2004). Association of the DTNBP1 Locus with schizophrenia in a US population. *The American Journal of Human Genetics, 75*(5), 891–898.

Fusar-Poli, P., Rubia, K., Rossi, G., Sartori, G., & Balottin, U. (2012). Striatal dopamine transporter alterations in ADHD: Pathophysiology or adaptation to psychostimulants? A meta-analysis. *American Journal of Psychiatry, 169*(3), 264–272.

Gabbard, G. O. (2005). Mind, brain, and personality disorders. *American Journal of Psychiatry, 162*(4), 648–655.

Galderisi, S., Quarantelli, M., & Volpe, U. (2008). Patterns of structural MRI abnormalities in deficit and non-deficit schizophrenia. *Schizophrenia Bulletin, 34*, 393–401.

Garavan, H., Brennan, K., Hester, R., & Whelan, R. (2013). The neurobiology of successful abstinence. *Current Opinion in Neurobiology, 23*(4), 468–474.

Garriock, H. A., & Moreno, F. A. (2011). Genetics of depression: Implications for clinical practice yet? *Clinical Neuropsychology, 8*(1), 37–46. www.genecards.org

Gennarelli, T. A., & Graham, D. I. (1998). Neuropathology of the head injuries. *Seminars in Clinical Neuropsychiatry, 3*(3), 160–175.

Gennarelli, T. A., Thibault, L. E., & Graham, D. I. (1998). Diffuse axonal injury: An important form of traumatic brain damage. *The Neuroscientist, 4*(3), 202–215.

George, D. T., Rawlings, R. R., Williams, W. A., Phillips, M. J., Fong, G., Kerich, M., . . . Hommer, D. (2004). A select group of perpetrators of domestic violence: Evidence of decreased metabolism in the right hypothalamus and reduced relationships between cortical/subcortical brain structures in position emission tomography. *Psychiatry Research: Neuroimaging, 130*(1), 11–25.

Getz, G. E., Fleck, D. E., Shear, P. K., & Strakowski, S. M. (2003). Positive emotion labeling bias in mania. *Journal of International Neuropsychological Society, 9* abstract, 327.

Getz, G. E., Shear, P. K., & Strakowski, S. M. (2003). Facial affect recognition deficits in bipolar disorder. *Journal of the International Neuropsychological Society, 9*, 623–632.

Giedd, J. N. (2004). Structural magnetic resonance imaging of the adolescent brain. *Annals of the New York Academy of Sciences, 1021*(1), 77–85.

Gold, S., Arndt, S., Nopoulos, P., O'Leary, D. S., & Andreasen, N. C. (1999). Longitudinal study of cognitive function in first-episode and recent-onset schizophrenia. *American Journal of Psychiatry, 156*(9), 1342–1348.

Goldapple, K., Zindel, S., Garson, C., Lau, M., Bieling, P., Kennedy, S., & Mayberg, H. (2004). Modulation of cortical-limbic pathways in major depression: Treatment-specific effects of cognitive behavior therapy. *Archives of General Psychiatry, 61*(4), 34–41.

Goldberg, T. E., Weinberger, D. R., Pliskin, N. H., Berman, K. F., & Podd, M. H. (1989). Recall memory deficit in schizophrenia: A possible manifestation of prefrontal dysfunction. *Schizophrenia Research, 2*(3), 251–257.

Goodman, A. (2008). Neurobiology of addiction: An integrative review. *Biochemical Pharmacology, 75*(1), 266–322.

Goodwin, F. K., & Jamison, K. R. (2007). *Manic-depressive illness: Bipolar disorders and recurrent depression* (2nd ed.). New York, NY: Oxford University Press.

Gould, E., Reeves, A. J., Fallah, M., Tanapat, P., Gross, C. G., & Fuchs, E. (1999). Hippocampal neurogenesis in adult Old World primates. *Proceedings of the National Academy of Sciences of the United States of America 96*(9), 5263–5267.

Gouzoulis-Mayfrank, E. (2008). Dual diagnosis psychosis and substance use disorders: Theoretical foundations and treatment. *Zeitschrift fur Kinder- und Jugendpsychiatrie und Psychotherapie, 36*(4), 245–253.

Graham, D. I., Horsburgh, K., Nicoll, J. A. R., & Teasdale, G. M. (1999). Apolipoprotein E and the response of the brain to injury. In A. Baethmann, N. Plesnila, F. Ringel, & J. Eriskat (Eds.), *Current progress in the understanding of secondary brain damage from trauma and ischemia* (pp. 89–92). Vienna: Springer.

Grant, B. F., Hasin, D. S., Stinson, F. S., Dawson, D. A., Chou, S. P., Ruan, W. J., & Pickering, R. P. (2004). Prevalence, correlates, and disability of personality disorders in the United States: Results from the National Epidemiologic Survey on Alcohol and Related Conditions. *The Journal of Clinical Psychiatry, 65*(7), 948–958.

Greek, R., & Greek, J. (2000). Animal research and human disease. *JAMA: The Journal of the American Medical Association, 283*(6), 743–744.

Green, M. F. (1996). What are the functional consequences of neurocognitive deficits in schizophrenia? *The American Journal of Psychiatry, 153*(3), 321–330.

Greenberg, J. J. Validation of the pain anxiety symptoms scale using psychophysiological and behavioral variables. AAI3061532, *Dissertation Abstracts International: Section B: The Sciences and Engineering*, 3914.

Greenwald, B. D., Cifu, D. X., Marwitz, J. H., Enders, L. J., Brown, A. W., Englander, J. S., & Zafonte, R. D. (2001). Factors associated with balance deficits on admission to rehabilitation after traumatic brain injury: A multicenter analysis. *The Journal of Head Trauma Rehabilitation, 16*(3), 238–252.

Greenwood, T. A., Light, G. A., Swerdlow, N. R., Radant, A. D., & Braff, D. L. (2012). Association analysis of 94 candidate genes and schizophrenia-related endophenotypes. *Plos One, 7*(1). doi:10.1371/journal.pone.0029630

Grisso, T., & Appelbaum, P. S. (1995) MacArthur Treatment Competence Study: I. Mental illness and competence to consent to treatment. *Law and Human Behaviour, 19,* 105–126.

Grunnert, B. K., Smucker, M. R., Weis, J. M., & Rusch, M. D. (2003). When prolonged exposure fails: Adding an imagery-based cognitive restructuring component in the treatment of industrial accident victims suffering from PTSD. *Cognitive and Behavioral Practice, 10*(4), 333–346.

Gruzelier, J., Seymour, K., Wilson, L., Jolley, A., & Hirsch, S. (1988). Impairments on neuropsychologic tests of temporohippocampal and frontohippocampal functions and word fluency in remitting schizophrenia and affective disorders. *Archives of General Psychiatry, 45*(7), 623.

Gualtieri, C., Johnson, L. G., & Benedict, K. B. (2006). Neurocognition in depression: Patients on and off medication versus healthy comparison subjects. *The Journal of Neuropsychiatry and Clinical Neurosciences, 18*(2), 217–225.

Gunderson, J. G. (2009). Borderline personality disorder: Ontogeny of a diagnosis. *American Journal of Psychiatry, 166*(5), 530–539.

Gur, R. E., Maany, V., Mozley, P. D., Swanson, C., Bilker, W., & Gur, R. C. (1998). Subcortical MRI volumes in neuroleptic-naive and treated patients with schizophrenia. *American Journal of Psychiatry, 155*(12), 1711–1717.

Haber, J., Leach, A. M., Schudy, S. M., & Sideleau, B. F. (1982). *Comprehensive psychiatric nursing.* Philadelphia, PA: McGraw-Hill.

Haddad, G. G., Sun, Y., Wyman, R. J., & Xu, T. (1997). Genetic basis of tolerance to $O_2$ deprivation in *Drosophila melanogaster. Proceedings of the National Academy of Sciences, 94*(20), 10809–10812.

Hagen, C. (1984). Langauge disorders in head trauma. In A. Holland (Ed.), *Language disorders in adults.* San Diego, CA: College-Hill.

Hajek, T., Kopecek, M., Kozeny, J., Gunde, E., Alda, M., & Höschl, C. (2009). Amygdala volumes in mood disorders: Meta-analysis of magnetic resonance volumetry studies. *Journal of Affective Disorders, 115*(3), 395–410.

Hajek, T., Kozeny, J., Kopecek, M., Alda, M., & Höschl, C. (2008). Reduced subgenual cingulated volumes in mood disorders: Meta-analysis. *Journal of Psychiatry and Neuroscience, 33*(2), 91–99.

Hall, M. N., Friedman II, R. J., & Leach, L. (2008). Treatment of bulimia nervosa. *American Family Physician, 77*(11), 1588.

Hall, T. K. (1944). *One hundred years of American psychiatry.* New York, NY: Columbia University Press for the American Psychiatric Association.

Hamilton, J. P., Siemer, M. M., & Gotlib, I. H. (2008). Amygdala volume in major depressive disorder: A meta-analysis of magnetic resonance imaging studies. *Molecular Psychiatry, 13*(11), 993–1000.

Han, S. D., Drake, A. I., Cessante, L. M., Jak, A. J., Houston, W. S., Delis, D. C., . . . Bondi, M. W. (2007). Apolipoprotein E and traumatic brain injury in a military population: Evidence of a neuropsychological compensatory mechanism. *Journal of Neurology, Neurosurgery & Psychiatry, 78*(10), 1103–1108.

Hanson, J. L., Chung, M. K., Avants, B. B., Shirtcliff, E. A., Gee, J. C., Davidson, R. J., & Pollak, S. D. (2010). Early stress is associated with alterations in the orbitofrontal cortex: A tensor-based morphometry investigation of brain structure and behavioral risk. *The Journal of Neuroscience, 30*(22), 7466–7472.

Harmell, A. L., Palmer, B. W., & Jeste, D. V. (2012). Preliminary study of a web-based tool for enhancing the informed consent process in schizophrenia research. *Schizophrenia Research, 141*(2–3), 247–250.

Hart, H., & Rubia, K. (2012). Neuroimaging of child abuse: A critical review. *Frontiers in Human Neuroscience, 6,* 1–24.

Hay, P. J., Bacaltchuk, J., & Stefano, S. (2004). Psychotherapy for bulimia nervosa and binging: Review. *Cochrane Database of Systematic Reviews, 3*, 1–122.

Headache Classification Committee of The International Headache Society. (2004). The international classification of headache disorders (2nd ed.), *Cephalalgia 2004, 24*, 1–160.

Heckers, S., Rauch, S., Goff, D., Savage, C., Schacter, D., Fischman, A., & Alpert, N. (1998). Impaired recruitment of the hippocampus during conscious recollection in schizophrenia. *Nature Neuroscience, 1*(4), 318–323.

Heim, C., & Nemeroff, & C. B. (2002). Neurobiology of early life stress: Clinical studies. *Seminars in Clinical Neuropsychiatry, 7*(2), 147–159.

Heimer, L. (2003). A new anatomical framework for neuropsychiatric disorders and drug abuse. *American Journal of Psychiatry, 160*(10), 1726–1739.

Herbener, E. S., & Harrow, M. (2004). Are negative symptoms associated with functioning deficits in both schizophrenia and nonschizophrenia patients? A 10-year longitudinal analysis. *Schizophrenia Bulletin, 30*(4), 813–825.

Herpertz, S. C., Dietrich, T. M., Wenning, B., Krings, T., Erberich, S. G., Willmes, K., ... Sass, H. (2001). Evidence of abnormal amygdala functioning in borderline personality disorder: A functional MRI study. *Biological Psychiatry, 50*(4), 292–298.

Herrmann, N., Black, S. E., Chow, T., Cappell, J., Tang-Wai, D. F., & Lanctôt, K. L. (2012). Serotonergic function and treatment of behavioral and psychological symptoms of frontotemporal dementia. *The American Journal of Geriatric Psychiatry, 20*(9), 789–797.

Hiatt, K. D., Schmitt, W. A., & Newman, J. P. (2004). Stroop tasks reveal abnormal selective attention among psychopathic offenders. *Neuropsychology, 18*(1), 50–59.

Hiekkanen, H., Kurki, T., Brandstack, N., Kairisto, V., & Tenovuo, O. (2009). Associations of injury severity, MRI-results and ApoE genotype with 1-year outcome in mainly mild TBI: A preliminary study. *Brain Injury, 23*(5), 396–402.

Hill, D. E., Yeo, R. A., Campbell, R. A., Hart, B., Vigil, J., Brooks, W. (2003). Magnetic resonance imaging correlates of attention-deficit/hyperactivity disorder in children. *Neuropsychology, 17*, 496–506.

Hill, K., Mann, L., Laws, K. R., Stephenson, C. M. E., Nimmo-Smith, I., & McKenna, P. J. (2004). Hypofrontality in schizophrenia: A meta-analysis of functional imaging studies. *Acta Psychiatrica Scandinavica, 110*(4), 243–256.

Hinz, M., Stein, A., Trachte, G., & Uncini, T. (2010). Neurotransmitter testing of the urine: A comprehensive analysis. *Open Access Journal of Urology, 2*, 177–183.

Hirschfeld, R. M., Calabrese, J. R., Weissman, M. M., Reed, M., Davies, M. A., Frye, M. A., ... Wagner, K. D. (2003). Screening for bipolar disorder in the community. *The Journal of Clinical Psychiatry, 64*(1), 53.

Hobson, J. A., Pace-Schott, E. F., & Stickgold, R. (2000). Dreaming and the brain: Toward a cognitive neuroscience of conscious states. *Behavioral and Brain Sciences, 23*(6), 793–842.

Hoff, P. (1992). Emil Kraepelin and philosophy: The implicit philosophical assumptions of Kraepelinian psychiatry. In M. Spitzer et al. (Eds.), *Phenomenology, language and schizophrenia* (pp. 115–125). New York, NY: Springer.

Hofman, P. A., Stapert, S. Z., van Kroonenburgh, M. J., Jolles, J., de Kruijk, J., & Wilmink, J. T. (2001). MR imaging, single-photon emission CT, and neurocognitive performance after mild traumatic brain injury. *American Journal of Neuroradiology, 22*(3), 441–449.

Holtz, J. L. (2010). *Applied clinical neuropsychology*. New York, NY: Springer Publishing Company.

Holtz, J. L. (2011). *Applied clinical neuropsychology: An introduction*. New York, NY: Springer Publishing Company.

Honea, R., Crow, T. J., Passingham, D., & Mackay, C. E. (2005). Regional deficits in brain volume in schizophrenia: A meta-analysis of voxel-based morphometry studies. *American Journal of Psychiatry, 162*(12), 2233–2245.

Hong, S. B., Tae, W. S., & Joo, E. Y. (2006). Cerebral perfusion changes during cataplexy in narcolepsy patients. *Neurology, 66*(11), 1747–1749.

Hornberger, M., Wong, S., Tan, R., Irish, M., Piguet, O., Kril, J., ... Halliday, G. (2012). In vivo and post-mortem memory circuit integrity in frontotemporal dementia and Alzheimer's disease. *Brain, 135*(10), 3015–3025.

Hshieh, T. T., Fong, T. G., Marcantonio, E. R., & Inouye, S. K. (2008). Cholinergic deficiency hypothesis in delirium: A synthesis of current evidence. *The Journals of Gerontology Series A: Biological Sciences and Medical Sciences, 63*(7), 764–772.

Hublin, C., Partinen, M., Koskenvuo, M., & Kaprio, J. (2011). Heritability and mortality risk of insomnia-related symptoms: a genetic epidemiologic study in a population-based twin cohort. *Sleep, 34*(7), 957.

Hudson, J. I., Hiripi, E., Pope Jr., H. G., & Kessler, R. C. (2007). The prevalence and correlates of eating disorders in the National Comorbidity Survey Replication. *Biological Psychiatry, 61*(3), 348–358.

Hughes, D. G., Jackson, A., Mason, D. L., Berry, E., Hollis, S., & Yates, D. W. (2004). Abnormalities on magnetic resonance imaging seen acutely following mild traumatic brain injury: Correlation with neuropsychological tests and delayed recovery. *Neuroradiology, 46*(7), 550–558.

Hulshoff Pol, H. E., Schnack, H. G., Mandl, R. C., Cahn, W., Collins, D. L., Evans, A. C., & Kahn, R. S. (2004). Focal white matter density changes in schizophrenia: reduced inter-hemispheric connectivity. *Neuroimage, 21*(1), 27–35.

Hulshoff Pol, H. E., Schnack, H. G., Mandl, R. C., van Haren, N. E., Koning, H., Collins, D. L., ... Kahn, R. S. (2001). Focal gray matter density changes in schizophrenia. *Archives of General Psychiatry, 58*(12), 1118.

Hyman, S. E. (2005). Neurotransmitters. *Current Biology: CB, 15*(5), R154–R158.

Inouye, S. K. (2006). Delirium in older persons. *New England Journal of Medicine, 354*(11), 1157–1165.

International League Against Epilepsy Commission report: A proposed diagnostic scheme for people with epileptic seizures and with epilepsy. Report of the ILAE Task Force on Classification and Terminology. (2001). 42, 796–803.

Ip, R. R., Hesch, P. P., Brandys, C. C., Dornan, J. J., & Schentag, C. C. (2000). Traumatic brain injury: Causes, severity, and outcome. *Brain and Cognition, 44*(1), 42–44.

Isaacson, R. L. (1982). *The Limbic System*. New York, NY: Plenum Press.

Isoniemi, H., Kurki, T., Tenovuo, O., Kairisto, V., & Portin, R. (2006). Hippocampal volume, brain atrophy, and APOE genotype after traumatic brain injury. *Neurology, 67*(5), 756–760.

Iverson, G. L. (2006). Complicated vs uncomplicated mild traumatic brain injury: Acute neuropsychological outcome. *Brain Injury, 20*(13–14), 1335–1344.

Iverson, G. L. (2006). Misdiagnosis of the persistent postconcussion syndrome in patients with depression. *Archives of Clinical Neuropsychology, 21*(4), 303–310.

Iverson, G. L. (2007). Predicting slow recovery from sport-related concussion: The new simple-complex distinction. *Clinical Journal of Sport Medicine, 17*(1), 31–37.

Jack, C. R., Lowe, V. J., Weigand, S. D., Wiste, H. J., Senjem, M. L., Knopman, D. S., ... Petersen, R. C. (2009). Serial PIB and MRI in normal, mild cognitive impairment and Alzheimer's disease: Implications for sequence of pathological events in Alzheimer's disease. *Brain, 132*(5), 1355–1365.

Jacobi, C., Abascal, L., & Taylor, C. B. (2004). Screening for eating disorders and high-risk behavior: Caution. *International Journal of Eating Disorders, 36*(3), 280–295.

Jaffee, M. S., Helmick, K. M., Girard, P. D., Meyers, K. S., Dinegar, K., & George, K. (2009). Acute clinical care and care coordination for traumatic brain injury within Department of Defense. *Journal of Rehabilitation Research & Development, 46*(6), 655–666.

Jain, G., Chakrabarti, S., & Kulhara, P. (2011). Symptoms of delirium: An exploratory factor analytic study among referred patients. *General Hospital Psychiatry, 33*(4), 377–385.

Jan, J. E., Reiter, R. J., Wasdell, M. B., & Bax, M. (2009). The role of the thalamus in sleep, pineal melatonin production, and circadian rhythm sleep disorders. *Journal of Pineal Research, 46*(1), 1–7.

Jeste, D. V., Palmer, B. W., Golshan, S., Eyler, L. T., Dunn, L. B., Meeks, T., . . . Appelbaum, P. S. (2009). Multimedia consent for research in people with schizophrenia and normal subjects: A randomized controlled trial. *Schizophrenia Bulletin, 35*(4), 719–729.

Johnstone, E., Frith, C. D., Crow, T. J., Husband, J., & Kreel, L. (1976). Cerebral ventricular size and cognitive impairment in chronic schizophrenia. *The Lancet, 308*(7992), 924–926.

Joiner Jr., T. E., Brown, J. S., & Wingate, L. R. (2005). The psychology and neurobiology of suicidal behavior. *Annual Review of Psychology, 56,* 287–314.

Jollant, F., Buresi, C., Guillaume, S., Jaussent, I., Bellivier, F., Leboyer, M., . . . Courtet, P. (2007). The influence of four serotonin-related genes on decision-making in suicide attempters. *American Journal of Medical Genetics Part B: Neuropsychiatric Genetics, 144B*(5), 615–624.

Jonson-Reid, M., Kohl, P. L., & Drake, B. (2012). Child and adult outcomes of chronic child maltreatment. *Pediatrics, 129*(5), 839–845.

Judd, L. L. (1997). The clinical course of unipolar major depressive disorders. *Archives of General Psychiatry, 54*(11), 989–991.

Judd, L. L., & Akiskal, H. S. (2003). The prevalence and disability of bipolar spectrum disorders in the US population: Re-analysis of the ECA database taking into account subthreshold cases. *Journal of Affective Disorders, 73*(1–2), 123–131.

Kadir, A., Almkvist, O., Forsberg, A., Wall, A., Engler, H., Långström, B., & Nordberg, A. (2012). Dynamic changes in PET amyloid and FDG imaging at different stages of Alzheimer's disease. *Neurobiology of Aging, 33*(1), 198.e1–198.e14.

Kamphuis, J., Meerlo, P., Koolhaas, J. M., & Lancel, M. (2012). Poor sleep as a potential causal factor in aggression and violence. *Sleep Medicine, 13*(4), 327–334.

Kanbayashi, T., Kodama, T., Kondo, H., Satoh, S., Inoue, Y., Chiba, S., . . . Nishino, S. (2009). CSF histamine contents in narcolepsy, idiopathic hypersomnia and obstructive sleep apnea syndrome. *Sleep, 32*(2), 181–187.

Kapur, S., Mizrahi, R., & Li, M. (2005). From dopamine to salience to psychosis—linking biology, pharmacology and phenomenology of psychosis. *Schizophrenia Research, 79*(1), 59–68.

Kapur, S., Zipursky, R., Jones, C., Shammi, C. S., Remington, G., & Seeman, P. (2000). A positron emission tomography study of quetiapine in schizophrenia: A preliminary finding of an antipsychotic effect with only transiently high dopamine D2 receptor occupancy. *Archives of General Psychiatry, 57*(6), 553.

Karlsgodt, K. H., Ellman, L. M., Sun, D., Mittal, V., & Cannon, T. D. (2012). The neurodevelopmental hypothesis of schizophrenia. In A. S. David, P. McGuffin, & S. Kapur (Eds.), *Schizophrenia: The final frontier—A Festschrift for Robin M. Murray* (p. 1) New York, NY: Psychology Press.

Kaufmann, C., Schuld, A., Pollmächer, T., & Auer, D. P. (2002). Reduced cortical gray matter in narcolepsy: Preliminary findings with voxel-based morphometry. *Neurology, 58*(12), 1852–1855.

Kaufmann, C., Wehrle, R., Wetter, T. C., Holsboer, F., Auer, D. P., Pollmächer, T., & Czisch, M. (2006). Brain activation and hypothalamic functional connectivity during human non-rapid eye movement sleep: An EEG/fMRI study. *Brain, 129*(3), 655–667.

Kaye, W. (2007). Neurobiology of anorexia and bulimia nervosa. *Physiology & Behavior, 94*(1):121–135.

Keefe, R. S., Bilder, R. M., Davis, S. M., Harvey, P. D., Palmer, B. W., Gold, J. M., & Lieberman, J. A. (2007). Neurocognitive effects of antipsychotic medications in patients with chronic schizophrenia in the CATIE Trial. *Archives of General Psychiatry, 64*(6), 633.

Keenan, H. T., Runyan, D., K., Marshall, S. W., Nocera, M. A., Merten, D. F., & Sinal, S. H. (2003). A population-based study of inflicted traumatic brain injury in young children. *JAMA: The Journal of the American Medical Association, 290*(5), 621–626.

Kegeles, L. S., Abi-Dargham, A., Zea-Ponce, Y., Rodenhiser-Hill, J., Mann, J. J., Van Heertum, R. L., . . . Laruelle, M. (2000). Modulation of amphetamine-induced striatal dopamine release by ketamine in humans: Implications for schizophrenia. *Biological Psychiatry, 48*(7), 627–640.

Kelley, A. E., Schiltz, C. A., & Landry, C. F. (2005). Neural systems recruited by drug- and food-related cues: Studies of gene activation in corticolimbic regions: Ingestive mechanisms in obesity, substance abuse and mental disorders. *Physiology & Behavior, 86*(1–2), 11–14.

Kemps, E., & Wilsdon, A. (2010). Preliminary evidence for a role for impulsivity in cognitive disinhibition in bulimia nervosa. *Journal of Clinical and Experimental Neuropsychology, 32*(5), 515–521.

Kempton, M. J., Salvador, Z., Munafò, M. R., Geddes, J. R., Simmons, A., Frangou, S., & Willams, S. C. R. (2011). Structural neuroimaging: studies in major depressive disorder: Meta-analysis and comparison with bipolar disorder. *Archives of General Psychiatry, 68*(7), 675–690.

Kendler, K. S., Gatz, M., Gardner, S. O., & Pederson, N. L. (2006). A Swedish national twin study of lifetime major depression. *American Journal of Psychiatry, 163*(1), 109–114.

Kendler, K. S., Karkowski, L. M., & Prescott, C. A. (1999). Fears and phobias: Reliability and heritability. *Psychological Medicine, 29*(3), 539–553.

Kennan, H. T., Runyan, D. K., Marshall, S. W., Nocera, M. A., Merten, D. F., & Sinal, S. H. (2003). A population-based study on inflicted traumatic brain injury in young children. *JAMA: The Journal of the American Medical Association, 290*(5), 621–626

Kenny, J. T., & Friedman, L. (2002). Cognitive impairment in early-stage schizophrenia. In R. B. Zipursky & S. C. Schulz (Eds.), *The early stages of schizophrenia* (pp. 205–231). Washington, DC: American Psychiatric Publishing.

Keshavan, M S., Kulkarni, S., Bhojral, T., Francis, A., Diwadkar, V., Montrose, D. M., . . . Sweeney J (2007). Premorbid cognitive deficits in young relatives of schizophrenia patients. *Frontiers in Human Neuroscience, 3,* 62.

Kessler, R. C., Aguilar-Gaxiola, S., Alonso, J., Chatterji, S., Lee, S., Ormel, J., . . . Wang, P. S. (2009). The global burden of mental disorders: An update from WHO World Mental Health Surveys. *Epidemiological Psychiatry Society, 18*(1), 23–33.

Kessler, R. C., Berglund, P., Demler, O., Jin, R., Koretz, D., Merikangas, K., . . . Wang, P. S. (2003). The epidemiology of major depressive disorder. *JAMA: The Journal of the American Medical Association, 289*(23), 3095–3105.

Kessler, R. C., Dupont, R., Wittchen, H., & Berglund, P. A. (2000). Dr. Kessler and colleagues reply. *The American Journal of Psychiatry, 157*(12), 2061–2061.

Kety, S. S. (1987). The significance of genetic factors in the etiology of schizophrenia: Results from the national study of adoptees in Denmark. *Journal of Psychiatric Research, 21*(4), 423–429.

Khalil, R. B. (2013). Dopamine D3 receptor antagonists in pathologic gambling. *Journal of Clinical Psychopharmacology, 33*(1), 146–148.

Kim, S. J., Lyoo, I. K., Hwang, J., Sung, Y. H., Lee, H. Y., Lee, D. S., & Jeong, D. (2005). Frontal glucose hypometabolism in abstinent methamphetamine users. *Neuropsychopharmacology, 30,* 1383–1391.

Kim-Cohen, J., Caspi, A., Taylor, A., Williams, B., Newcombe, R., Craig, I. W., & Moffitt, T. E. (2006). MAOA, maltreatment, and gene–environment interaction predicting children's mental health: New evidence and a meta-analysis. *Molecular Psychiatry, 11*(10), 903–913.

Kirov, R., & Brand, S. (2012). The memory, cognitive and psychological functions of sleep: Update from electroencephalographic and neuroimaging studies. *Neuroimaging: Cognitive and Clinical Neuroscience. InTech, Rijeca,* 155–180.

Kjaer, T. W., Law, I., Wiltschiotz, G., Pauson, O. B., & Madsen, P. L. (2002). Regional cerebral blood flow during light sleep: An $H_2^{15}O$-PET study. *Journal of Sleep Research, 11*(3), 201–207.

Knopik, V. S., Smith, S. D., Cardon, L., Pennington, B., Gayan, J., Olson, R. K., & DeFries, J. C. (2002). Differential genetic etiology of reading component processes as a function of IQ. *Behavior Genetics, 32*(3), 181–198.

Kolb, B., & Whishaw, I. Q. (2008). *Fundamentals of human neuropsychology* (6th ed.). New York, NY: Worth Publishers.

Krabbendam, L., Arts, B., van Os, J., & Aleman, A. (2005). Cognitive functioning in patients with schizophrenia and bipolar disorder: A quantitative review. *Schizophrenia Research, 80*(2), 137–149.

Kraus, J. F., Morgenstern, H., Fife, D., Conroy, C., & Nourjah, P. (1989). Blood alcohol tests, prevalence of involvement, and outcomes following brain injury. *American Journal of Public Health, 79*(3), 294–299.

Krieg, J.-C., Lauer, C., & Pirke, K. M. (1987). Hormonal and metabolic mechanisms in the development of cerebral pseudoatrophy in eating disorders. *Psychotherapy and Psychosomatics, 48*, 176–180.

Krieg, J. C., Pirke, K. M., Lauer, C., & Herbert, B. (1988). Endocrine, metabolic, and cranial computed tomographie findings in anorexia nervosa. *Biological Psychiatry, 23*(4), 377–387.

Krieger, I. (1982). *Pediatric disorders of feeding, nutrition, and metabolism*. Hoboken, NJ: John Wiley & Sons.

Kumari, V., Das, M., Taylor, P. J., Barkataki, I., Andrew, C., Sumich, A., . . . Ffytche, D. H. (2009). Neural and behavioural responses to threat in men with a history of serious violence and schizophrenia or antisocial personality disorder. *Schizophrenia Research, 110*, 47–58.

Kupfer, D. J., Frank, E., & Phillips, M. L. (2012). Major depressive disorder: New clinical, neurobiological, and treatment perspectives. *The Lancet, 379*(9820), 1045–1055.

Labuschagne, I., Phan, K. L., Angstadt, A., Chua, P., Heinrichs, M., . . . Nathan, P. J. (2010). Medial frontal hyperactivity to sad faces in generalized social anxiety disorder and modulation by oxytocin. *The International Journal of Neuropsychophramacology, 15*, 883–896.

Ladd, C. O., Huot, R. L., Thrivikraman, K. V., Nemeroff, C. B., Meaney, M. J., & Plotsky, P. M. (1999). Long-term behavioral and neuroendocrine adaptations to adverse early experience. *Progress in Brain Research, 122*, 81–103.

Lambe, E. K., Katzman, D. K., Mikulis, D. J., Kennedy, S. H., & Zipursky, R. B. (1997). Cerebral gray matter volume deficits after weight recovery from anorexia nervosa. *Archives of General Psychiatry, 54*(6), 537.

Langlois, J. A., Rutland-Brown, W., & Wald, M. M. (2006). The epidemiology and impact of traumatic brain injury: A brief overview. *The Journal of Head Trauma Rehabilitation, 21*(5), 375–378.

LaRue, B. L., & Padilla, P. A. (2011). Environmental and genetic preconditioning for long-term anoxia responses requires AMPK in *Caenorhabditis elegans*. *PloS One, 6*(2), e16790.

Lauer, M., Senitz, D., & Beckmann, H. (2001). Increased volume of the nucleus accumbens in schizophrenia. *Journal of Neural Transmission, 108*(6), 645–660.

Lawrie, S. M., Whalley, H. C., Abukmeil, S. S., Kestelman, J. N., Miller, P., Best, J. J., . . . Johnstone, E. C. (2002). Temporal lobe volume changes in people at high risk of schizophrenia with psychotic symptoms. *The British Journal of Psychiatry, 181*(2), 138–143.

Lebedev, A., V., Westman, E., Beyer, M., K., Kramberger, M., G., Aguilar, C., Pirtosek, Z., & Aarsland, D. (2012). Multivariate classification of patients with Alzheimer's and dementia with Lewy bodies using high-dimensional cortical thickness measurements: An MRI surface-based morphometric study. *Journal of Neurology, 260*(4), 1104–1115.

Lee, B. H., Shofer, J. L., & Koppelman, F. S. (2005). Bicycle safety helmet legislation and bicycle-related non-fatal injuries in California. *Accident Analysis & Prevention, 37*(1), 93–102.

Leeb, R. T. (2008). *Child maltreatment surveillance: Uniform definitions for public health and recommended data elements*. Atlanta, GA: Centers for Disease Control and Prevention, National Center for Injury Prevention and Control.

le Grange, D., Crosby, R. D., Rathouz, P. J., & Leventhal, B. L. (2007). A randomized controlled comparison of family-based treatment and supportive psychotherapy for adolescent bulimia nervosa. *Archives of General Psychiatry, 64*(9), 1049.

LeGris, J., & van Reekum, R. (2006). The neuropsychological correlates of borderline personality disorder and suicidal behaviour. *The Canadian Journal of Psychiatry, 51*(3), 131–142.

Lenze, E. J., Mulsant, B. H., Mohlman, J., Shear, M. K., Dew, M. A., Schulz, R., & Reynolds III, C. F. (2005). Generalized anxiety disorder in late life: Lifetime course and comorbidity with major depressive disorder. *American Journal of Geriatric Psychiatry, 13*(1), 77–80.

Lenzenweger, M. F., Lane, M. C., Loranger, A. W., & Kessler, R. C. (2007). DSM-IV personality disorders in the National Comorbidity Survey Replication. *Biological Psychiatry, 62*(6), 553–564.

Lencz, T., Morgan, T. V., Athanasiou, M., Dain, B., Reed, C. R., Kane, J. M., & Malhotra, A. K. (2007). Converging evidence for a pseudoautosomal cytokine receptor gene locus in schizophrenia. *Molecular Psychiatry, 12*(6), 572–580.

Levin, H. S., O'Donnell, V. M., & Grossman, R. G. (1979). The Galveston Orientation and Amnesia Test: A practical scale to assess cognition after head injury. *The Journal of Nervous and Mental Disease, 167*(11), 675–684.

Levine, J., Goldberger, I., Rapaport, A., Schwartz, M., Schield, C., Elizur, A., & Agam, G. (1994). CSF inositol in schizophrenia and high-dose inositol treatment of schizophrenia. *European Neuropsychopharmacology, 4*(4), 487–490.

Levinson, D. F. (2006) The genetics of depression: A review. *Biological Psychiatry, 60*(2), 84–92.

Lew, H. L., Otis, J. D., Tun, C., Kerns, R. D., Clark, M. E., & Cifu, D. X. (2009). Prevalence of chronic pain, posttraumatic stress disorder, and persistent postconcussive symptoms in OIF/OEF veterans: Polytrauma clinical triad. *Journal of Rehabilitation Research & Development, 46*(6), 697–702.

Lewis, C. M., Levinson, D. F., Wise, L. H., DeLisi, L. E., Straub, R. E., Hovatta, I., . . . Petursson, H. (2003). Genome scan meta-analysis of schizophrenia and bipolar disorder, part II: Schizophrenia. *The American Journal of Human Genetics, 73*(1), 34–48.

Lewis, D. A., Volk, D. W., & Hashimoto, T. (2004). Selective alterations in prefrontal cortical GABA neurotransmission in schizophrenia: A novel target for the treatment of working memory dysfunction. *Psychopharmacology, 174*(1), 143–150.

Lezak, M. D., Howieson, D. B., & Loring, D. W. (2004). *Neuropsychological assessment* (4th ed.). New York, NY: Oxford University Press.

Li, M., Chen, Z., Deng, W., He, Z., Wweng, Q., Jiang, L., . . . Li, T. (2012). Volume increases in putamen associated with positive symptom reduction in previously drug-naive schizophrenia after 6 weeks antipsychotic treatment. *Psychological Medicine, 42*(7), 1475–1483.

Liao, P. C., Uher, R., Lawrence, N., Treasure, J., Schmidt, U., Campbell, I. C., . . . Tchanturia, K. (2009). An examination of decision making in bulimia nervosa. *Journal of Clinical and Experimental Neuropsychology, 31*(4), 455–461.

Liao, Y., Tang, J., Ma, M., Wu, Z., Yang, M., Wang, X., . . . Hao, W. (2010). Frontal white matter abnormalities following chronic ketamine use: A diffusion tensor imaging study. *Brain, 133*(7), 2115–2122.

Liddle, P. F. (2000). Cognitive impairment in schizophrenia: Its impact on social functioning. *Acta Psychiatrica Scandinavica, 101*, 11–16.

Lieberman, J. A., Kane, J. M., & Alvir, J. (1987). Provocative tests with psychostimulant drugs in schizophrenia. *Psychopharmacology, 91*(4), 415–433.

Lieverse, R., Van Someren, E. W., Nielen, M. A., Uitdehaag, B. J., Smit, J. H., & Hoogendijk, W. G. (2011). Bright light treatment in elderly patients with nonseasonal major depressive disorder: a randomized placebo-controlled trial. *Archives of General Psychiatry, 68*(1), 61–70.

Lilenfeld, L. R., Kaye, W. H., Greeno, C. G., Merikangas, K. R., Plotnicov, K., Pollica, C., . . . Nagy, L. (1998). A controlled family study of anorexia nervosa and bulimia nervosa: Psychiatric disorders in first-degree relatives and effects of proband comorbidity. *Archives of General Psychiatry, 55*, 603–610.

Lin, F., Weng, S., Xie, B., Wu, G., & Lei, H. (2011). Abnormal frontal cortex white matter connections in bipolar disorder: A DTI tractography study. *Journal of Affective Disorders, 131*(1–3), 299–306.

Lin, T. Y., & Flak, B. (1988). Ventricular and sulcal size at the onset of psychosis. *American Journal of Psychiatry, 145*(7), 820–824.

Lincoln, T. M., Wilhelm, K., & Nestoriuc, Y. (2007). Effectiveness of psychoeducation for relapse, symptoms, knowledge, adherence and functioning in psychotic disorders: A meta-analysis. *Schizophrenia Research, 96*(1), 232–245.

Lipsey, J. R., Robinson, R. G., Pearlson, G. D., Rao, K., & Price, T. R. (1983). Mood change following bilateral hemisphere brain injury. *The British Journal of Psychiatry, 143*(3), 266–273.

Lis, E., Greenfield, B., Henry, M., Guilé, J. M., & Dougherty, G. (2007). Neuroimaging and genetics of borderline personality disorder: A review. *Journal of Psychiatry & Neuroscience, 32*(3), 162.

Lisdahl, K. M., Gilbert, E. R., Wright, N. E., & Shollenbarger, S. (2013). Dare to delay? The impacts of adolescent alcohol and marijuana use onset on cognition, brain structure and function. *Frontiers in Psychiatry, 4*, 1–19.

Liu, L., Foroud, T., Xuei, X., Berrettini, W., Byerley, W., Coryell, W., ... Nurnberger Jr., J. I. (2008). Evidence of association between brain-derived neurotrophic factor (BDNF) gene and bipolar disorder. *Psychiatric Genetics, 18*(6), 267.

Loman, M. M., Wiik, K. L., Frenn, K. A., Pollak, S. D., & Gunnar, M. R. (2009). Postinstitutionalized children's development: Growth, cognitive, and language outcomes. *Journal of Developmental and Behavioral Pediatrics: JDBP, 30*(5), 426.

Lopez, C., Tchanturia, K., Stahl, D., & Treasure, J. (2008). Central coherence in eating disorders: A systematic review. *Psychological Medicine, 38*(10), 1393–1404.

López Léon, S. (2008). *Genetic determinants of depression* (Doctoral dissertation). Erasmus University, Rotterdam.

Lu, L. H., Zhou, X., Fitzgerald, J., Keedy, S. K., Reilly, J. L., Passarotti, A. M., ... Pavuluri, M. (2012). Microstructural abnormalities of white matter differentiate pediatric and adult onset bipolar disorder. *Bipolar Disorders, 14*(6), 597–606.

Lu, S., Ahn, D., Johnson, G., Law, M., Zagzag, D., & Grossman, R. I. (2004) Diffusion-tensor MR imaging of intracranial neoplasia and associated peritumoral edema: Introduction of the tumor infiltration index. *Radiology, 232*, 221–228.

Lupien, S. J., McEwen, B. S., Gunnar, M. R. & Heim, C. (2009). Effects of stress throughout the lifespan on the brain, behaviour and cognition. *Nature, 10*, 434–445.

Lykken , D. T. (1995). *The antisocial personalities*. Hillsdale, NJ: Erlbaum.

MacLullich, A. M., Ferguson, K. J., Miller, T., de Rooij, S. E., & Cunningham, C. (2008). Unravelling the pathophysiology of delirium: A focus on the role of aberrant stress responses. *Journal of Psychosomatic Research, 65*(3), 229–238.

Maheu, F. S., Dozier, M., Guyer, A. E., Mandell, D., Peloso, E., Poeth, K., ... Ernst, M. (2010). A preliminary study of medial temporal lobe function in youths with a history of caregiver deprivation and emotional neglect. *Cognitive, Affective, & Behavioral Neuroscience, 10*(1), 34–49.

Mahone, E. M., Crocetti, D., Ranta, M. E., Gaddis, A., Cataldo, M., Slifer, K. J., ... Mostofsky, S. H. (2011). A preliminary neuroimaging study of preschool children with ADHD. *The Clinical Neuropsychologist, 25*(6), 1009–1028.

Mahoney, C. J., Beck, J., Rohrer, J. D., Lashley, T., Mok, K., Shakespeare, T., ... Warren, J. D. (2012). Frontotemporal dementia with the C9ORF72 hexanucleotide repeat expansion: Clinical, neuroanatomical and neuropathological features. *Brain, 135*(3), 736–750.

Mainz, V., Schulte-Rüther, M., Fink, G. R., Herpertz-Dahlmann, B., & Konrad, K. (2012). Structural brain abnormalities in adolescent anorexia nervosa before and after weight recovery and associated hormonal changes. *Psychosomatic Medicine, 74*(6), 574–582.

Mann, J., Malone, K. M., Sweeney, J. A., Brown, R. P., Linnoila, M., Stanley, B., & Stanley, M. (1996). Attempted suicide characteristics and cerebrospinal fluid amine metabolites in depressed inpatients. *Neuropsychopharmacology, 15*(6), 576–586.

Mann, J. J. (2003). Neurobiology of suicidal behaviour. *Nature Reviews Neuroscience, 4*, 819–828.

Maquet, P., Laureys, S., Peigneux, P., Fuchs, S., Petiau, C., Phillips, C., ... Cleeremans, A. (2000). Experience-dependent changes in cerebral activation during human REM sleep. *Nature Neuroscience, 3*(8), 831–836.

Marsh, R., Steinglass, J. E., Gerber, A. J., Graziano O'Leary, K., Wang, Z., Murphy, D., ... Peterson, B. S. (2009). Deficient activity in the neural systems that mediate self-regulatory control in bulimia nervosa. *Archives of General Psychiatry, 66*(1), 51.

Martin, E. I., Ressler, K. J., Binder, E., & Nemeroff, C. B. (2010). The neurobiology of anxiety-disorders: brain imaging, genetics, and psychoneuroendocrinology. *Clinics in Laboratory Medicine, 30*(4), 865.

Martínez-Arán, A. A., Vieta, E., Colom, F. F., Torrent, C. C., Sánchez-Moreno, J. J., Reinares, M. M., . . . Salamero, M. M. (2004). Cognitive impairment in euthymic bipolar patients: Implications for clinical and functional outcome. *Bipolar Disorders, 6*(3), 224–232.

Martland, H. S. (1928). Punch drunk. *JAMA: The Journal of the American Medical Association, 91*(15), 1103–1107.

Mathers, J., Sitch, A., Marsh, J. L., & Parry, J. (2011). Widening access to medical education for under-represented socioeconomic groups: Population-based cross-sectional analysis of UK data, 2002–6. *BMJ: British Medical Journal, 342,* 298–307.

Matsuo, K., Sanches, M., Brambilla, P., & Soares, J. C. (2012). Structural brain abnormalities in bipolar disorder. In S. M. Strakowski (Ed.), *The bipolar brain: Integrating neuroimaging and genetics* (pp. 17–141). New York, NY: Oxford University Press.

Mayberg, H. S., Lozano, A. M., Voon, V., McNeely, H. E., Seminowicz, D., Hamani, C., Schwalb, J. M., & Kennedy, S. H. (2005). Deep brain stimulation for treatment-resistant depression. *Neuron, 45*(5), 651–660.

McAdams, C. J., & Krawczyk, D. C. (2011). Impaired neural processing of social attribution in anorexia nervosa. *Psychiatry Research: Neuroimaging, 194*(1), 54–63.

McCormick, L. M., Keel, P. K., Brumm, M. C., Bowers, W., Swayze, V., Andersen, A., & Andreasen, N. (2008). Implications of starvation-induced change in right dorsal anterior cingulate volume in anorexia nervosa. *International Journal of Eating Disorders, 41*(7), 602–610.

McElroy, S. L., Guerdjikova, A. I., Winstanley, E. L., O'Melia, A. M., Mori, N., McCoy, J., . . . Hudson, J. I. (2011). Acamprosate in the treatment of binge eating disorder: A placebo-controlled trial. *International Journal of Eating Disorders, 44*(1), 81–90.

McFall, R. (2006). Doctoral training in clinical psychology. *Annual Review of Clinical Psychology, 2,* 21–49.

McGrath, J., Saha, S., Welham, J., El Saadi, O., MacCauley, C., & Chant, D. (2004). A systematic review of the incidence of schizophrenia: The distribution of rates and the influence of sex, urbanicity, migrant status and methodology. *BMC Medicine, 2*(1), 13.

McGrorry, P. D., Killackey, E., & Yung, A. (2008). Early intervention in psychosis: Concepts evidence and future directions. *World Psychiatry, 7,* 148–157.

McGurk, S., Twamley, E., Sitzer, D., McHugo, G., & Mueser, K. (2007). A meta-analysis of cognitive remediation in schizophrenia. *American Journal of Psychiatry, 164*(12), 1791–1802.

McHugh, R. K., Smits, J. A. J. & Otto, M. W. (2009). Empirically supported treatments for panic disorder. *Psychiatric Clinics of North America, 32*(3).

McKee, A. C., Gavett, B. E., Stern, R. A., Nowinski, C. J., Cantu, R. C., Kowall, N. W., . . . Budson, A. E. (2010). TDP-43 proteinopathy and motor neuron disease in chronic traumatic encephalopathy. *Journal of Neuropathology and Experimental Neurology, 69*(9), 918–929.

McKnight, R. F., & Park, R. J. (2010). Atypical antipsychotics and anorexia nervosa: A review. *European Eating Disorders Review, 18*(1), 10–21.

Meaney, M. J., & Szyf, M. (2005). Maternal care as a model for experience-dependent chromatin plasticity. *TRENDS in Neurosciences, 28*(9), 456–463.

Meares, R., Stevenson, J., & Gordon, E. (1999). A Jacksonian and biopsychosocial hypothesis concerning borderline and related phenomena. *Australian and New Zealand Journal of Psychiatry, 33*(6), 831–840.

Medina, K. L., Hansen, K. L., Schweinsberg, A. D., Cohen-Zion, M., Nagel, B. J., & Tapert, S. F. (2007). Neuropsychological functioning in adolescent marijuana users: Subtle deficits detectable after a month of abstinence. *Journal of the International Neuropsychological Society, 13*(5), 807–820.

Mellon, M. W., & McGrath, M. L. (2000). Empirically supported treatments in pediatric psychology: Nocturnal enuresis. *Journal of Pediatric Psychology, 25*(4), 193–214.

Mendoza, J. E., & Foundas, A. L. (2008). *Clinical neuroanatomy: A neurobehavioral approach.* New York, NY: Springer Publishing Company.

Mercer, D., Douglass, A. B., & Links, P. S. (2009). Meta-analyses of mood stabilizers, antidepressants and antipsychotics in the treatment of borderline personality disorder: Effectiveness for depression and anger symptoms. *Journal of Personality Disorders*, 23(2), 156–174.

Mercer, K. D., Selby, M. J., & McClung, J. (2005). The effects of psychopathy, violence and drug use on neuropsychological functioning. *American Journal of Forensic Psychology*, 23(3), 65–86.

Michaelides, M., Thanos, P. K., Volkow, N. D., & Wang, G. J. (2012). Dopamine-related frontostriatal abnormalities in obesity and binge-eating disorder: Emerging evidence for developmental psychopathology. *International Review of Psychiatry*, 24(3), 211–218.

Minzenberg, M. J., Fan, J., New, A. S., Tang, C. Y., & Siever, L. J. (2008). Frontolimbic structural changes in borderline personality disorder. *Journal of Psychiatric Research*, 42(9), 727–733.

Mitchell, D. G. V., Avny, S. B., & Blair, R. J. R. (2006). Divergent patterns of aggressive and neurocognitive characteristics in acquired versus developmental psychopathy. *Neurocase*, 12(3), 164–178.

Mitchell, J. E., Pyle, R. L., Eckert, E. D., Hatsukami, D., Pomeroy, C., & Zimmerman, R. (1990). A comparison study of antidepressants and structured intensive group psychotherapy in the treatment of bulimia nervosa. *Archives of General Psychiatry*, 47(2), 149.

Moeller, G. F., & Dougherty, D. M. (2006). Antisocial personality disorder, alcohol, and aggression. *Alcohol Research & Health*. National Institute on Alcohol Abuse and Alcoholism.

Moffit, M. T. (1993). *Bright light treatment of late-life depression* (Doctoral dissertation). Wright Institute, Berkeley, CA.

Mond, J. M., & Calogero, R. M. (2009). Excessive exercise in eating disorder patients and in healthy women. *Australasian Psychiatry*, 43(3), 227–234.

Mond, J. M., Latner, J. D., Hay, P. H., Owen, C., & Rodgers, B. (2010). Objective and subjective bulimic episodes in the classification of bulimic-type eating disorders: Another nail in the coffin of a problematic distinction. *Behaviour Research and Therapy*, 48(7), 661–669.

Moore, G. J., Cortese, B. M., Glitz, D. A., Zajac-Benitez, C., Quiroz, J. A., Uhde, T. W., . . . Manji, H. K. (2009). A longitudinal study of the effects of lithium treatment on prefrontal and subgenual prefrontal gray matter volume in treatment-responsive bipolar disorder patients. *The Journal of Clinical Psychiatry*, 70(5), 699–705.

Morrell, M. J., McRobbie, D. W., Quest, R. A., Cummin, A. R., Ghiassi, R., & Corfield, D. R. (2003). Changes in brain morphology associated with obstructive sleep apnea. *Sleep Medicine*, 4(5), 451–454.

Morrow, R. L., Garland, E. J., Wright, J. M., Maclure, M., Taylor, S., & Dormuth, C. R. (2012). Influence of relative age on diagnosis and treatment of attention-deficit/hyperactivity disorder in children. *Canadian Medical Association Journal*, 184(7), 755–762.

Moyer, D. D. (2011). Review article: terminal delirium in geriatric patients with cancer at end of life. *American Journal of Hospice and Palliative Medicine*, 28(1), 44–51.

Mrad, A., Mechri, A., Rouissi, K., Khiari, G., & Gaha, L. (2007). Clinical characteristics of bipolar I patients according to their family history of affective disorders. *Encephale*, 33(5), 762–767.

Murashita, J., Kato, T., Shioiri, T., Inubushi, T., & Kato, N. (2000). Altered brain energy metabolism in lithium-resistant bipolar disorder detected by photic stimulated 31 P-MR spectroscopy. *Psychological Medicine*, 30(1), 107–115.

Nasrallah, H. A., Coffman, J. A., & Olson, S. C. (1989). Structural brain-imaging findings in affective disorders: An overview. *The Journal of Neuropsychiatry and Clinical Neurosciences*, 1(1), 21–26.

National Institute of Mental Health. (2002). Hypericum Depression Trial Study Group. Effect of *Hypericum perforatum* (St. John's wort) in major depressive disorder: A randomized, controlled trial. *JAMA: The Journal of the American Medical Association*, 287, 1807–1814.

National Institute of Neurological Disorders and Stroke. (2007). Brain basics: Understanding sleep. *National Institute of Neurological Disorders and Stroke*. Retrieved May 30, 2013, from http://www.ninds.nih.gov/disorders/brain_basics/understanding_sleep.htm

Nehra, R., Chakrabarti, S., Pradhan, B. K., & Khehra, N. (2006). Comparison of cognitive functions between first- and multi-episode bipolar affective disorders. *Journal of Affective Disorders, 93*(1–3), 185–192.

Nelson, J. (1999). A review of the efficacy of serotonergic and noradrenergic reuptake inhibitors for treatment of major depression. *Biological Psychiatry, 46*(9), 1301–1308.

Newcomer, J. W., & Hennekens, C. H. (2007). Severe mental illness and risk of cardiovascular disease. *JAMA: The Journal of the American Medical Association, 298*(15), 1794–1796.

Niego, S. H., Pratt, E. M., & Agras, W. S. (1997). Subjective or objective binge: Is the distinction valid? *International Journal of Eating Disorders, 22*(3), 291–298.

NIH Consensus Development Panel on Rehabilitation of Persons With Traumatic Brain Injury. (1999). Rehabilitation of persons with traumatic brain injury. *JAMA: The Journal of American Medical Association, 282*, 974–983.

Nilsson, K. W., Sjöberg, R. L., Wargelius, H. L., Leppert, J., Lindström, L., & Oreland, L. (2007). The monoamine oxidase A (MAO-A) gene, family function and maltreatment as predictors of destructive behaviour during male adolescent alcohol consumption. *Addiction, 102*(3), 389–398.

Nobler, M. S., Oquendo, M. A., Kegeles, L. S., Malone, K. M., Campbell, C., Sackheim, H. A., & Mann, J. (2001). Decreased regional brain metabolism after ECT. *The American Journal of Psychiatry, 158*(2), 305–308.

Nolan, M., Carr, A., Fitzpatrick, C., O'Flaherty, A., Keary, K., Turner, R., . . . Tobin, G. (2002). A comparison of two programmes for victims of child sexual abuse: A treatment outcome study. *Child Abuse Review, 11*(2), 103–123.

Numakawa, T., Yagasaki, Y., Ishimoto, T., Okada, T., Suzuki, T., Iwata, N., & Hashimoto, R. (2004). Evidence of novel neuronal functions of dysbindin, a susceptibility gene for schizophrenia. *Human Molecular Genetics, 13*(21), 2699–2708.

Nussbaum, A., Thurstone, C., & Binswanger, I. (2011). Medical marijuana use and suicide attempt in a patient with major depressive disorder. *The American Journal of Psychiatry, 168*(8), 778–781.

Nutt, D. J. (2001). Neurobiological mechanisms in generalized anxiety disorder. *Journal of Clinical Psychiatry, 62*, 22–27.

Oberndorfer, T. A., Kaye, W. H., Simmons, A. N., Strigo, I. A., & Matthews, S. C. (2011). Demand-specific alteration of medial prefrontal cortex response during an inhibition task in recovered anorexic women. *International Journal of Eating Disorders, 44*(1), 1–8.

Oldham, J. M. (2006). Borderline personality disorder and suicidality. *American Journal of Psychiatry, 163*(1), 20–26.

O'Leary, K. M., Brouwers, P., Gardner, D. L., & Cowdry, R. W. (1991). Neuropsychological testing of patients with borderline personality disorder. *American Journal of Psychiatry, 148*(1), 106–111.

Olsen, K. A., & Rosenbaum, B. (2006). Prospective investigations of the prodromal state of schizophrenia: Assessment instruments. *Acta Psychiatrica Scandinavica, 113*(4), 273–282.

Owens, M. J., & Nemeroff, C. B. (1994). Role of serotonin in the pathophysiology of depression: Focus on the serotonin transporter. *Clinical Chemistry, 40*(2), 288–295.

Palmer, B. A., Pankratz, V. S., & Bostwick, J. M. (2005). The lifetime risk of suicide in schizophrenia: A reexamination. *Archives of General Psychiatry, 62*(3), 247–253.

Patel, J. K., & DeUgiannidis, K. M. (2010). Schizophrenia and schizoaffective disorder. In A. J. Rothsch Ud (Ed.), *Antipsychotic medications* (pp. 5–43). Washington, DC: American Psychiatric Association.

Pearlson, G. D., & Robinson, R. G. (1981). Suction lesions of the frontal cerebral cortex in the rat induce asymmetrical behavioral and catecholaminergic responses. *Brain Research, 218*(1), 233–242.

Pengas, G., Hodges, J. R., Watson, P., & Nestor, P. J. (2010). Focal posterior cingulate atrophy in incipient Alzheimer's disease. *Neurobiology of Aging, 31*(1), 25–33.

Pengas, G., Williams, G., B., Acosta-Cabronero, J., Ash, T., W., Hong, Y. T., Izquierdo-Garcia, D., . . . Nestor, P., J. (2012). The relationship of topographical memory performance to regional neurodegeneration in Alzheimer's disease. *Frontiers in Aging Neuroscience, 4*, 17.

Perdikouri, M., Rathbone, G., Huband, N., & Duggan, C. (2007). A comparison of adults with antisocial personality traits with and without childhood conduct disorder. *Annals of Clinical Psychiatry, 19*(1), 17–23.

Perkins, D. O., Gu, H., Boteva, K., & Lieberman, J. A. (2005). Relationship between duration of untreated psychosis and outcome in first-episode schizophrenia: A critical review and meta-analysis. *American Journal of Psychiatry, 162*(10), 1785–1804.

Perlick, D. A., Miklowitz, D. J., Lopez, N., Chou, J., Kalvin, C., Adzhiashvili, V., & Aronson, A. (2010). Family-focused treatment for caregivers of patients with bipolar disorder. *Bipolar Disorders, 12*(6), 627–637.

Peters, R., Peters, J., Warner, J., Beckett, N., & Bulpitt, C. (2008). Alcohol, dementia, and cognitive decline in the elderly: A systemic review. *Oxford Journals Age and Ageing, 37*(5), 505–512.

Peterson, B. S., Warner, V., Bansal, R., Zhu, H., Hao, X., Liu, J., . . . Weissman, M. M. (2009). Cortical thinning in persons at increased familial risk for major depression. *Proceedings of the National Academy of Sciences of the United States of America, 106*(15), 6273–6278.

Phillips, L. J., Mcgorry, P. D., Garner, B., Thompson, K. N., Pantelis, C., Wood, S. J., & Berger, G. (2006). Stress, the hippocampus and the hypothalamic–pituitary–adrenal axis: Implications for the development of psychotic disorders. *Australian and New Zealand Journal of Psychiatry, 40*(9), 725–741.

Phillips, L. J., Rantz, M., & Petroski, G. F. (2011). Indicators of a new depression diagnosis in nursing home residents. *Journal of Gerontological Nursing, 37*(1), 42–52.

Picard, H., Amado, I., Mouchet-Mages, S., Olié, J. P., & Krebs, M. O. (2008). The role of the cerebellum in schizophrenia: An update of clinical, cognitive, and functional evidences. *Schizophrenia Bulletin, 34*(1), 155–172.

Pilling, S., Bebbington, P., Kuipers, E., Garety, P., Geddes, J., Martindale, B., . . . Morgan, C. (2002). Psychological treatments in schizophrenia: II. Meta-analyses of randomized controlled trials of social skills training and cognitive remediation. *Psychological Medicine, 32*(5), 783–791.

Pinheiro, A. P., Root, T., & Bulik, C. M. (2009). The genetics of anorexia nervosa: Current findings and future perspectives. *International Journal of Child and Adolescent Health, 2*(2), 153.

Pollak, S. D., Nelson, C. A., Schlaak, M. F., Roeber, B. J., Wewerka, S. S., Wiik, K. L., . . . Gunnar, M. R. (2010). Neurodevelopmental effects of early deprivation in postinstitutionalized children. *Child Development, 81*(1), 224–236.

Pollak, S. D., & Tolley-Schell, S. A. (2003). Selective attention to facial emotion in physically abused children. *Journal of Abnormal Psychology, 112*(3), 323–338.

Pollock, L. R. & Williams, J. R. G. (1998). Problem solving and suicidal behavior. *Suicide and Life-Threatening Behavior, 28*(4), 375–387.

Porter, R. J., Phipps, A. J., Gallagher, P., Scott, A., Stevenson, P. S., & O'Brien, J. T. (2005). Effects of acute tryptophan depletion on mood and cognitive functioning in older recovered depressed subjects. *The American Journal of Geriatric Psychiatry, 13*(7), 607–615.

Price, T. R. P., Goetz, K. L., & Lovell, M. R. (2010). Neuropsychiatric aspects of brain tumors. In C. Yudofsky & R. E. Hales (Eds.). *Essentials of neuropsychiatry and behavioral neurosciences* (2nd ed.) (pp. 323–347). Washington, DC: American Psychiatric Association.

Qiu, A., Crocetti, D., Adler, M., Mahone, E. M., Denckla, M., Miller, M. I., & Mostofsky, S. H. (2009). Basal ganglia volume and shape in children with attention deficit hyperactivity disorder. *American Journal of Psychiatry, 166*, 74–82.

Qureshi, S. U., Long, M. E., Bradshaw, M. R., Pyne, J. M., Magruder, K. M., Kimbrell, T., . . . Kunik, M. E. (2011, December 1). Does PTSD impair cognition beyond the effect of trauma? *The Journal of Neuropsychiatry, 23*(1), 16–28.

Rabinowitz, J., Levine, S. Z., Garibaldi, G., Bugarski-Kirola, D., Berardo, C. G., & Kapur, S. (2012). Negative symptoms have greater impact on functioning than positive symptoms in schizophrenia: Analysis of CATIE data. *Schizophrenia Research, 137*(1), 147–150.

Raine, A., & Benishay, D. (1995). The SPQ-B: A brief screening instrument for schizotypal personality disorder. *Journal of Personality Disorders, 9*(4), 346–355.

Raine, A., Ishikawa, S. S., Arce, E., Lencz, T., Knuth, K. H., Bihrle, S., . . . Colletti, P. (2004). Hippocampal structural asymmetry in unsuccessful psychopaths. *Biological Psychiatry, 55*(2), 185–191.

Raine, A., Lee, L., Yang, Y., & Colletti, P. (2010). Neurodevelopmental marker for limbic maldevelopment in antisocial personality disorder and psychopathy. *The British Journal of Psychiatry, 197*(3), 186–192.

Raine, A., Lencz, T., & Mednick, S. A. (1995). *Schizotypal personality*. Cambridge, UK: Cambridge University Press.

Raine, A., & Yang, Y. (2006). Neural foundations to moral reasoning and antisocial behavior. *Social Cognitive and Affective Neuroscience, 1*(3), 203–213.

Raney, T. J., Thornton, L. M., Berrettini, W., Brandt, H., Crawford, S., Fichter, M. M., . . . Bulik, C. M. (2008). Influence of overanxious disorder of childhood on the expression of anorexia nervosa. *International Journal of Eating Disorders, 41*(4), 326–332.

Ray, L. A., & Hutchinson, K. E. (2007). Effects of naltrexone on alcohol sensitivity and genetic moderators of medication response. *Archives of General Psychiatry, 64*(9), 1069–1077.

Reas, D. L., & Grilo, C. M. (2008). Review and meta-analysis of pharmacotherapy for binge-eating disorder. *Obesity, 16*(9), 2024–2038.

Reddy, P. L., Khanna, S., Subhash, M. N., Channabasavanna, S. M., & Sridhara Rama Rao, B. S. (1992). CSF amine metabolites in depression. *Biological Psychiatry, 31*(2), 112–118.

Regier, D. A., Narrow, W. E., Rae, D. S., Manderscheid, R. W., Locke, B. Z., & Goodwin, F. K. (1993). The de facto mental and addictive disorders service system. Epidemiologic Catchment Area prospective 1-year prevalence rates of disorders and services. *Archives of General Psychiatry, 50*(2), 85–94.

Reichenberg, A., Caspi, A., Harrington, H., Houts, R., Keefe, R. S., Murray, R. M., . . . Moffitt, T. E. (2010). Static and dynamic cognitive deficits in childhood preceding adult schizophrenia: A 30-year study. *The American Journal of Psychiatry, 167*(2), 160.

Retz, W., Retz-Junginger, P., Supprian, T., Thome, J., & Rösler, M. (2004). Association of serotonin transporter promoter gene polymorphism with violence: Relation with personality disorders, impulsivity, and childhood ADHD psychopathology. *Behavioral Sciences & the Law, 22*(3), 415–425.

Reynolds, G. P., & Mason, S. L. (1995). Absence of detectable striatal dopamine D4 receptors in drug-treated schizophrenia. *European Journal of Pharmacology, 281*(2), 5–6.

Ridley, R. M., Frith, C. D., Crow, T. J., & Conneally, P. M. (1988). Anticipation in Huntington's disease is inherited through the male line but may originate in the female. *Journal of Medical Genetics, 25*(9), 589–595.

Riemann, D., Voderholzer, U., Spiegelhalder, K., Hornyak, M., Buysse, D. J., Nissen, C., . . . Feige, B. (2007). Chronic insomnia and MRI-measured hippocampal volumes: A pilot study. *Sleep, 30*(8), 955.

Rivkin, M. J., Davis, P. E., Lemaster, J. L., Cabral, H. J., Warfield, S. K., Mulkern, R. V., & Robson, C. D. (2008). Volumetric MRI study of brain in children with intrauterine exposure to cocaine, alcohol, tobacco, and marijuana. *Pediatrics, 121*(4), 741–750.

Roberts, M. E., Tchanturia, K., Stahl, D., Southgate, L., & Treasure, J. (2007). A systematic review and meta-analysis of set shifting ability in eating disorders. *Psychological Medicine, 37*(8), 1075–1084

Roca, J., Fuentes, L. J., Marotta, A., López-Ramón, M. F., Castro, C., Lupiáñez, J., & Martella, D. (2012). The effects of sleep deprivation on the attentional functions and vigilance. *Acta Psychologica, 140*(2), 164–176.

Rocca, P., Montemagni, C., Castagna, F., Giugiario, M., Scalese, M., & Bogetto, F. (2009). Relative contribution of antipsychotics, negative symptoms and executive functions to social functioning in stable schizophrenia. *Progress in Neuro-Psychopharmacology and Biological Psychiatry, 33*(2), 373–379.

Rosoklija, G., Toomayan, G., Ellis, S. P., Keilp, J., Mann, J. J., Latov, N., . . . Dwork, A. J. (2000). Structural abnormalities of subicular dendrites in subjects with schizophrenia and mood disorders: Preliminary findings. *Archives of General Psychiatry, 57*(4), 349–356.

Ross, C. A., & Tabrizi, S. J. (2011). Huntington's disease: From molecular pathogenesis to clinical treatment. *The Lancet Neurology, 10*(1), 83–98.

Rosso, I. M., Bearden, C. E., Hollister, J. M., Gasperoni, T. L., Sanchez, L. E., . . . Cannon, T. D. (2000). Childhood neuromotor dysfunction in schizophrenia patients and their unaffected siblings: A prospective cohort study. *Schizophrenia Bulletin, 26*(2), 367–378.

Rosval, L., Steiger, H., Bruce, K., Israël, M., Richardson, J., & Aubut, M. (2006). Impulsivity in women with eating disorders: Problem of response inhibition, planning, or attention? *International Journal of Eating Disorders, 39*(7), 590–593.

Rowland, A. S., Lesesne, C. A., & Abramowitz, A. J. (2002). The epidemiology of attention-deficit/hyperactivity disorder (ADHD): A public health view. *Mental Retardation and Developmental Disabilities Research Reviews, 8*(3), 162–170.

Rund, B. R., Melle, I., Friis, S., Johannessen, J. O., Larsen, T. K., Midbøe, L. J., . . . McGlashan, T. (2007). The course of neurocognitive functioning in first-episode psychosis and its relation to premorbid adjustment, duration of untreated psychosis, and relapse. *Schizophrenia Research, 91*(1–3), 132–140.

Ruocco, A. C. (2005). The neuropsychology of borderline personality disorder: A meta-analysis and review. *Psychiatry Research, 137*(3), 191–202.

Ryan, R. E., Prictor, M. J., McLaughlin, K. J., & Hill, S. J. (2008). Audio-visual presentation of information for informed consent for participation in clinical trials. *Cochrane Database Syst Rev, 1*(1).

Saccone, S. F., Hinrichs, A. L., Saccone, N. L., Chase, G. A., Konvicka, K., Madden, P. A., & Breslau, N. (2006). Cholinergic nicotinic receptor genes implicated in a nicotine dependence association study targeting 348 candidate genes with 3713 SNPs. *Human Molecular Genetics, 16*(1), 36–49.

Saha, S., Chant, D., Welham, J., & McGrath, J. (2005). A systematic review of the prevalence of schizophrenia. *PLoS Medicine, 2*(5), e141.

Samaraweera, Y., & Abeysena, C. (2010). Maternal sleep deprivation, sedentary lifestyle and cooking smoke: Risk factors for miscarriage. A case control study. *Australian and New Zealand Journal of Obstetrics and Gynecology, 50*(4), 352–357.

Sanfilipo, M., Lafargue, T., Rusinek, H., Arena, L., Loneragan, C., Lautin, A., . . . Wolkin, A. (2000). Volumetric measure of the frontal and temporal lobe regions in schizophrenia: Relationship to negative symptoms. *Archives of General Psychiatry, 57*(5), 471–480.

Saxena, S., & Lawley, D. (2009). Delirium in the elderly: A clinical review. *Postgraduate Medical Journal, 85*(1006), 405–413.

Schabus, M., Hödlmoser, K., Gruber, G., Sauter, C., Anderer, P., Klösch, G., . . . Zeitlhofer, J. (2006). Sleep spindle–related activity in the human EEG and its relation to general cognitive and learning abilities. *European Journal of Neuroscience, 23*(7), 1738–1746.

Schabus, M., Hoedlmoser, K., Pecherstorfer, T., Anderer, P., Gruber, G., Parapatics, S., . . . Zeitlhofer, J. (2008). Interindividual sleep spindle differences and their relation to learning-related enhancements. *Brain Research, 1191*, 127–135.

Schäfer, A., Vaitl, D., & Schienle, A. (2010). Regional grey matter volume abnormalities in bulimia nervosa and binge-eating disorder. *Neuroimage, 50*(2), 639–643.

Schienle, A., Schäfer, A., Hermann, A., & Vaitl, D. (2009). Binge-eating disorder: Reward sensitivity and brain activation to images of food. *Biological Psychiatry, 65*(8), 654–661.

Schmahl, C., & Bremner, J. D. (2006). Neuroimaging in borderline personality disorder. *Journal of Psychiatric Research, 40*(5), 419–427.

Schmidt, H., Freudenberger, P., Seiler, S., & Schmidt, R. (2012). Genetics of subcortical vascular dementia. *Experimental Gerontology, 47*(11), 873–877.

Schmidt, M., Morgan, J. F., & Yousaf, F. (2008). Treatment adherence and the care programme approach in individuals with eating disorders. *Psychiatric Bulletin, 32*(11), 426–430.

Schmidt, U., Lee, S., Perkins, S., Eisler, I., Treasure, J., Beecham, J., . . . Yi, I. (2008). Do adolescents with eating disorder not otherwise specified or full-syndrome bulimia nervosa differ in clinical severity, comorbidity, risk factors, treatment outcome or cost? *International Journal of Eating Disorders, 41*(6), 498–504.

Schmidt, U., & Treasure, J. (2006). Anorexia nervosa: Valued and visible. A cognitive-interpersonal maintenance model and its implications for research and practice. *British Journal of Clinical Psychology, 45*(3), 343–366.

Schneider, C., Fulda, S., & Schulz, H. (2004). Daytime variation in performance and tiredness/sleepiness ratings in patients with insomnia, narcolepsy, sleep apnea and normal controls. *Journal of Sleep Research, 13*(4), 373–383.

Schrijvers, E. M., Schürmann, B., Koudstaal, P. J., van den Bussche, H., Van Duijn, C. M., Hentschel, F., . . . Ikram, M. A. (2012). Genome-wide association study of vascular dementia. *Stroke, 43*(2), 315–319.

Schulkin, J. (2007). Autism and the amygdala: An endocrine hypothesis. *Brain and Cognition, 65*(1), 87–99.

Schultz, R. T., & Robins, D. L. (2005). Functional neuroimaging studies of autism spectrum disorders. In F. R. Volkmar, R. Paul, A. Klin, & D. J. Cohen (Eds.), *Handbook of autism and pervasive developmental disorders: Vol. 1* (3rd ed., pp. 515–533) New York, NY: John Wiley.

Schwartz, S., Ponz, A., Poryazova, R., Werth, E., Boesiger, P., Khatami, R., & Bassetti, C. L. (2008). Abnormal activity in hypothalamus and amygdala during humour processing in human narcolepsy with cataplexy. *Brain, 131*(2), 514–522.

Schweizer, T. A., & Vogel-Sprott, M. (2008). Alcohol-impaired speed and accuracy of cognitive functions: A review of acute tolerance and recovery of cognitive performance. *Environmental and Clinical Psychopharmacology, 16*(3), 240–250.

Seeman, P., Guan, H. C., & Van Tol, H. H. (1993). Dopamine D4 receptors elevated in schizophrenia. *Nature, 365*, 441–445.

Seiger, B. H. (2005). *An exploratory study of social workers' attitudes towards harm reduction with substance abusing individuals utilizing the Substance Abuse Treatment Survey (SATS)* (Doctoral dissertation). New York University, School of Social Work.

Seres, I., Unoka, Z., Bódi, N., Aspán, N., & Kéri, S. (2009). The neuropsychology of borderline personality disorder: Relationship with clinical dimensions and comparison with other personality disorders. *Journal of Personality Disorders, 23*(6), 555–562.

Shackman, J. E., Shackman, A. J., & Pollak, S. D. (2007). Physical abuse amplifies attention to threat and increases anxiety in children. *Emotion, 7*(4), 838–852.

Shang, J., Fu, Q., Dienes, Z., Shao, C., & Fu, X. (2013). Negative affect reduces performance in implicit sequence learning. *PLoS One, 8*(1).

Shapiro, J. R., Bauer, S., Andrews, E., Pisetsky, E., Bulik-Sullivan, B., Hamer, R. M., & Bulik, C. M. (2010). Mobile therapy: Use of text-messaging in the treatment of bulimia nervosa. *International Journal of Eating Disorders, 43*(6), 513–519.

Sheedy, J., Geffen, G., Donnelly, J., & Faux, S. (2006). Emergency department assessment of mild traumatic brain injury and prediction of post-concussion symptoms at one month post injury. *Journal of Clinical and Experimental Neuropsychology, 28*(5), 755–772.

Sheline, Y. I., Mokhtar, H. G., & Kraemer, H. C. (2003). Untreated depression and hippocampal volume loss. *The American Journal of Psychiatry, 160*(8), 1516–1518.

Shen, J., Chung, S. A., Kayumov, L., Moller, H., Hossain, N., Wang, X., . . . Shapiro, C. M. (2006). Polysomnographic and symptomatological analyses of major depressive disorder patients treated with mirtazapine. *Canadian Journal of Psychiatry, 51*(1), 27–34.

Shenton, M. E., Dickey, C. C., Frumin, M., & McCarley, R. W. (2001). A review of MRI findings in schizophrenia. *Schizophrenia Research, 49*(1–2), 1–52.

Shiles, M. (2009). Discriminatory referrals: Uncovering a potential ethical dilemma facing practitioners. *Ethics & Behavior, 19*(2), 142–155.

Shores, E. A., Lammel, A., Hullick, C., Sheedy, J., Flynn, M., Levick, W., & Batchelor, J. (2008). The diagnostic accuracy of the Revised Westmead PTA Scale as an adjunct to the Glasgow Coma Scale in the early identification of cognitive impairment in patients with mild traumatic brain injury. *Journal of Neurology, Neurosurgery, and Psychiatry, 79*(10), 1100–1106.

Sidor, M. M., & MacQueen, G. M. (2011). Antidepressants in acute bipolar depression: A systematic review and meta-analysis. *Journal of Clinical Psychiatry, 72*(2), 156–167.

Silk, K. R. (2008). Augmenting psychotherapy for borderline personality disorder: The STEPPS program. *American Journal of Psychiatry, 165*(4), 413–415.

Silver, J. M., Hales, R. E., Yudofsky, S. C., Yudofsky, S. C., & Hales, R. E. (2004). Neuropsychiatric aspects of traumatic brain injury in essentials of neuropsychiatric and clinical neurosciences. *American Psychiatric Publishing*, 241–291.

Silverman, D. H. (2004). Brain 18F-FDG PET in the diagnosis of neurodegenerative dementias: Comparison with perfusion SPECT and with clinical evaluations lacking nuclear imaging. *Journal of Nuclear Medicine, 45*(4), 594–607.

Simon, S. L., Domier, C., Carnell, J., Brethen, P., Rawson, R., & Ling, W. (2000). Cognitive impairment in individuals currently using methamphetamine. *The American Journal on Addictions, 9*(3), 222–231.

Skodol, A. E., Bender, D. S., Pagano, M. E., Shea, M. T., Yen, S., Sanislow, C. A., . . . Gunderson, J. G. (2007). Positive childhood experiences: Resilience and recovery from personality disorder in early adulthood. *Journal of Clinical Psychiatry, 68*(7), 1102–1108.

Skodol, A. E., Gunderson, J. G., Shea, M. T., McGlashan, T. H., Morey, L. C., Sanislow, C. A, . . . Stout, R. L. The Collaborative Longitudinal Personality Disorders Study (CLPS): Overview and implications. *Journal of Personality Disorders, 2005, 19*(5), 487–504.

Smith, M. T., Perlis, M. L., Chengazi, V. U., Pennington, J., Soeffing, J., Ryan, J. M., & Giles, D. E. (2002). Neuroimaging of NREM sleep in primary insomnia: A Tc-99-HMPAO single photon emission computed tomography study. *Sleep, 25*(3), 325–335.

Smoller, J. W., & Finn, C. T. (2003). Family, twin, and adoption studies of bipolar disorder. *American Journal of Medical Genetics Part C: Seminars in Medical Genetics, 123C*(1), 48–58.

Smoller, J. W., Gardner-Schuster, E., & Covino, J. (2008). The genetic basis of panic and phobic anxiety disorders. *American Journal of Medical Genetics, 148C*(2), 118–126.

Snell, F. I., Halter, M. J. (2010). A signature wound of war: Mild traumatic brain injury. *Journal of Psychosocial Nursing and Mental Health Services, 48*(2), 22–28.

Snook, L., Paulson, L. A., Roy, D., Phillips, L., & Beaulieu, C. (2005). Diffusion tensor imaging of neurodevelopment in children and young adults. *Neuroimage, 26*(4), 1164–1173.

Society for Neuroscience. (2012). *Brain facts: A primer on the brain and nervous system* (6th ed.). Washington, DC: Meadows Design Office Incorporated.

Södersten, P., Nergårdh, R., Bergh, C., Zandian, M., & Scheurink, A. (2008). Behavioral neuroendocrinology and treatment of anorexia nervosa. *Frontiers in Neuroendocrinology, 29*(4), 445.

Solms, M. (1997). *The neuropsychology of dreams: A clinico-anatomical study*. Hillsdale, NJ: Lawrence Erlbaum Associates.

Soloff, P. H., George, A., Nathan, R. S., Schulz, P. M., Cornelius, J. R., Herring, J., & Perel, J. M. (1989). Amitriptyline versus haloperidol in borderlines: Final outcomes and predictors of response. *Journal of Clinical Psychopharmacology, 9*(4), 238.

Spence, S. A., Brooks, D. J., Hirsch, S. R., Liddle, P. F., Meehan, J., & Grasby, P. M. (1997). A PET study of voluntary movement in schizophrenic patients experiencing passivity phenomena (delusions of alien control). *Brain, 120*(11), 1997–2011.

Steen, R. G., Mull, C., Mcclure, R., Hamer, R. M., & Lieberman, J. A. (2006). Brain volume in first-episode schizophrenia: systematic review and meta-analysis of magnetic resonance imaging studies. *The British Journal of Psychiatry, 188*(6), 510–518.

Steinhausen, H. C., & Weber, S. (2009). The outcome of bulimia nervosa: Findings from one-quarter century of research. *American Journal of Psychiatry, 166*(12), 1331–1341.

Stergiakouli, E., Hamshere, M., Holmans, P., Langley, K., Zaharieva, I., Hawi, Z., & Thapar, A. (2012). Investigating the contribution of common genetic variants to the risk and pathogenesis of ADHD. *American Journal of Psychiatry, 169*(2), 186–194.

Strakowski, S. M. (2012). Integration and consolidation: A neurophysiological model of bipolar disorder. In S. M. Strakowski (Ed.), *The bipolar brain: Integrating neuroimaging and genetics* (pp. 253–274). New York, NY: Oxford University Press.

Strakowski, S. M., Adler, C. M., Almeida, J., Altshuler, L. L., Blumberg, H. P., Chang, K. D., . . . Townsend, J. D. (2012). The functional neuroanatomy of bipolar disorder: A consensus model. *Bipolar Disorders, 14*, 313–325.

Striegel-Moore, R. H., & Bulik, C. M. (2007). Risk factors for eating disorders. *American Psychologist, 62*(3), 181.

Stroup, T. S., Appelbaum, P. S., Gu, H., Hays, S., Swartz, M. S., Keefe, R. S., . . . Lieberman, J. A. (2011). Longitudinal consent-related abilities among research participants with schizophrenia: Results from the CATIE study. *Schizophrenia Research, 130*(1), 47–52.

Sullivan, P. F. (1995). Mortality in anorexia nervosa. *American Journal of Psychiatry, 152*(7), 1073–1074.

Sullivan, P. F., Neale, M. C., & Kendler, K. S. (2000). Genetic epidemiology of major depression: Review and meta-analysis. *The American Journal of Psychiatry, 157*(10).

Sutterby, S. R., & Bedwell, J. S. (2012). Lack of neuropsychological deficits in generalized social phobia. *PLoS One, 7*(8), e42675.

Swanson, J., Holzer, C., Ganju, V. K., & Jano, R. T. (1990). Violence and psychiatric disorder in the community: Evidence from the Epidemiologic Catchment Area surveys. *Hospital and Community Psychiatry. 41*(7), 761–770.

Sysko, R., Hildebrandt, T., Wilson, G. T., Wilfley, D. E., & Agras, W. S. (2010). Heterogeneity moderates treatment response among patients with binge eating disorder. *Journal of Consulting and Clinical Psychology, 78*(5), 681–690.

Szentagotai, A., & David, D. (2010). The efficacy of cognitive–behavioral therapy in bipolar disorder: A quantitative meta-analysis. *Journal of Clinical Psychiatry, 71*(1), 66–72.

Tadic, A., Elsaber, A., Victor, A., von Cube, R., Baakaya, O., Wagner, S., & Dahmen, N. (2009). Association analysis of serotonin receptor IB (HTRIB) and brain-denved neurotrophic factor gene polymorphisms in borderline personality disorder. *Journal of Neural Transmission, 116,* 1185–1188.

Taki, Y., Hashizume, H., Thyreau, B., Sassa, Y., Takeuchi, H., Wu, K., . . . Kawashima, R. (2012). Sleep duration during weekdays affects hippocampal gray matter volume in healthy children. *NeuroImage, 60*(1), 471–475.

Tamminga, C. A. (1998). Schizophrenia and glutamatergic transmission. *Critical Reviews in Neurobiology, 12*(1–2), 21–36.

Tamminga, C. A., Shad, M. U., & Ghose, S. (2008). Neuropsychiatric aspects of schizophrenia. In S. C. Yudofsky & R. E. Hales (Eds.). *The American Psychiatric Publishing textbook of neuropsychiatric and behavioral neurosciences* (5th ed.). Washington, DC: American Psychiatric Association.

Tamminga, C. A., Thaker, G. K., Buchanan, R., Kirkpatrick, B., Alphs, L. D., Chase, T. N., & Carpenter, W. T. (1992). Limbic system abnormalities identified in schizophrenia using positron emission tomography with fluorodeoxyglucose and neocortical alterations with deficit syndrome. *Archives of General Psychiatry, 49*(7), 522–530.

Tandon, R., Belmaker, R. H., Gattaz, W. F., Lopez-Ibor Jr., J. J., Okasha, A., Singh, B., & Moeller, H. J. (2008). World Psychiatric Association Pharmacopsychiatry Section statement on comparative effectiveness of antipsychotics in the treatment of schizophrenia. *Schizophrenia Research, 100*(1–3), 20–38.

Tandon, R., Keshavan, M. S., & Nasrallah, H. A. (2008). Schizophrenia, "just the facts": What we know in 2008: Part 1: Overview. *Schizophrenia Research, 100*(1), 4–19.

Tapert, S. F., Cheung, E. H., Brown, G. G., Frank, L. R., Paulus, M. P., Schweinsburg, A. D., & Meloy, M. J. (2003). Neural response to alcohol stimuli in adolescents with alcohol use disorder. *Archives of General Psychiatry, 60*(7), 727–735.

Tapert, S. F., Schweinsburg, A. D., Drummond, S. P., Paulus, M. P., Brown, S. A., Yang, T. T., . . . . Meloy, M. J. (2007). Functional MRI of inhibitory processing in abstinent adolescent marijuana users. *Psychopharmacology (Berl.) 194,* 173–183.

Taylor, M., Stefanis, N., Frangou, S., Yakeley, J., Sharma, T., O'Connell, P., . . . Murray, R. (1999). Hippocampal volume reduction in schizophrenia: Effects of genetic risk and pregnancy and birth complications. *Biological Psychiatry, 46*(5), 697–702.

Tchanturia, K., Davies, H., & Campbell, I. C. (2007). Cognitive remediation therapy for patients with anorexia nervosa: Preliminary findings. *Annals of General Psychiatry, 6*(1), 14.

Teasdale, G., & Jennett, B. (1974). Assessment of coma and impaired consciousness: A practical scale. *The Lancet, 304*(7872), 81–84.

Tebartz van Elst, L., Hesslinger, B., & Thiel, T. (2003). Frontolimbic brain abnormalities in patients with borderline personality disorder: A volumetric magnetic resonance imaging study. *Biological Psychiatry, 54, 163*–171.

Teicher, M. H., Samson, J. A., Sheu, Y. S., Polcari, A., & McGreenery, C. E. (2010). Hurtful words: Exposure to peer verbal aggression is associated with elevated psychiatric symptom scores and corpus callosum abnormalities. *The American Journal of Psychiatry, 167*(12), 1464.

Teicher, M. H., Tomoda, A., & Andersen, S. L. (2006). Neurobiological consequences of early stress and childhood maltreatment: Are results from human and animal studies comparable? *Annals of the New York Academy of Sciences, 1071*(1), 313–323.

Teodorczuk, A., Reynish, E., & Milisen, K. (2012). Improving recognition of delirium in clinical practice: A call for action. *BMC Geriatrics, 12*(1),1–5.

Thompson, P. M., Vidal, C., Giedd, J. N., Gochman, P., Blumenthal, J., Nicolson, R., & Rapoport, J. L. (2001). Mapping adolescent brain change reveals dynamic wave of accelerated gray matter loss in very early-onset schizophrenia. *Proceedings of the National Academy of Sciences, 98*(20), 11650–11655.

Thorpe, C. M., Floresco, S. B., Carr, J. A., & Wilkie, D. M. (2002). Alterations in time–place learning induced by lesions to the rat medial prefrontal cortex. *Behavioural Processes, 59*(2), 87–100.

Titone, D., Holzman, P. S., & Levy, D. L. (2002). Idiom processing in schizophrenia: Literal implausibility saves the day for idiom priming. *Journal of Abnormal Psychology, 111*(2), 313–320.

Torgersen, S., Kringlen, E., & Cramer, V. (2001). The prevalence of personality disorders in a community sample. *Archives of General Psychiatry, 58*(6), 590.

Toschlog, E. A., MacElligot, J., Sagraves, S. G., Schenarts, P. J., Bard, M. R., Goettler, C. E., . . . Swanson, M. S. (2003). The relationship of Injury Severity Score and Glasgow Come Score to rehabilitative potential in patients suffering traumatic brain injury. *The American Surgeon, 69*(6), 491–497.

Townsend, J., & Altshuler, L. L (2012). Emotion processing and regulation in bipolar disorder: A review. *Bipolar Disorders, 14*(4), 326–339.

Tranøy, J., & Blomberg, W. (2005). Lobotomy in Norwegian psychiatry. *History of Psychiatry, 16*(1), 107–110.

Tuulio-Henriksson, A., Partonen, T., Suvisaari, J., Haukka, J., & Lönnqvist, J. (2004). Age at onset and cognitive functioning in schizophrenia. *The British Journal of Psychiatry, 185*(3), 215–219.

Uher, R., Murphy, T., Brammer, M. J., Dalgleish, T., Phillips, M. L., Ng, V. W., . . . Treasure, J., (2004). Medial prefrontal cortex activity associated with symptom provocation in eating disorder. *American Journal of Psychiatry 161*(7), 1238–1246.

Uher, R., & Rutter, M. (2012). Classification of feeding and eating disorders: Review of evidence and proposals for ICD-11. *World Psychiatry, 11*(2), 80–92.

U.S. Department of Health and Human Services: Interagency Head Injury Task Force Report. (1989). Washington, DC: Author.

Valdes, I. H., Steinberg, J. L., Narayana, P. A., Kramer, L. A., Dougherty, D. M., & Swann, A. C. (2006). Impulsivity and BOLD fMRI activation in MDMA users and healthy control subjects. *Psychiatry Research, 147*(2–3), 239–242.

Van den Eynde, F., Claudino, A. M., Mogg, A., Horrell, L., Stahl, D., Ribeiro, W., . . . Schmidt, U. (2010). Repetitive transcranial magnetic stimulation reduces cue-induced food craving in bulimic disorders. *Biological Psychiatry, 67*(8), 793–795.

Van den Eynde, F., Guillaume, S., Broadbent, H., Stahl, D., Campbell, I. C., Schmidt, U., & Tchanturia, K. (2011). Neurocognition in bulimic eating disorders: A systematic review. *Acta Psychiatrica Scandinavica, 124*(2), 120–140.

Van Dijk, E. J., Vermeer, S. E., De Groot, J. C., Van De Minkelis, J., Prins, N. D., Oudkerk, M., . . . Breteler, M. M. B. (2004). Arterial oxygen saturation, COPD, and cerebral small vessel disease. *Journal of Neurology, Neurosurgery & Psychiatry, 75*(5), 733–736.

Van Hoeken, D., Veling, W., Sinke, S., Mitchell, J. E., & Hoek, H. W. (2009). The validity and utility of subtyping bulimia nervosa. *International Journal of Eating Disorders, 42*(7), 595–602.

Van Praag, H. M. (1983). CSF 5-HIAA and suicide in non-depressed schizophrenics. *The Lancet, 322*(8356), 977–978.

Velligan, D. I., & Bow-Thomas, C. C. (1999, January). Executive function in schizophrenia. *Seminars in Clinical Neuropsychiatry, 4*(1), 24–33.

Videbach, P. (2000). PET measurements of brain glucose metabolism and blood flow in major depressive disorder: A critical review. *Acta Psychiatrica Scandinavica, 101*, 11–20.

Vonsattel, J. P. G., Keller, C., & Amaya, M. P. (2008). Neuropathology of Huntington's disease. *Handbook of Clinical Neurology, 89*, 599–618.

Vos, P. E., Battistin, L., Birbamer, G., Gerstenbrand, F., Potapov, A., Prevec, T., . . . Wild, K. V. (2002). EFNS guideline on mild traumatic brain injury: Report of an EFNS task force. *European Journal of Neurology, 9*(3), 207–219.

Voyer, P., Richard, S., McCusker, J., Cole, M. G., Monette, J., Champoux, N., . . . Belzile, E. (2012). Detection of delirium and its symptoms by nurses working in a long-term care facility. *Journal of the American Medical Directors Association, 13*(3), 264–271.

Wade, S. L., Wolfe, C., Brown, T. M., & Pestian, J. P. (2005). Putting the pieces together: Preliminary efficacy of a web-based family intervention for children with traumatic brain injury. *Journal of Pediatric Psychology, 30*(5), 437–442.

Wagner, A., Aizenstein, H., Venkatraman, V., Fudge, J., May, J., Mazurkewicz, L., . . . Kaye, W. (2007). Altered reward processing in women recovered from anorexia nervosa. *American Journal of Psychiatry, 164*(12), 1842–1849.

Wagner, A. K., Sasser, H. C., Hammond, F. M., Wiercisiewski, D., & Alexander, J. (2000). Intentional traumatic brain injury: Epidemiology, risk factors, and associations with injury severity and mortality. *The Journal of Trauma and Acute Care Surgery, 49*(3), 404–410.

Wagner, G., Koch, K., Schachtzabel, C., Schultz, C. C., Sauer, H., & Schlosser, R. G. (2010). Structural brain alterations in patients with major depressive disorder and high risk for suicide: Evidence for a distinct neurobiological entity? *Neuroimage, 54*, 1607–1614.

Wagner, S., Baskaya, O., Lieb, K., Dahmen, K., & Tadec, K. (2009). The 5 HTTLPR polymorphism modulates the association of serious life events and impulsivity in patients with Borderline Personality Disorder. *Journal of Psychiatric Research, 43*(13), 1067–1072.

Walpoth, M., Hoertnagl, C., Mangweth-Matzek, B., Kemmler, G., Hinterhozl, J., Conca, A., & Hausmann, A. (2008). Repetitive transcranial magnetic stimulation in bulimia nervosa: Preliminary results of a single-centre randomized, double blind, sham controlled trial in female outpatients. *Psychotherapy Psychosomatics, 77*, 57–60.

Walsh, B. T., Kaplan, A. S., Attia, E., Olmsted, M., Parides, M., Carter, J. C., . . . Rockert, W. (2006). Fluoxetine after weight restoration in anorexia nervosa. *JAMA: The Journal of the American Medical Association, 295*(22), 2605–2612.

Wang, G. J., Volkow, N. D., Chang, L., Miller, E., Sedler, M., Hitzemann, R., . . . Fowler, J. S. (2004). Partial recovery of brain metabolism in methamphetamine abusers after protracted abstinence. *American Journal of Psychiatry, 161*, 242–248.

Wang, L., Hosakere, M., Trein, C. L., Miller, A., Ratnanather, J. T., Barch, D. M., . . . Csernansky, J. G. (2007). Abnormalities of cingulate gyrus neuroanatomy in schizophrenia. *Schizophrenia Research, 93*(1–3), 66–78.

Watts, S., Mackenzie, A., Thomas, C., Griskaitis, A., Mewton, L., Williams, A., & Andrews, G. (2013). CBT for depression: A pilot RCT comparing mobile phone vs. computer. *BMC Psychiatry, 13*, 13–49.

Waxman, S. E. (2009). A systematic review of impulsivity in eating disorders. *European Eating Disorders Review, 17*(6), 408–425.

Weinberg, S. M., Jenkins, E. A., Marazita, M. L., & Maher, B. S. (2007). Minor physical anomalies in schizophrenia: A meta-analysis. *Schizophrenia Research, 89*(1), 72–85.

Weiss, R. P. (2000). Memory and learning. *Training & Development, 54*(10), 46–50.

Wertz, D. C., Fanos, J. H., & Reilly, P. R. (1994). Genetic testing for children and adolescents. *JAMA: The Journal of the American Medical Association, 272*(11), 875–881.

Wesnes, K. A., McKeith, I. G., Ferrara, R., Emre, M., Del Ser, T., Spano, P. F., . . . Spiegel, R. (2002). Effects of rivastigmine on cognitive function in dementia with Lewy bodies: A randomised placebo-controlled international study using the cognitive drug research computerised assessment system. *Dementia and Geriatric Cognitive Disorders, 13*(3), 183–192.

Weygandt, M., Schaefer, A., Schienle, A., & Haynes, J. D. (2012). Diagnosing different binge-eating disorders based on reward-related brain activation patterns. *Human Brain Mapping, 33*(9), 2135–2146.

White, T., Nelson, M., & Lim, K. O. (2008). Diffusion tensor imaging in psychiatric disorders. *Topics in Magnetic Resonance Imaging, 19*(2), 97–109.

Wible, C. G., Anderson, J., Shenton, M. E., Kricun, A., Hirayasu, Y., Tanaka, S., & McCarley, R. W. (2001). Prefrontal cortex, negative symptoms, and schizophrenia: An MRI study. *Psychiatry Research, 108*(2), 65–78.

Williams, D. H., Levin, H. S., & Eisenberg, H. M. (1990). Mild head injury classification. *Neurosurgery, 27*(3), 422–428.

Williams, N. M., Franke, B., Mick, E., Anney, R. J. L., Freitag, C. M., Thapar, A., & Faranoe, S. V. (2012). Genome-wide analysis of copy number variants in attention deficit hyperactivity disorder: The role of rare variants and duplications at 15q13.3. *American Journal of Psychiatry, 169*, 195–204.

Wilson, S. J., Nutt, D. J., Alford, C., Argyropoulos, S. V., Baldwin, D. S., Bateson, A. N., . . . Wade, A. G. (2010). British Association for Psychopharmacology consensus statement on evidence-based treatment of insomnia, parasomnias, and circadian rhythm disorders. *Journal of Psychopharmacology, 24*(11), 1577–1601.

Wilson, W., Mathew, R., Turkington, T., Hawk, T., Coleman, R. E., & Provenzale, J. (2000). Brain morphological changes and early marijuana use: A magnetic resonance and positron emission tomography study. *Journal of Addictive Diseases, 19*, 1–22.

Winkelman, J., Lin, L., Schormair, B., Kornum, B. R., Faraco, J., Plazzi, G., . . . Mignot, E. (2012). Mutations in DNMT1 cause autosomal dominant cerebellar ataxia, deafness and narcolepsy. *Human Molecular Genetics, 21*, 2205–2210.

Winterer, G., Coppola, R., Goldberg, T. E., Egan, M. F., Jones, D. W., Sanchez, C. E., & Weinberger, D. R. (2004). Prefrontal broadband noise, working memory, and genetic risk for schizophrenia. *American Journal of Psychiatry, 161*(3), 490–500.

Wolfe, B. E., Baker, C. W., Smith, A. T., & Kelly-Weeder, S. (2009). Validity and utility of the current definition of binge eating. *International Journal of Eating Disorders, 42*(8), 674–686.

Wolfson, A. R., & Carskadon, M. A. (1998). Sleep schedules and daytime functioning in adolescents. *Child Development, 69*(4), 875–887.

Wood, L. R. (2004). Understanding the "miserable minority": A diathesis-stress paradigm for post-concussional syndrome. *Brain Injury, 18*(11), 1135–1153.

Wood, P. K., Sher, K. J., & Bartholow, B. D. (2002). Alcohol use and cognitive abilities in young adulthood: A prospective study. *Journal of Consulting and Clinical Psychology, 70*(4), 897–907.

Woods, S. P., Rippeth, J. D., Conover, E., Gongvatana, A., Gonzalez, R., Carey, C. L., . . . Grant, I. (2005). Deficient strategic control of verbal encoding and retrieval in individuals with methamphetamine dependence. *Neuropsychology, 19*(1), 35–43.

Woon, F. L., & Hedges, D. W. (2008). Hippocampal and amygdala volumes in children and adults with childhood maltreatment-related posttraumatic stress disorder: A meta-analysis. *Hippocampus, 18*(8), 729–736.

Wright, I. C., Rabe-Hesketh, S., Woodruff, P. W., David, A. S., Murray, R. M., & Bullmore, E. T. (2000). Meta-analysis of regional brain volumes in schizophrenia. *American Journal of Psychiatry, 157*(1), 16–25.

Wykes, T., Brammer, M., Mellers, J., Bray, P., Reeder, C., Williams, C., & Corner, J. (2002). Effects on the brain of a psychological treatment: Cognitive remediation therapy functional magnetic resonance imaging in schizophrenia. *The British Journal of Psychiatry, 181*(2), 144–152.

Wykes, T., Reeder, C., Corner, J., Williams, C., & Everitt, B. (1999). The effects of neurocognitive remediation on executive processing in patients with schizophrenia. *Schizophrenia Bulletin, 25*(2), 291–307.

Wykes, T., Reeder, C., Williams, C., Corner, J., Rice, C., & Everitt, B. (2003). Are the effects of cognitive remediation therapy (CRT) durable? Results from an exploratory trial in schizophrenia. *Schizophrenia Research, 61*(2), 163–174.

Xia, J., & Li, C. (2007). Problem solving skills for schizophrenia. *Cochrane Database of Systematic Reviews, 18*(2).

Yager, J. (2008). Binge eating disorder: The search for better treatments. *American Journal of Psychiatry, 165*(1), 4–6.

Yeo, R. A., Hill, D. E., Campbell, R. A., Vigil, J., Petropoulos, H., Hart, B., . . . Brooks, W. M. (2003). Proton magnetic resonance spectroscopy investigation of the right frontal lobe in children with attention-deficit/hyperactivity disorder. *Journal of the American Academy of Child and Adolescent Psychiatry, 42*(3), 303–310.

Yoo, S. S., Hu, P. T., Gujar, N., Jolesz, F. A., & Walker, M. P. (2007). A deficit in the ability to form new human memories without sleep. *Nature Neuroscience, 10*(3), 385–392.

Yudofsky, S. C., & Hales, R. E. (2008). *The American Psychiatric Publishing textbook of neuropsychiatry and behavioral neurosciences.* Washington, DC: American Psychiatric Association.

Yuii, K., Suzuki, M., & Kurachi, M. (2007). Stress sensitization in schizophrenia. *Annals of the New York Academy of Sciences, 1113*(1), 276–290.

Zanello, A., Curtis, L., Badan, B. M., & Merlo, M. C. (2009). Working memory impairments in first-episode psychosis and chronic schizophrenia. *Psychiatry Research, 165*(1), 10–18.

Zastrow, A., Kaiser, S., Stippich, C., Walther, S., Herzog, W., Tchanturia, K., . . . Friederich, H. C. (2009). Neural correlates of impaired cognitive-behavioral flexibility in anorexia nervosa. *American Journal of Psychiatry, 166*(5), 608–616.

Zeeck, A., Weber, S., Sandholz, A., Wetzler-Burmeister, E., Wirsching, M., & Hartmann, A. (2009). Inpatient versus day clinic treatment for bulimia nervosa: A randomized trial. *Psychotherapy and Psychosomatics, 78*(3), 152–160.

Zelkowitz, P., Paris, J., Guzder, I., & Feldman, K. (2001). Diathesis and stressors in borderline pathology of childhood: The role of neuropsychological risk and trauma. *Journal of the American Academy of Child and Adolescent Psychiatry, 40,* 100–105.

Zhao, X., Shi, Y., Tang, J., Tang, R., Yu, L., Gu, N., . . . He, L. (2004). A case control and family based association study of the neuregulin1 gene and schizophrenia. *Journal of Medical Genetics, 41*(1), 31–34.

Zilbovicius, M., Meresse, I., & Boddaert, N. (2006). Autism: Neuroimaging. *Revista Brasileira de Psiquiatria, 28,* s21–s28.

Zito, J. M., Safer, D. J., Sai, D., Gardner, J. F., Thomas, D., Coombes, P., . . . Mendez-Lewis, M. (2008). Psychotropic medication patterns among youth in foster care. *Pediatrics, 121*(1), e157–e163.

# Index

ablation 14–15
acetylcholine, 33–34
ADHD. *See* attention deficit hyperactivity disorder
adolescents, alcohol abuse, 166
agoraphobia, 120
alcohol abuse
  in adolescents, 166
  cognitive impairment, 165–166
  introduction, 164–165
  mechanism, 165
  treatment, 166
Alzheimer's disease
  cause, 178
  and genetics, 177–178
  and imaging, 178–179
  introduction, 176–177
  lifestyle risks, 177
  mood, 177
  treatment, 179–180
amphetamines, 171–172
amygdala, 49
AN. *See* anorexia nervosa
anorexia nervosa (AN)
  and cognition, 132–133
  and genetics, 131
  introduction, 130–131
  and neuroimaging, 131–132
  treatment, 133–134, 136
    medication, 133–134
    psychotherapy, 134, 136
anoxia and hypoxia, 190–191
antisocial personality disorder (ASPD)
  and cognition, 212
  and genetics, 211
  introduction, 210
  and neuroimaging, 211–212
  treatment, 212–213
anxiety disorders
  agoraphobia, 120
  fight or flight response
    introduction, 115–116
    SNS activation, 116–118
  generalized anxiety disorder (GAD), 126–128
    and cognition, 127–128
    and genetics, 127
    and neurochemistry, 127
    and neuroimaging, 127
    treatment, 128
  introduction, 113, 115
  panic disorder (PD), 118–120
    and cognition, 119
    and neurochemistry, 119
    and neuroimaging, 119
    treatment, 119–120
  phobias, specific, 122–123
    and genetics, 123
    and neuroimaging, 123
    treatment, 123
  posttraumatic stress disorder (PTSD), 123–126
    and cognition, 126
    and genetics, 124
    introduction, 122–123
    and neurochemistry, 124–125
    and neuroimaging, 125
    treatment, 126

social anxiety, 121–122
   and cognition, 122
   and genetics, 121
   and neurochemistry, 121
   and neuroimaging, 121–122
   treatment, 122
  stress in clinical and research environments, 114–115
ASPD. *See* antisocial personality disorder
attention deficit hyperactivity disorder (ADHD)
  characteristics, 65
  and cognition, 67
  and genetics, 65
  introduction, 65
  and neuroimaging, 66–67
  and neurotransmitters, 66
  treatment, 67–68
autism spectrum disorders (ASD)
  and cognition, 71
  and genetics, 70
  introduction, 68–70
  and neuroimaging, 70–71
  and neurotransmitters, 70
  treatment, 71–72

basal ganglia
  introduction, 51
  globus pallidus, 52
  role, 51–52
  striatum, 52
BED. *See* binge eating disorder
behavior genetic testing, 25
binge eating disorder (BED)
  and cognition, 142
  introduction, 139, 141
  and neurochemistry, 141
  and neuroimaging, 141–142
  treatment, 142–143
    medication, 142
    psychotherapy, 142–143
bipolar disorder (BD)
  and cognition, 107–108
  dual-role and relationships with, 214
  and genetics, 104–105
  introduction, 104
  and neurochemistry, 106–107
  and neuroimaging, 105–106
  treatment, 108
    psychotherapy, 109
    psychotropic medications, 108–109
BN. *See* bulimia nervosa
borderline personality disorder (BPD)
  and genetics, 215
  introduction, 213, 215
  and neurocognition, 216–217
  and neuroimaging, 215–216
  treatment, 217–218
BPD. *See* borderline personality disorder
brain division
  cerebellum, 37
  corpus callosum, 38
  glial cells, 38
  gray versus white matter, 36–37
  introduction, 36
  left and right hemispheres, 37
  meninges and cerebral spinal fluid (CSF), 38
brain lobes
  frontal lobe, 40, 42
  introduction, 39
  occipital lobe, 39
  parietal lobe, 40
  temporal lobe, 40
brain systems
  basal ganglia
    globus pallidus, 52
    introduction, 51
    role, 51–52
    striatum, 52
  cingulate system, 54
  fusiform gyrus, 54–55
  introduction, 45–48
  limbic system
    amygdala, 49
    hippocampus, 49–50
    hypothalamic–pituitary–adrenal (HPA) axis, 51
    mesolimbic system, 51
    role, 48–49
    thalamus, 50–51
  prefrontal system, 53–54
  secondary systems
    Circle of Willis, 56–57
    cranial nerves, 55
    introduction, 55
    mirror system, 56
bulimia nervosa (BN)
  and cognition, 137–138
  introduction, 136–137
  and neuroimaging, 137
  treatment, 138–139
    medication, 138
    psychotherapy, 139
    repetitive transcranial magnetic stimulation, 139

caffeine and nicotine abuse, 170
cannabis abuse, 167
catecholamines, 34
central nervous system
  development of, 35–36
  firing of the neuron, 32–33
  neuron, 31

properties of the neuron, 31–32
cerebellum, 37
cerebral spinal fluid (CSF), 38
childhood disorders
   attention deficit hyperactivity disorder (ADHD)
      characteristics, 65
      and cognition, 67
      and genetics, 65
      introduction, 65
      and neuroimaging, 66–67
      and neurotransmitters, 66
      treatment, 67–68
   autism spectrum disorders (ASD)
      and cognition, 71
      and genetics, 70
      introduction, 68–70
      and neuroimaging, 70–71
      and neurotransmitters, 70
      treatment, 71–72
   childhood maltreatment
      and cognition, 76–77
      cognitive behavioral therapy, 77
      and genetics, 75
      introduction, 74
      and neuroimaging, 75–76
      and neurotransmitters, 75
      treatment, 77
   childhood trauma, 74
   introduction, 61, 63
   intellectual disability
      characteristics of, 72–73
      dyslexia, 73–74
      introduction, 72
      learning disorders, 73
   neurodevelopment overview
      embryogenesis, 63
      neuronal migration, 63–64
      stages, 63
      synaptic pruning, 64
   psychotropic medication in children, 62–63
childhood maltreatment
   and cognition, 76–77
   cognitive behavioral therapy, 77
   and genetics, 75
   introduction, 74
   and neuroimaging, 75–76
   and neurotransmitters, 75
   treatment, 77
childhood trauma, 74
circadian rhythms, 150
cingulate system, 54
clinical case reports, neurobiological history
   current case examinations, 6
   H.M., 4–6
   introduction, 2
   Mr. Tan, 3–4
   Phineas Gage, 2–3
cocaine abuse, 168–169
cognitive behavioral therapy
   and binge eating disorder, 142–143
   for childhood maltreatment, 77
   with major depressive disorder (MDD), 99
   for persistent postconcussion syndrome (PPCS), 200
   for panic disorder, 119–120
   for traumatic brain injury, 203
   treatment technique, history of, 10–11
competency and neurobiology, 28
computed tomography (CT), 7, 17–18
concussion, 200
confidentiality, 205
continuous seizure disorder (status epilepticus), 184
corpus callosum, 38
cranial nerves, 55
CSF. *See* cerebral spinal fluid
CT. *See* computed tomography

decision-making capacity, 5
deep brain stimulation (DBS), 101
delirium, 174–176
   causes, 175
   introduction, 174–175
   and neurobiology, 175–176
   symptoms, 175
dementia of Lewy body (DLB), 181–182
dementias, introduction, 176. *See also* Alzheimer's disease, vascular dementia, frontal lobe dementia, dementia of Lewy body
depression
   clinical symptoms, 94–95
   introduction, 94
   subtypes, 95
diffusion tensor imaging (DTI), 20–21
dopamine, 34–35
dreaming, 149–150
driving safety, medical disorders, 174
DTI. *See* diffusion tensor imaging
dyslexia, 73–74

eating disorders
   anorexia nervosa (AN)
      and cognition, 132–133
      and genetics, 131
      introduction, 130–131
      and neuroimaging, 131–132
      treatment, 133–134, 136
   binge eating disorder (BED)
      and cognition, 142
      introduction, 139, 141
      and neurochemistry, 141

and neuroimaging, 141–142
treatment, 142–143
bulimia nervosa (BN)
and cognition, 137–138
introduction, 136–137
and neuroimaging, 137
treatment, 138–139
introduction, 129–130
obesity, and therapist role, 140–141
safety and, 135
ECT. *See* electroconvulsive therapy
EEG. *See* electroencephalogram
electroconvulsive therapy (ECT), 9, 101–102
electroencephalogram (EEG), 16
electrophysiology recordings
electroencephalogram (EEG), 16
event-related potentials, 17
single-cell technique, 15–16
embryogenesis, 63
epinephrine, 35
ethics
amphetamines as performance
enhancers, 171–172
animal research in neuroscience, 15
competency and neurobiology, 28
collaboration and cooperation, 47
confidentiality, 205
decision-making capacity in schizophrenia,
89–90
dual-role and relationships with BPD, 214
falsifying neurobiological data, 69
genetic testing, 187
safety and eating disorders, 135
safety of patients, 9
sleep and clinician self-care, 153
stress in clinical and research environments,
114–115
suicidal behavior and confidentiality,
110–111
event-related potentials, 17

fight or flight response
introduction, 115–116
SNS activation, 116–118
fMRI. *See* functional magnetic resonance
imaging
focal (partial) seizure disorder, 184
fright, enjoyment of, 117
frontal lobe, 40, 42
frontal lobe dementia, 180–181
functional magnetic resonance imaging (fMRI),
22–23
functional neuroimaging
functional magnetic resonance imaging (fMRI),
22–23
magnetoencephalography (MEG), 23–24

positron emission tomography (PET), 22
single-photon emission computed tomography
(SPECT), 21–22
fusiform gyrus, 54–55

GAD. *See* generalized anxiety disorder
gamma amino butric acid (GABA), 35
generalized anxiety disorder (GAD), 126–128
and cognition, 127–128
and genetics, 127
introduction, 126
and neurochemistry, 127
and neuroimaging, 127
treatment, 128
generalized seizure disorder, 183–184
genetics
antisocial personality disorder (ASPD), 211
attention deficit hyperactivity disorder
(ADHD), 65
Alzheimer's disease, 177–178
anorexia nervosa (AN), 131
autism, 70
bipolar disorder (BD), 104–105
borderline personality disorder (BPD), 215
childhood maltreatment, 75
Huntington's disease (HD), 186
posttraumatic stress disorder (PTSD), 124
seizure disorders, 185
substance abuse, 163–164
testing, ethics of, 187
glial cells, 38
globus pallidus, 52
glutamate, 35
gray versus white matter, 36–37

hallucinogen abuse, 169
hippocampus, 49–50
H.M., case of, 4–6
HPA axis. *See* hypothalamic–pituitary–adrenal
axis
Huntington's disease (HD)
characteristics, 186
and cognition, 187–188
and genetics, 186
introduction, 186
mechanism, 186
treatment, 188
hypersomnia, 154–156
hypothalamic–pituitary–adrenal (HPA) axis, 51

insomnia, 152–154
and neuroimaging, 153
treatment, 153–154
intellectual disability
characteristics of, 72–73
dyslexia, 73–74

introduction, 72
learning disorders, 73

law and neuroscience, 208–209
learning disorders, 73
left and right hemispheres of the brain, 37
life span and sleep, 150–151
limbic system
   amygdala, 49
   hippocampus, 49–50
   hypothalamic–pituitary–adrenal (HPA) axis, 51
   mesolimbic system, 51
   thalamus, 50–51
lobotomy, 8–9

magnetic resonance spectroscopy (MRS), 19
magnetic resonance imaging (MRI), 7–8, 18–19
   errors in analysis of, 41–42
   safety, 20
magnetoencephalography (MEG), 23–24
major depressive disorder (MDD)
   and cognition, 98
   and genetics, 96
   introduction, 95
   and neuroimaging, 96–98
   predictors of success, 103–104
   treatment, 98–99
      cognitive behavioral therapy, 99
      deep brain stimulation (DBS), 101
      electroconvulsive therapy, 101–102
      nonmedical techniques, 102
      psychotropic medications, 99–100
      transcranial magnetic stimulation, 100
      vagus nerve stimulation, 101
marijuana abuse. *See* cannabis abuse
marijuana, medical, 160
MDD. *See* major depressive disorder
medical disorders
   Alzheimer's disease
      cause, 178
      and genetics, 177–178
      and imaging, 178–179
      introduction, 176–177
      lifestyle risks, 177
      mood, 177
      treatment, 179–180
   anoxia and hypoxia, 190–191
   delirium, 174–176
      causes, 175
      introduction, 174–175
      and neurobiology, 175–176
      symptoms, 175
   dementia of Lewy body (DLB), 181–182
   dementias, introduction, 176
   frontal lobe dementia, 180–181
   Huntington's disease (HD)
      characteristics, 186
      and cognition, 187–188
      and genetics, 186
      mechanism, 186
      treatment, 188
   introduction, 173
   Parkinson's disease (PD), 188–189
   seizure disorders
      and cognition, 185
      continuous (status epilepticus), 184–185
      focal (partial), 184
      generalized, 183–184
      and genetics, 185
      introduction, 182–183
      mechanism, 185
      and neurochemistry, 185
      and neuroimaging, 185
      treatment, 186
   tumors, 189–190
   vascular dementia, 180
medical marijuana, 160
medication
   treatment technique, history of, 10
   for anorexia nervosa (AN), 133–134
   for binge eating disorder (BED), 142
   for bipolar disorder (BD), 108–109
   for bulimia nervosa (BN), 138
   for major depressive disorder (MDD), 99–100
   for schizophrenia, 88, 90
   for traumatic brain injury (TBI), 204–205
MEG. *See* magnetoencephalography
meninges, 38
mesolimbic system, 51
methamphetamine abuse, 167–168
methodological advancements, neurobiology
   computed tomography (CT), 7
   introduction, 6–7
   magnetic resonance imaging (MRI), 7–8
   neuropsychology testing, 7
methods, research and clinical
   ablation 14–15
   behavior genetic testing, 25
   electrophysiology recordings
      electroencephalogram (EEG), 16
      single-cell technique, 15–16
      event-related potentials, 17
   functional neuroimaging
      functional magnetic resonance imaging (fMRI), 22–23
      magnetoencephalography (MEG), 23–24
      positron emission tomography (PET), 22
      single-photon emission computed tomography (SPECT), 21–22
   introduction, 13–14
   microscopic evaluations, 14
   neuropsychological evaluations, 24–25

structural neuroimaging
    computed tomographic imaging (CT), 17–18
    diffusion tensor imaging (DTI), 20–21
    magnetic resonance imaging (MRI), 18–19
    magnetic resonance spectroscopy (MRS), 19
microscopic evaluations, 14
mirror system, 56
mood disorders
    bipolar disorder (BD)
        and genetics, 104–105
        introduction, 104
        treatment, 108
    depression
        clinical symptoms, 94–95
        introduction, 94
        subtypes, 95
    introduction, 93–94
    major depressive disorder (MDD)
        introduction, 95
        predictors of success, 103–104
movement disorders, 186–189
MRI. See magnetic resonance imaging
MRI scanner, safety and, 20
MRS. See magnetic resonance spectroscopy
Mr. Tan, case of, 3–4

neurobiology, history of
    clinical case reports
        current case examinations, 6
        H.M., 4–6
        introduction, 2
        Mr. Tan, 3–4
        Phineas Gage, 2–3
    introduction, 1–2
    methodological advancements
        computed tomography (CT), 7
        introduction, 6–7
        magnetic resonance imaging (MRI), 7–8
        neuropsychology testing, 7
    neuroscientists, 11
    treatment technique
        cognitive behavioral therapy, 10–11
        lobotomy and shock treatment, 8–9
        medication, 10
        psychotherapy, 10
neurodevelopment overview
    embryogenesis, 63
    neuronal migration, 63–64
    stages, 63
    synaptic pruning, 64
neuronal migration, 63–64
neurons
    definition, 32
    firing, 32–33
    properties, 32
neuroplasticity, 58

neuropsychological evaluations, 24–25
neuropsychology testing, 7
neuroscience and law, 208–209
neuroscientists, 11
neurotransmitters and psychiatry
    acetylcholine, 33–34
    catecholamines, 34
    dopamine, 34–35
    epinephrine, 35
    gamma amino butric acid (GABA), 35
    glutamate, 35
    introduction, 33
    norepinephrine, 35
    serotonin, 34
nocturnal enuresis, 157
norepinephrine, 35

obesity, and therapist role, 140–141
occipital lobe, 39
opiate abuse, 169–170

panic disorder (PD), 118–120
    and cognition, 119
    and neurochemistry, 119
    and neuroimaging, 119
    treatment, 119–120
parasomnia, 156–157
parietal lobe, 40
partial (focal) seizure, 184
PD. See panic disorder
peripheral nervous system (PNS), 29–31
persistent postconcussion syndrome (PPCS), 200
personality disorders
    antisocial personality disorder (ASPD)
        and cognition, 212
        and genetics, 211
        introduction, 210
        and neuroimaging, 211–212
        treatment, 212–213
    borderline personality disorder (BPD)
        and genetics, 215
        introduction, 213, 215
        and neurocognition, 216–217
        and neuroimaging, 215–216
        treatment, 217–218
    introduction, 207, 209–210
PET. See positron emission tomography
Phineas Gage, case of, 2–3
positron emission tomography (PET), 22
posttraumatic stress disorder (PTSD)
    and cognition, 126
    and genetics, 124
    introduction, 123–124
    and neurochemistry, 124–125
    and neuroimaging, 125
    treatment, 126

PPCS. *See* persistent postconcussion syndrome
prefrontal system, 53–54
psychotherapy
   for anorexia nervosa (AN), 134, 136
   for binge eating disorder (BED), 142–143
   for bipolar disorder (BD), 109
   for bulimia nervosa (BN), 139
   for traumatic brain injury, 203
   treatment technique, history of, 10
psychotropic medication
   for bipolar disorder (BD), 108–109
   in children, 62–63
   for MDD, 99–100
PTSD. *See* posttraumatic stress disorder

REM sleep, 148–149

safety and eating disorders, 135
safety of patients, 9
self-care, and sleep, 153
secondary neurobiological brain systems
   Circle of Willis, 56–57
   cranial nerves, 55
   introduction, 55
   mirror system, 56
schizophrenia
   clinical symptoms, 80–81
   and cognition, 87
   and decision-making capacity, 89–90
   and genetics, 82–83
   introduction, 79–80
   and neurochemistry, 83
   and neuroimaging, 83–86
   onset and course, 81–82
   treatment, 88–91
      medication, 88, 90
      therapy, 90–91
seizure disorders
   and cognition, 185
   continuous (status epilepticus), 184–185
   focal (partial), 184
   generalized, 183–184
   and genetics, 185
   introduction, 182–183
   and neurochemistry, 185
   and neuroimaging, 185
   treatment, 185
serotonin, 34
single-cell technique, 15–16
single-photon emission computed tomography (SPECT), 21–22
sleep
   circadian rhythms, 150
   clinical disorders
      hypersomnia, 154–156
      insomnia, 152–154
      parasomnia, 156–157
   dreaming, 149–150
   introduction, 145
   life span and sleep, 150–151
   nocturnal enuresis, 157
   normal sleep
      cycle of stages, 149
      introduction, 146
      REM sleep, 148–149
      stage I sleep, 147
      stage II sleep, 147–148
      stage III sleep, 148
      stage IV sleep, 148
      wakefulness, 146–147
   purpose of sleep, 145–146
   sleep deprivation
      and cognition, 151–152
      introduction, 151
      and neuroimaging, 152
social anxiety, 121–122
   and cognition, 122
   and genetics, 121
   and neurochemistry, 121
   and neuroimaging, 121–122
   treatment, 122
SPECT. *See* single-photon emission computed tomography
stage I sleep, 147
stage II sleep, 147–148
stage III sleep, 148
stage IV sleep, 148
status epilepticus (continuous seizure), 184–185
striatum, 52
stress in clinical and research environments, 114–115
structural neuroimaging
   computed tomographic imaging (CT), 17–18
   diffusion tensor imaging (DTI), 20–21
   magnetic resonance spectroscopy (MRS), 19
   magnetic resonance imaging (MRI), 18–19
substance abuse disorders
   alcohol abuse
      in adolescents, 166
      cognitive impairment, 165–166
      introduction, 164–165
      mechanism, 165
      treatment, 166
   amphetamines, 171–172
   behavioral and neurobiological response, 161–163
   caffeine and nicotine abuse, 170
   cannabis abuse, 167
   cocaine abuse, 168–169
   comorbidity, psychiatric disorders, 163
   and genetics, 163–164
   hallucinogen abuse, 169

introduction, 159–161
medical marijuana, 160
methamphetamine abuse, 167–168
opiate abuse, 169–170
prenatal exposure, 164
suicidal behavior and confidentiality, 110–111
synaptic pruning, 64

TBI. *See* traumatic brain injury
temporal lobe, 40
thalamus, 50–51
transcranial magnetic stimulation (TMS), 100
traumatic brain injury (TBI)
    definition, 194–195
    introduction, 193–194
    malingering issues, 201–202
    mild
        concussion, 200
        definition and symptoms, 199–200
        persistent postconcussion syndrome (PPCS), 200
    and neurobiology, 195–197
    and neuroimaging, 197–198
    outcome, 202
    risk factors, 194
    severity classification, 198–199
    sports-related, 201
    treatment, 202–206
        cognitive rehabilitation, 202–203
        diet/nutrition/exercise, 205–206
        for headache, 204
        medication, 204–205
        neuropsychological assessment, 203
        psychotherapy, 203
        sensory evaluations, 204
        speech and language therapy, 203–204
treatment technique, history of
    cognitive behavioral therapy, 10–11
    lobotomy and shock treatment, 8–9
    medication, 10
    psychotherapy, 10
tumors, 189

vagus nerve stimulation, 101
vascular dementia, 180

wakefulness, 146–147

www.ingramcontent.com/pod-product-compliance
Ingram Content Group UK Ltd.
Pitfield, Milton Keynes, MK11 3LW, UK
UKHW050042210326
4879IPUK00006B/106